SOTHIS PRESS

Sothis Press aims to completely reimagine the public domain book with an emphasis on universal access to captivatingly designed print editions. We believe that technology will enable every single book ever published to be perpetually accessible in beautiful print form. Sothis is beauty in ubiquity.

www.sothis.press
1 Waterhouse Square, London, EC1N 2ST

The Normal Union Arithmetic

Designed for Common Schools, Normal Schools, High Schools, Academies, Etc

Edward Brooks

SOTHIS PRESS

ISBN 978-1-0365-2611-5

The Normal Union Arithmetic
is in the public domain.

THE

NORMAL

UNION ARITHMETIC,

DESIGNED FOR

COMMON SCHOOLS, NORMAL SCHOOLS, HIGH SCHOOLS,
ACADEMIES, ETC.

BY

EDWARD BROOKS, A.M., PH. D.,

PRINCIPAL AND PROFESSOR OF MATHEMATICS IN PENNSYLVANIA STATE NORMAL SCHOOL, AND AUTHOR OF "THE NORMAL SERIES OF ARITHMETICS," "NORMAL ELEMENTARY ALGEBRA," "NORMAL GEOMETRY AND TRIGONOMETRY," "PHILOSOPHY OF ARITHMETIC," ETC.

"Analysis and Synthesis;—the magic wand whose touch unfolds the mysterious and beautiful relations of numbers."

PHILADELPHIA:
SOWER, POTTS & COMPANY,
530 MARKET ST., AND 523 MINOR ST.

CONTENTS.

PREFACE.

EDUCATION is progressive. The development of the popular mind is becoming the transcendent question of the day. Improvements are being made in every department, dull routine is giving way to intellectual activity, instruction is becoming a science, and teaching a profession.

This advance in education has been nowhere more noticeable than in the improvements of text-books upon Arithmetic. A few years ago an unpretending little work, Colburn's Intellectual Arithmetic, was presented to the public. That little work touched Arithmetic as with the wand of an enchantress, and transformed it from a dry collection of mechanical processes to a thing of interest and beauty. It laid the foundation of that system of Mental Arithmetic which has infused a new spirit into the science of numbers, and has done more than any other influence to vitalize the methods of common school instruction in this country.

In presenting a new work upon the subject, I desire to acknowledge my obligations to this and other works which have followed it. Bringing to the task the reflection and experience of many years of educational labor, I hope to be able to present a text-book upon Arithmetic which will take an honorable position among the many valuable works upon the subject which are doing so much for the educational interests of the country. Some of the general and special features of this work will be briefly noticed.

METHOD OF TREATMENT.—The method of treatment is both Inductive and Deductive, embracing Analysis and Synthesis. In some cases both of these methods are employed in the development of the same subject; in other cases they are combined in the same solution or explanation, and such combination is characteristic of the entire work. I have endeavored to meet the wants of both teacher and pupil, by preparing a work convenient for instruction, adapted to the natural and logical development of the mind of the pupil in the study of numbers, and containing such applications as will prepare students for the business relations of life.

ARRANGEMENT.—The arrangement of the work is believed to be strictly *logical* and at the same time, *practical*, being adapted to the natural mental growth and development of the pupil. The mottoes have been,—*from the easy to the difficult, from the simple to the complex, from the known to the unknown.* Care has been taken to present the simpler and more practical subjects first, and not to anticipate any principles or processes before the pupil is prepared for them. Thus, I have placed Compound Numbers after Fractions, Percentage before Ratio and Proportion, Equation of Payments after Proportion, and other arrangements have been determined by the same principle.

THE REASONING.—All reasoning is *comparison*. A comparison requires a standard, and this standard is the *fixed*, the *axiomatic*, the *known*. The law of correct reasoning, therefore, is to compare the *complex* with the *simple*, the *theoretic* with the *axiomatic*, the *unknown* with the *known*. This law is kept prominently before the mind in the development of this work, and upon it are based its definitions, solutions and explanations, etc. As an illustration, notice the definitions of Ratio, Proportion, etc., the method of stating a proportion, etc.

SOLUTIONS.—The solutions and demonstrations are so simple and clear, that they may be understood by very young pupils, yet they are expressed in language concise and logically accurate, and *in the form which the pupil should be required to use at recitation*. A solution may be too concise to be readily understood, and it may also be too prolix, the idea being smothered or concealed in a multiplicity of words. Both of these errors I have endeavored to avoid, remembering that *the highest science is the greatest simplicity*.

RULES.—The rules or methods of operation are expressed in brief and simple language, and are given as the results of solutions and explanations. I have endeavored to lead the pupil to see the reason for the different processes, thus enabling him to derive his own method of operation based upon such reasoning. The object has been to develop mind as well as the power of computation—to make thinkers rather than arithmetical machines.

APPLICATIONS.—One of the most prominent features of the work is its *practical character*. The applications of the science are not the thought of the scholar as what business may be, but represent the *actual business of the day*. Many of the problems and processes are derived from *actual business transactions*. Our *Bills* and *Accounts* came out of the stores; our *Taxes*, *Banking*, *Exchange*, etc., have been submitted to and endorsed by those connected with the business; several of the problems on *Duties* are out of the *Custom House; Insurance* has been examined by experts in the business; the subject of *Building Associations*, for the first time introduced into an arithmetic, was partly prepared by one practically familiar with the subject; etc.

UNION OF MENTAL AND WRITTEN.—Another leading feature of the work is the union of mental and written arithmetic in one book. Many who recognize the importance of Mental Arithmetic think that it takes *too much time for* the pupil to study two separate books—one on Mental and the other on Written Arithmetic—and hold that these two subjects should be embraced in *one book*. To meet this demand I have made a complete and harmonious combination of the two subjects, introducing many of those forms of analysis that have given such popularity to my Mental Arithmetic. It is this combination that gives the work its name, *The Union Arithmetic;* and this union will be found to be not a mere nominal thing, but a reality. In the study of the work the pupil can obtain quite a thorough course in *arithmetical analysis* while he is becoming familiar with the art of computation and the application of the art to business. These mental exercises are so arranged and printed that any teacher who prefers to omit them can do so without any inconvenience to either the pupil or the teacher.

SPECIAL FEATURES.—There are several special features peculiar to this work, to which we desire to call attention.

1st. Many new definitions, as of Fraction, Least Common Multiple, Percentage, Ratio, etc.

2d. New and concise method of explaining Greatest Common Divisor, and a method of Least Common Multiple not usually given.

3d. The two distinct methods of the development of Fractions, the *relation* of fractions, the method of stating a problem in Simple Proportion and reason for it, and the development of Compound Proportion.

4th. The Analytic and Synthetic methods of developing Involution and Evolution, the greater attention to Involution as a preparation to Evolution, a new method of cube root, etc.

5. Great number and variety of problems, especially after the Fundamental Rules, Fractions, etc., and at the close of the book. Other features also important, will present themselves upon a careful examination.

It should be stated that this work was first published in 1863, and that some of the definitions and processes which were then new have since been introduced into other works. The present edition is thoroughly revised, and brought up to the very latest methods of business calculations.

Thanking my friends for the cordial reception given to my previous labors, I send forth this new volume, with the earnest desire that it may meet their approbation, and aid in the development and diffusion of a deeper interest in the beautiful science of numbers—a science which practically lies at the foundation of all science and all thought, and one which is doing so much to promote the cause of popular education.

EDWARD BROOKS.

STATE NORMAL SCHOOL,
May 10, 1877.

THE

NORMAL

UNION ARITHMETIC

SECTION I.

ARITHMETICAL LANGUAGE.

1. Arithmetic is the science of numbers and the art of computing with them.

2. A **Unit** is a single thing or *one*. A thing is a *concrete unit; one* is an *abstract unit.*

3. A **Number** is a unit or a collection of units. Numbers are *concrete* and *abstract.*

4. A **Concrete Number** is one in which the kind of unit is named; as, two *yards,* five *books.*

5. An **Abstract Number** is one in which the kind of unit is not named; as, *two, four,* etc.

6. Similar Numbers are those in which the units are alike; as, two *boys* and four *boys.*

7. Dissimilar Numbers are those in which the units are unlike; as, two *boys* and four *books.*

8. A **Problem** is a question requiring some unknown result from that which is known.

9. A **Solution** of a problem is a process of obtaining the required result.

10. A **Rule** is a statement of the method of solving a problem.

11. Mental Arithmetic treats of performing arithmetical operations without the aid of written characters.

12. Written Arithmetic treats of performing arithmetical operations with written characters.

13. Arithmetical Language is the method of expressing numbers.

14. Arithmetical Language is of two kinds, *Oral* and *Written.* The former is called *Numeration* and the latter is called *Notation.*

NOTE.—A number is really the *how many* of the collection instead of the *collection;* but the definition given, which is a modification of Euclid's, is simpler and sufficiently accurate.

NUMERATION.

15. Numeration is the method of naming numbers, and of reading them when expressed by characters. It is the *oral expression* of numbers.

16. Since it would require too many words to give each number a separate name, numbers are named according to the following simple principle:

Principle.—*We name a few of the first numbers, and then form groups or collections, name these groups, and use the names of the first numbers to number these groups.*

17. A single thing is named *one:* one and one more are named *two;* two and one more, *three;* three and one more, *four;* and thus we obtain the simple names,

One, two, three, four, five, six, seven, eight, nine, ten.

18. Now, regarding the collection *ten* as a single thing, we might count *one and ten, two and ten, three and ten,* etc., as far as *ten and ten,* which we would call *two tens.* By this principle were obtained the following numbers:

Eleven, twelve, thirteen, fourteen, fifteen, sixteen, seventeen, eighteen, nineteen, twenty.

19. Proceeding in the same way, we would have *two tens and one, two tens and two, two tens and three,* etc. By this principle were obtained the following numbers:

Twenty-one, twenty-two, twenty-three, twenty-four, twenty five, twenty-six, twenty-seven, twenty-eight, twenty-nine.

20. Continuing in the same manner, we would have *three tens, four-tens, five-tens,* etc. By this principle were derived the following ordinary names:

Thirty, forty, fifty, sixty, seventy, eighty, ninety.

21. A group of *ten tens* is called a *hundred;* a group of *ten hundreds,* a *thousand;* the next group receiving a new name consists of a *thousand thousands,* called a *million;* the next group of a *thousand millions,* called a *billion,* etc.

22. After a *thousand,* the two intermediate groups between those having a distinct name, are numbered by *tens* and *hundreds,* as *ten thousand* and *hundred thousand.*

NOTES.—1. The above shows the *principle* by which the numbers were named. The names, however, were not derived from the particular expressions given, but originated in the Saxon language.

2. *Eleven* is from the Saxon *endlefen,* or Gothic *ainlif* (*ain, one,* and *lif, ten*); *twelve* is from the Saxon *twelif,* or Gothic *tvalif* (*tva, two,* and *lif, ten*). Some have supposed that *eleven* meant *one left after ten,* and *twelve, two left after ten.*

3. *Twenty* is from the Saxon *twentig* (*twegen, two,* and *tig,* a *ten*); *thirty* is from the Saxon *thritig* (*thri, three,* and *tig,* a *ten*), etc.

4. *Hundred* is a primitive word; *thousand* is from the Saxon *thusend,* or Gothic *thusundi,* (*thus, ten,* and *hund, hundred*); *million, billion,* etc., are from the Latin.

NOTATION.

23. Notation is the method of writing numbers. Numbers may be written in three ways:

1st. By *words,* or common language.

2d. By *figures,* called the *Arabic Method.*

3d. By *letters,* called the *Roman Method.*

NOTE.—The method by words is that of ordinary written language and needs no explanation.

ARABIC NOTATION.

24. The **Arabic System** of Notation is the method of expressing numbers by characters called *figures.*

25. In this system numbers are expressed according to the following principle:

Principle.—*We employ characters to represent the first nine numbers, and then use these characters to number the groups, the group numbered being indicated by the position of the character.*

26. Figures.—Figures are characters used in expressing numbers. There are ten figures used, as follows:

FIGURES.	1,	2,	3,	4,	5,	6,	7,	8,	9,	0.
NAMES AND VALUES.	one,	two,	three,	four,	five,	six,	seven,	eight,	nine,	naught, cipher or zero.

27. By the combination of these figures all numbers may be expressed; hence they are appropriately called the *alphabet* of *arithmetic*.

28. Combination.—These figures are combined according to the following principle:

1. *A figure standing alone, or in the first place at the right of other figures, expresses* UNITS *or* ONES.

2. *A figure standing in the second place, counting from the right, expresses* TENS; *in the third place,* HUNDREDS; *in the fourth place,* THOUSANDS, etc.; thus,

10 is 1 ten, or ten.	100 is 1 hundred.
20 " 2 tens, or twenty.	200 " 2 hundred.
30 " 3 tens, or thirty.	300 " 3 hundred.
40 " 4 tens, or forty.	400 " 4 hundred.
50 " 5 tens, or fifty.	1000 " 1 thousand.
60 " 6 tens, or sixty.	2000 " 2 thousand.
90 " 9 tens, or ninety.	4000 " 4 thousand.

29. The name of each of the first twenty-one places is represented by the following

NUMERATION TABLE.

	Hundred-quintillions.	Ten-quintillions.	Quintillions.	Hundred-quadrillions.	Ten-quadrillions.	Quadrillions.	Hundred-trillions.	Ten-trillions.	Trillions.	Hundred-billions.	Ten-billions.	Billions.	Hundred-millions.	Ten-millions.	Millions.	Hundred-thousands.	Ten-thousands.	Thousands.	Hundreds.	Tens.	Units.
	8	8	8,	8	8	8,	8	8	8	8	8	8,	8	8	8,	8	8	8,	8	8	8.
PLACES.	21st,	20th.	19th.	18th,	17th,	16th	15th,	14th,	13th,	12th,	11th,	10th,	9th,	8th,	7th,	6th,	5th,	4th,	3d,	2d,	1st,
PERIODS.	7th.			6th.			5th.			4th.			3d.			2d.			1st		

30. Periods.—For convenience in writing and reading numbers, the figures are arranged in *periods* of three places each, as shown in the table. The first three places constitute the *first*, or *units period;* the second three places constitute the *second*, or *thousands period*, etc.

1. Required the names of the following places:

First; third; second; sixth; fourth; eighth; tenth; ninth; twelfth; fifth; seventh; eleventh; thirteenth; seventeenth; fourteenth; sixteenth; eighteenth; fifteenth; nineteenth; twenty-first; twentieth.

2. Required the places of the following:

Tens; hundreds; thousands; millions; ten-thousands; hundred thousands; ten-millions; billions; hundred-millions; hundred-billions; units; ten-billions; trillions; quadrillions; hundred-quintillions; ten-trillions; ten-quintillions; hundred-quadrillions; quintillions; hundred-trillions; ten-quadrillions.

3. Required the names of the following periods:

1. First.	3. Second.	5. Fourth.
2. Third.	4. Fifth.	6. Seventh.

4. Required the period and place of the following:

Thousands; millions; ten-thousands; hundred-millions; billions; hundred-trillions; trillions; ten-trillions; quadrillions; ten-quadrillions; hundred-trillions; quintillions; hundred-quintillions; hundred-thousands; ten-millions.

31. The combination of figures to express a number forms a *numerical expression.* Thus, 25 is a numerical expression which denotes the same as the common word *twenty-five.*

32. The different figures of a numerical expression are called *terms. Terms* are also used to indicate the *numbers* represented by the figures.

NOTE.—The use of the word *term*, to indicate both the figures and the numbers represented by them, enables us to avoid the error of using the word *figure* for the word *number.*

EXERCISES IN NUMERATION.

33. The pupils are now prepared to learn to *read* numbers when expressed by *figures.* From the preceding explanations we have the following *rule for numeration:*

Rule.—I. *Begin at the right hand, and separate the numerical expression into periods of three figures each.*

II. *Then begin at the left hand and read each period in succession, giving the name of each period except the last.*

NOTE.—The name of the last period is usually omitted, because it is understood.

1. What number is expressed by 6325478?

SOLUTION.—Separating the numerical expression into periods of three figures each, beginning at the right hand, we have 6,325,478. The third period is 6 *millions*, the second period is 325 *thousands*, and the first is 478 *units*; hence the number is 6 *millions*, 325 *thousands*, 478.

OPERATION.
6,325,478

Read the following numerical expressions:

2.	3876	**10.**	468217	**18.**	80305072
3.	2185	**11.**	654879	**19.**	65073058
4.	3072	**12.**	803006	**20.**	484378513
5.	5678	**13.**	1234567	**21.**	123456789
6.	12630	**14.**	3078560	**22.**	854327031
7.	70851	**15.**	8507032	**23.**	80735468579
8.	32468	**16.**	54372568	**24.**	20650708462067
9.	507035	**17.**	87072135	**25.**	798653013285678521

NOTE.—After pupils are familiar with reading by dividing into periods, the division may be omitted or performed mentally.

EXERCISES IN NOTATION.

34. Having learned to *read* numerical expressions, we are now prepared to *write* them. From the principles which have been explained, we derive the following rule:

Rule.—I. *Begin at the left and write the hundreds, tens, and units of each period in their proper order.*

II. *When there are vacant places, fill them with ciphers*

1. Express in figures the number *four thousand three hundred and four.*

SOLUTION.—We write the 4 thousands in the 4th place, 3 hundreds in the 3d place, a cipher in the 2d place, there being no tens, and 4 units in the 1st place, and we have 4304.

OPERATION.
4,304

Express the following numbers in figures:

2. One hundred and six.

3. One hundred and ten.

4. Two hundred and forty.

5. Two hundred and sixty-five.

6. Two hundred and nine.

7. Three hundred and twelve

8. Three hundred and fifty seven.

9. Four hundred and twenty eight.

10. Seven hundred and eighty-four.

11. Nine hundred and thirty-seven.

12. Eight hundred and ninety-nine.

13. Four thousand and seven.

14. Five thousand two hundred and thirty-six.

15. Six thousand and eighty-five.

16. Twenty-three thousand six hundred and forty-seven.

17. One hundred and forty-five thousand seven hundred and six.

18. Three hundred and eight thousand three hundred and eight.

19. Six hundred and four thousand three hundred and sixty-eight.

20. Eight hundred and seventy-four thousand one hundred and twenty.

21. Seven hundred and seven thousand seven hundred and seven.

22. One million.

23. Two million and six.

24. Three million and twelve.

25. Forty-five million and twenty-four.

26. Six million forty-seven thousand.

27. Eight million two thousand and sixty-seven.

28. Five million two hundred and ninety-six thousand.

29. Seventy million, one thousand and forty-five.

30. One million, two hundred and thirty thousand, four hundred and fifty-six.

31. Four million, three hundred and seven thousand, four hundred and nine.

32. Fifteen million, four hundred and eight thousand, and eighty-four.

33. Twenty-eight million, five hundred and ninety-four thousand, seven hundred and nine.

34. Forty-seven million, thirty-eight thousand, two hundred and eight.

35. Two hundred million, forty-nine hundred and twenty-eight.

36. Six billion, seven hundred and five million, thirty-five thousand and six.

37. Forty-nine trillion, fifty-eight thousand, seven hundred and ninety-eight.

35. Orders.—Since we may have 2 *tens*, 3 *tens*, etc., 2 *hundreds*, 3 *hundreds*, etc., the same as 2 *apples*, 3 *apples*, 2 *books*, 3 *books*, etc., these different groups may be regarded as *units* of different *orders;* thus,

UNITS	are	called	Units of the 1*st order*.
TENS	"	"	Units of the 2*d order*.
HUNDREDS	"	"	Units of the 3*d order*.
THOUSANDS	"	"	Units of the 4*th order*.
TEN-THOUSANDS	"	"	Units of the 5*th order*.
etc.,			etc.

36. From this it is seen that *ten* units of a lower order make one unit of the next higher order; the system of notation is therefore called the *Decimal System*, from the Latin, *decem*, ten.

WRITTEN EXERCISES.

1. Write and read two units of the second order, and four of the first.

2. Write and read four units of the third order, and six of the second.

3. Write and read nine units of the third order, and three of the first.

4. Write and read three units of the fifth order, six of the third, and seven of the first.

5. Write and read five units of the sixth order, four of the third, and eight of the second.

6. How many units in one ten? In 2 tens? In 3 tens? In 4 hundreds?

7. How many tens in 1 hundred? In two hundreds? 3 hundreds? 2 thousands?

8. How many hundreds in 3 thousands? In 5 thousands? In 60 tens? In 3 millions?

9. What is the result of changing a figure one place to the left? Two places? Three places?

10. What is the result of changing a figure one place to the right? Two places? Three places?

11. What is the result of placing one naught to the right of one or more figures? Two naughts? Three naughts?

12. What place is occupied by tens figure? Hundreds figure? Ten-thousands? Millions? Billions?

13. What is the denomination of a figure in third place? In fifth place? In fourth place? In seventh place? In ninth place? In sixth place? In eighth place?

37. The table which has been given enables us to read a numerical expression consisting of twenty-one figures; the periods which follow them in order are as follows:

Sextillions, Septillions, Octillions, Nonillions, Decillions, Undecillions. Duodecillions, Tertio-decillions, Quarto-decillions, Quinto-decillions, Sexto-decillions, Septo-decillions, Octo-decillions, Nono-decillions, Vigillions. With these, and those already given, we can write and read a numerical expression consisting of sixty-six places.

NOTES.—1. The first of the nine Arabic characters are called *digits*, from the Latin word *digitus*, a finger, owing to the fact that the ancients reckoned by counting the fingers. They are also called *significant* figures, because they always indicate a definite number of units. The character 0, called *zero*, *cipher*, or *naught*, always indicates an absence of units.

2. The Arabic Notation is named from the Arabs, who introduced it into Europe by their conquest of Spain during the 11th century. The Arabs obtained it from the Hindoos, by whom it was probably invented more than 2000 years ago.

3. There are three theories for the origin of the Arabic characters, for which see *Brooks's Philosophy of Arithmetic*.

THE DECIMAL SCALE.

38. Since figures express different orders of units in different places, they are said to have two kinds of values; a *simple value* and a *local value.*

39. The **Simple Value** of a figure is its value when it stands alone or in units place. The *Local Value* of a figure is its value when in some other place than units.

40. The value of a figure is *increased tenfold* for every place it is moved from *right to left;* and is *decreased tenfold* for every place it is moved from *left to right.* This law is called the *scale* of the system.

41. Since the value of terms decreases from left to right at a tenfold rate, if we fix the place of units by a point (.), we may extend the decimal scale to the right of units.

42. The first place on the right of the point will be one-tenth of units or *tenths*, the second place one-tenth of tenths, or *hundredths*, the third place, *thousandths*, etc.

43. Such terms are called *decimals*, and the point is called the *decimal point.* The expression 48.37 is read 48 units, 3 tenths, and 7 hundredths, or 48 and 37 hundredths.

44. The **Money** of the United States is expressed by the decimal system. The *dollar* is the *unit*, and is indicated by the symbol $. The first place at the right of the decimal point is called *dimes;* the second place, *cents*, etc.

45. Dimes and **Cents** are usually read as a *number of cents.* Thus, $4.65 is read 4 dollars and 65 cents; and $72.48 is read 72 dollars **and** 48 cents.

The Decimal system of numeration had its origin in the practice, common to all nations, of counting by groups of tens.

EXAMPLES FOR PRACTICE.

Read the following:

1.	12.5	**5.**	46.375	**9.**	$125.45
2.	36.25	**6.**	89.625	**10.**	$437.05
3.	47.75	**7.**	$16.25	**11.**	$548.475
4.	86.50	**8.**	$85.35	**12.**	$768.605

Write the following:

1. Fourteen and five tenths.
2. Eighty-four and twenty-five hundredths.
3. Two hundred and three, and sixty-seven hundredths.
4. Seven hundred and ninety-six, and eight hundred seventy-five thousandths.
5. Sixty dollars and seven cents.
6. Eighty seven dollars and twenty-five cents.
7. Four hundred and fifty dollars, and fifty cents.
8. Eight hundred and sixty-four dollars, thirty-seven cents and five mills.

ENGLISH METHOD OF NUMERATION.

46. The method of numeration by dividing numbers into periods of three figures each, is called the *French Method.* There is also another method called the *English Method.*

47. The **English Method** uses periods of six figures each, calling the first period *units*, the second *millions*, the third *billions*, the fourth *trillions*, etc.

48. The places in each period are *units, tens, hundreds, thousands, tens of thousands, hundreds of thousands.* The method is represented in the following table:

Hund. of Thou. of Trillions.	Tens of Thou. of Trillions.	Thousands of Trillions.	Hundreds of Trillions.	Tens of Trillions.	Trillions.
6	6	6	6	6	6,

4th or Trillions Period.

Hund. of Thou. of Billions.	Tens of Thou. of Billions.	Thousands of Billions.	Hundreds of Billions.	Tens of Billions.	Billions.
6	6	6	6	6	6,

3d or Billions Period.

Hund. of Thou. of Millions.	Tens. of Thou. of Millions.	Thousands of Millions.	Hundreds of Millions.	Tens of Millions.	Millions.
6	6	6	6	6	6,

2d or Millions Period.

Hundreds of Thousands.	Tens of Thousands.	Thousands.	Hundreds.	Tens.	Units.
6	6	6	6	6	6,

1st or Units Period

NOTE.—The French Method is used throughout the United States, France, etc., and being much more convenient. is very likely to supersede the other method in England.

WRITTEN EXERCISES.

Write the following by the English method.

1. One million.
2. *One billion.*
3. One trillion.
4. One quadrillion.

ROMAN NOTATION.

49. The **Roman Method** of Notation employs seven letters of the Roman alphabet. Thus, I represents *one ;* V, *five ;* X, *ten ;* L, *fifty ;* C, *one hundred ;* D, *five hundred ,* M, *one thousand.*

50. To express other numbers these characters are combined according to the following principles :

1. *Every time a letter is repeated its value is repeated.*

2 *When a letter is placed before one of greater value, the* DIFFERENCE *of their value is the number represented.*

3. *When a letter is placed after one of a greater value, the* SUM *of their values is the number represented*

4. *A dash placed over an expression increases its value a thousand fold.* Thus, $\overline{\text{VII}}$ denotes seven thousand.

51. These principles are exhibited in the following table, which the pupil will examine carefully :

ROMAN TABLE.

I	One.	XXX	Thirty.
II	Two.	XL	Forty.
III	Three.	L	Fifty.
IV	Four.	LX	Sixty.
V	Five.	LXX	Seventy.
VI	Six.	XC	Ninety.
VII	Seven.	C	One hundred.
VIII	Eight.	CC	Two hundred.
IX	Nine.	D	Five hundred.
X	Ten.	DC	Six hundred.
XI	Eleven.	DCCCC	Nine hundred.
XIV	Fourteen.	M	One thousand.
XV	Fifteen.	MM	Two thousand
XIX	Nineteen.	MCLX	One thousand one hundred [and sixty
XX	Twenty.	MDCCCLXXVI,	1876

52. The **Roman Method** is named from the Romans, who invented and used it. It is now only employed to denote the chapters and sections of books, pages of preface and introduction, and in other places for prominence and distinction.

WRITTEN EXERCISES.

Express the following numbers by the Roman method: 1. Twenty-seven. 2. Seventy-seven. 3. Two hundred and one. 4. Six hundred and fifty-six. 5. One thousand seven hundred and seventy-six. 6. Four thousand seven hundred and fifty-seven. 7. 25007. 8. 206484.

Read the following numbers:

1. LXXVII. 2. MXC. 3. MDCCCLXXVI. 4. MMMCCCXXXIII. 5. $\overline{\text{XV}}$DCCXLIV. 6. clxxxviii. 7. xlix. 8. xcix.

LUMBERMEN'S NOTATION.

53. Lumbermen in marking lumber employ a modification of the Roman Method of Notation. The first four characters are like the Roman. The others are as follows:

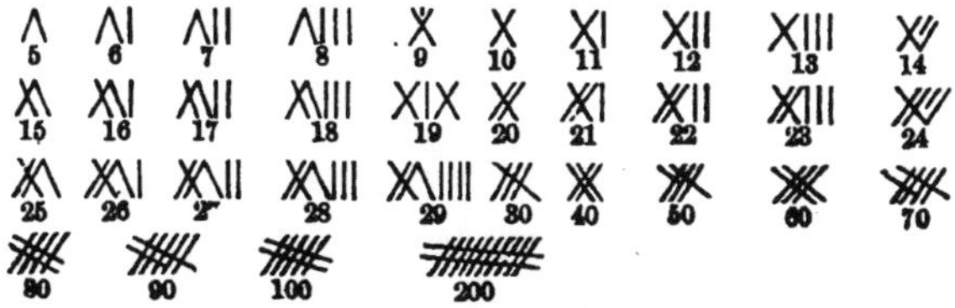

NOTE.—For a full discussion of Arithmetical Language, see the author's *Philosophy of Arithmetic.*

INTRODUCTION TO ADDITION.

MENTAL EXERCISES.

1. If there are 7 boys in one class and 6 in another class, how many boys in both classes?

SOLUTION.—If there are 7 boys in one class and 6 in another class, in both classes there are 7 boys plus 6 boys, which are 13 boys.

2. In a garden there are 9 apple-trees and 10 pear-trees; how many trees in the garden?

3. Anna bought a bonnet for 15 dollars and a cloak for 24 dollars; how much did both cost her?

4. Mary gave 19 apples to one of her schoolmates and 15 to another; how many did she give to both?

5. If a boy gave 75 cents for a pair of skates and 25 cents for a ball, what did they both cost him?

6. A lady gave 10 cents for needles, 14 cents for thread, and 15 cents for muslin; what did she give for all?

7. A farmer sold some wheat for 15 dollars, some corn for 25 dollars, and some oats for 30 dollars; what did he receive?

8. How many are 6 and 21? 9 and 22? 8 and 31? 10 and 17? 4 and 18? 7 and 27? 9 and 28? 12 and 11? 13 and 11?

9. How many are 6 and 17? 8 and 16? 9 and 18? 8 and 19? 7 and 12? 10 and 21? 12 and 20? 14 and 18? 11 and 21?

10. How many are 3 and 14? 4 and 17? 5 and 15? 7 and 17? 6 and 16? 8 and 18? 9 and 19? 11 and 21? 10 and 20?

11. Count by 2's from 2 to 20; from 20 to 40; from 40 to 60; from 1 to 21; from 21 to 41; from 41 to 61.

12. Count by 3's from 3 to 21; from 21 to 42; from 1 to 22; from 22 to 43; from 2 to 23; from 23 to 44.

13. Count by 4's from 4 to 40; from 40 to 80; from 1 to 25; from 25 to 41; from 2 to 30; from 30 to 50; from 3 to 35; from 35 to 55.

14. Count by 5's from 5 to 60; from 1 to 61; from 2 to 62; from 3 to 63; from 4 to 64.

15. Count by 6's, 7's, 8's, 9's, 10's, 11's, and 12's, as in the previous problems.

The uniting of two or more numbers into one sum is called *Addition*. The sign of addition is +, and is read *plus*.

Find the sum of

3+5+4	6+8+5	8+7+6	10+5+7	13+ 6+ 8
4+3+6	3+9+6	7+9+8	11+6+5	14+ 5+ 9
6+7+5	5+7+8	8+9+7	12+7+8	15+ 8+ 7
5+8+2	4+9+7	9+7+8	12+8+9	16+17+18

SECTION II.

FUNDAMENTAL OPERATIONS.

ADDITION.

54. Addition is the process of finding the *sum* of two or more numbers.

55. The **Sum** of several numbers is a number which contains as many units as the numbers added.

56. The **Sign of Addition** is +, and is read *plus*. It denotes that the numbers between which it is placed are to be added.

57. The **Sign of Equality** is =, and is read *equals*. Thus, $10=4+6$ is read 10 *equals* 4 *plus* 6.

NOTES.—1. The *Sign of Addition* consists of two short lines bisecting each other, the one in, and the other perpendicular to, the line of writing.
2. The symbol + was introduced by *Stifelius*, a German mathematician, in a work published in 1544.

PRINCIPLES.

1. *The numbers added must be similar.*
2. *The sum is a number similar to the numbers added.*

CASE I.

58. ***To add when the sum of no column exceeds nine units of that column.***

1. What is the sum of 34, 23, and 32?

OPERATION.
34
23
32
89

SOLUTION.—We write the numbers so that figures of the same order stand in the same column, draw a line beneath, and begin at the right to add. 2 units and 3 units are 5 units, and 4 units are 9 units, which we write under the column of units; 3 tens and 2 tens are 5 tens, and 3 tens are 8 tens, which we write under the column of tens. Hence the sum of 34, 23, and 32 is 8 tens and 9 units, or eighty-nine. Hence the following

Rule.—I. *Write the numbers to be added so that terms of the same order stand in the same column.*

II. *Begin at the right, add each column separately, and write each sum under the column which produced it.*

WRITTEN EXERCISES.

(2)	(3)	(4)	(5)	(6)
526	232	328	437	423
243	456	671	312	536
769				

(7)	(8)	(9)	(10)	(11)
$1.23	$5.42	$12.20	$20.06	$132.25
4.31	1.05	23.25	31.20	245.03
2.03	3.50	30.03	46.52	400.50
7.57				

12. What is the sum of $21.35, $15.02, and $50.61? *Ans.* $86.98.

13. What is the sum of $32.16, $21.50, and $24.23? *Ans.* $77.89.

14. What is the sum of 2131236, 2314010, and 4520652? *Ans.* 8965898.

15. A man gave his son $13.56 and his daughter $46.21; how many dollars did he give to both? *Ans.* $59.77

16. Marie's book contains 3516 words, and Jennie's book 4271 words; how many words in both? *Ans.* 7787.

17. John takes 3250 steps in going to school, and Henry takes 2518 steps; how many do both take? *Ans.* 5768.

18. If Mr. Martin rides 3164 miles in a year and walks 513 miles; how far does he travel in a year? *Ans.* 3677.

19. A house cost $4160, a barn $1702, and a store $2027; required the cost of all. *Ans.* $7889.

20. A book cost $2.25, a kite $1.10 and a ball 62 cents; what was the cost of all? *Ans.* $3.97.

21. A little girl's pulse beats 105216 times in a day, and her mother's pulse beats 93470 times; how many beats do both make in a day? *Ans.* 198686.

22. John bought a Higher Arithmetic for $1.38, a Peterson's Familiar Science for $1.60, and a Loomis's Algebra for $2.00; how much did he pay for all? *Ans.* $4.98.

23. An army consisting of 24,360 men received at one time a reinforcement of 10,220, and at another time of 5400; of how many men did it then consist? *Ans.* 39,980.

CASE II.

59. ***To add when the sum of a column exceeds nine units of that column.***

1. What is the sum of 475, 384, and 896 ?

OPERATION.
475
384
896
1755

SOLUTION.—We write the numbers so that terms of the same order stand in the same column, and begin at the right to add. 6 units and 4 units are 10 units, and 5 units are 15 units, which equal 1 *ten* and 5 *units*; we write the 5 units under the column of units, and add the 1 ten to the column of tens: 1 ten and 9 tens are 10 tens, and 8 tens are 18 tens, and 7 tens are 25 tens, which equal 2 *hundreds* and 5 *tens*; we write the 5 tens under the column of tens, and add the 2 hundreds to the column of hundreds: 2 hundreds and 8 hundreds are 10 hundreds, and 3 hundreds are 13 hundreds, and 4 hundreds are 17 hundreds, which equal 1 *thousand* and 7 *hundreds*; we write the 7 hundreds under the column of hundreds, and place the 1 thousand on the left in the place of thousands. Hence, the sum of 475, 384, and 896 is 1755.

Rule.—I. *Write the numbers to be added so that terms of the same order stand in the same column, and draw a line beneath.*

II. *Begin at the right, add the terms of each column separately, and write the sum underneath, if less than ten.*

III. *When the sum of any column is ten or more than ten, write the units figure only, and add the tens to the next column.*

IV. *Write the entire sum of the last column.*

NOTE.—In adding dollars and cents, dollars must be written under dollars and cents under cents, so that the points may be in a vertical line.

Proof.—Begin at the top and add the columns downward and if the work is correct the two sums will be equal.

SECOND METHOD. Separate the number into two or more parts, add these parts, and then add the sum of these parts; and if the work is correct the two results will be equal.

NOTES.—1. We write figures of the same order in the same column for *convenience* of adding, since only units of the same order can be directly added.

2. We begin at the *right* to add for *convenience*, so that when the sum of any column exceeds 9 we may add the left hand term of such sum to the next column.

WRITTEN EXERCISES.

2. What is the sum of 4326+7428+4675+3857+1231 ?

Ans. 21517.

3. What is the sum of 3213+3562+4572+4207+3208?
Ans. 18762.

4. What is the sum of 2573+3865+2101+5213+1202?
Ans. 14954.

5. What is the sum of 3216+3105+21678+36102?
Ans. 64101.

6. What is the sum of 3120+1356+2013+206+12345?
Ans. 19040.

7 Find the sum of 3123+204+36512+30567+123457.
Ans. 193863.

8. What is the sum of 21650+1456+287657+3608+146075? *Ans.* 460446.

9. Find the sum of 60758+12473+10572+876543+752075+875432. *Ans.* 2587853.

10. What is the sum of 67086+1257321+8752072+516203+2010507? *Ans.* 12603189.

11. What is the sum of 124607+576078+1555+6753209+12460729? *Ans.* 19916178.

12. Find the sum of 623405+328721+6075+2870572+395. *Ans.* 3829168.

13. What is the sum of 351670+8037508+367+12578+246+312? *Ans.* 8402681.

14. Required the sum of 2187587+3057218+207562+17473+312578. *Ans.* 5782418.

15. Find the sum of 875672+3408+13767+870823+138721+85432785. *Ans.* 87335176.

16. What is the sum of 8243+13750+372+82134+28793+876789? *Ans.* 1010081.

17. Required the sum of 3207+28543+2856+28732+87256+387128. *Ans.* 537722.

18. Find the sum of 8723+285768+36124+28765+37124+856078. *Ans.* 1252582.

19. What is the sum of 23456+7891011+121314+151617+1819202? *Ans.* 10006600.

20. Required the sum of 28567+385602+13728+38562+308029+2085703. *Ans.* 2860191.

21. Required the sum of 3142+2813+93104+160532++210356+213165. *Ans.* 683112.

22. Find the sum of 36875+28764+3128+6785+3876+38521+46212. *Ans.* 164161.

23. Find the sum of 41287, 32104, 80010, 678910, 136710, 896343, 257856. *Ans* 2123220

24. Find the sum of 67851, 21856, 38214, 621857, 36724, 128705, 1250641. *Ans.* 2165848.

25. Find the sum of 20875, 318764, 287656, 318567, 208725, 387124, 411403. *Ans.* 1953114.

26. Find the sum of 218765, 865432, 896, 32481, 2879823, 8128765, 1759824. *Ans.* 13885986.

27. Find the sum of 61857, 30876, 14687, 318572, 856, 30876, 12345, 678910, 328764. *Ans.* 1477743.

MENTAL EXERCISES.

1. Rose gave 15 cents to her brother and 8 cents to her sister; how much did she give to both?

SOLUTION.—If Rose gave 15 cents to her brother and 8 cents to her sister, to both she gave 15 cents plus 8 cents, which are 23 cents.

2. What is the cost of a suit of clothes if the material cost $25 and the making cost $10?

3. If a boy has 12 cents and then earns 15 cents, how many cents will he then have?

4. The head of a fish is 6 inches long, the tail 8 inches, and the body 12 inches; what is the length of the fish?

5. How many are 18 and 6? 17 and 8? 21 and 7? 25 and 8? 36 and 8? 42 and 7? 48 and 6? 57 and 7?

6. How many are 8 and 24? 9 and 26? 7 and 35? 6 and 48? 5 and 47? 4 and 48? 3 and 68?

7. How many are 105 and 6? 108 and 7? 115 and 8? 124 and 6? 135 and 5? 140 and 7? 155 and 8?

8. How many are 6 and 135? 5 and 206? 7 and 208? 4 and 212? 7 and 234? 8 and 326? 300 and 400?

9. Howard shot 11 robins, Howell shot 12 robins, and Hoyt shot 13 robins; how many robins did they all shoot?

10. A boy bought a robin for 10 cents, a jay for 20 cents, and a wren for 30 cents; what did they all cost?

11. A lady gave 6 cents for needles, 12 cents for thread, and 25 cents for some muslin; what was the entire cost?

12. What is the sum of 28 and 37?

SOLUTION.—28 and 30 are 58, and 7 are 65.

13. What is the sum of 25 and 38? 27 and 36? 45 and 28? 47 and 65? 67 and 48? 73 and 48?

14. What is the sum of 29 and 46? 28 and 38? 76 and 46? 85 and 91? 83 and 48? 96 and 69? 125 and 43? 173 and 205?

15. Add by 7's from 7 to 98; from 1 to 99; from 2 to 100; from 3 to 101. Add by 8's from 8 to 120; from 2 to 122; from 3 to 123.

16. Add by 9's from 0 to 108; from 1 to 109; from 2 to 110; from 3 to 111. Add by 11's from 0 to 132; from 1 to 133; from 2 to 134.

17. Add by 12's from 0 to 144; from 1 to 145; from 2 to 146; from 3 to 147; by 13's from 13 to 130; from 1 to 131; etc.

18. Add alternately by 4 and 5 till the sum exceeds 100; by 5 and 6; by 6 and 7; by 8 and 9; by 9 and 10; by 11 and 12.

19. A merchant sold some rice for $15, some sugar for $17, and some flour for $22; what was the amount of the sale?

WRITTEN EXERCISES.

1. A farmer raised 857 bushels of wheat, 372 bushels of rye, 720 bushels of oats, and 876 bushels of corn; how many bushels of grain did he raise?

SOLUTION.—If he raised 857 bushels of wheat, 372 bushels of rye, 720 bushels of oats, and 876 bushels of corn, he raised in all the sum of 857 bushels, 372 bushels, 720 bushels, and 876 bushels, which we find by adding, to be 2825 bushels.

OPERATION

$$\begin{array}{r} 857 \\ 372 \\ 720 \\ 876 \\ \hline 2825 \end{array}$$

2. A lady owns a farm worth $8326, a house worth $6575, a barn worth $3818, and $7584 in State bonds; what is she worth? *Ans.* $26303.

3. A grocer paid $897 for flour and $759 for sugar; for what must both be sold that the grocer may gain $125 on the transaction? *Ans.* $1781.

4. In a graded school there are 29 pupils in the first section, 42 in the second, 38 in the third, 45 in the fourth, 51 in the fifth, and 63 in the sixth; how many pupils are there in the whole school? *Ans.* 268 pupils.

5. A merchant "checked" out of bank $2560 one day, $3065 the next, $2780 the next, and then had $5760 in bank; how much had he in bank at first? *Ans.* $14165.

6. A dealer bought 450 barrels of wheat flour for $2700, and 286 barrels of rye flour for $1416; how many barrels did he buy *and what* was the whole cost? Ans. $4116

7. A gentleman dying left his eldest son $7286, the next younger son $9111, and the youngest as much as both the others; what was the fortune? *Ans.* $32794.

8. Bought calico for a dress for $2, a neck-tie for $0.45, a parasol for $3.50, a pair of boots for $4.25, and a pair of gloves for $1.75; what was the whole cost? *Ans.* $11.95.

9. A speculator bought three city lots for $5750 each; for what must he sell them that he may gain $2367 by the investment? *Ans.* $19617.

10. Mr. Bowman paid $628 for flour, $382 for sugar, and $125 for potatoes; how much did he receive for all if he sold them at a gain of $324? *Ans.* $1459.

11. A boy bought an arithmetic for 95 cents, a grammar for 65 cents, a history for 75 cents, a geography for $1.88, a piece of India-rubber for 5 cents, and a pencil for 3 cents; what was the whole cost? *Ans.* $4.31.

12. A merchant pays his book-keeper $1500 a year, three salesmen $750 each, a porter $450, and a boy $200; what amount of salaries does he pay? *Ans.* $4400.

13. A young man paid $450 for a horse, $115 for a sleigh, $137.50 for a harness and bells, and $33.35 for a robe; what was the cost of the whole outfit? *Ans.* $735.85.

14. At an auction a woman bid off a marble-topped bureau at $19.50, a dozen cane-seat chairs at $15.00, a carpet at $31, a small rug at $1.50, and a dozen china dinner plates at $1.75; what was her bill? *Ans.* $68.75.

15. A schooner sailed from Milwaukee to Erie, having on board 12391 bushels new No. 3 Milwaukee wheat, 11420 bushels Mixed Spring Minnesota wheat, and 1239 bushels common Winter Red Western wheat; of how many bushels did her cargo consist? *Ans.* 25050 bushels.

16. A dry goods merchant, in purchasing his stock, paid $650 for silks, $875 for woolens, $345.75 for cotton dress goods, and $795.45 for other goods; what did his stock cost? *Ans.* $2666.20.

17. Mr. Anthony's balance in bank on Monday morning *was $1546.75;* that day he deposited $500; Tuesday he deposited $350; Wednesday, $764.75; Thursday, $250; Friday, $640; Saturday, $54; how much did he have on deposit on the next Monday morning? *Ans.* $4105.50.

18. A builder bought a lot for $450, built upon it a house costing $3545, and a barn and carriage-house costing $847.50, fenced it at a cost of $127.50, and graded it at a cost of $77.25; for what must he sell the property in order to gain upon it $540? *Ans.* $5587.25.

19. A Normal school paid, in 1875, $18,622.17 for salaries of professors and teachers, $1350 for salaries of other officers, $34,721.03 for board, washing, etc., $11,806.95 for servants' hire, and $7159.15 for other expenses; what were the expenditures for the year? *Ans.* $73,659.30

20. In erecting an academy, the trustees paid $250 for digging the cellar, $580 for laying the foundation walls. $12,575 for the brickwork, and $10,650 for the woodwork, plastering, etc.; what was the entire cost? *Ans.* $24,055.

21. Mr. Johnson's house cost for brick $450; for lumber, $780; other materials, $350; digging the cellar, $87.50; masons' and bricklayers' work, $425; carpenters' work, $789.75; painting, glazing, etc., $350; what was the whole cost of the house? *Ans.* $3232.25.

22. In the year ending June 1, 1875, Pennsylvania expended for school-houses, $2,059,464.83; for teachers' wages, $4,746,875.52; for fuel and contingent expenses, $2,448,315 78; for other expenses, $109,270.94; $77.324.32 extra appropriation to city of Pittsburg; $85,815.84 for Normal schools; and $423,693.76 for Soldiers' Orphan schools; what was the whole amount expended that year for education by the State? *Ans.* $9,950,760.99.

23. The cash value of farms in New York in 1870 was $1,272,857,766; in Pennsylvania, $1,043,481,582; in New Jersey, $257,523,376; in Ohio, $1,054,465,226; in Indiana, $634,804,189; in Illinois, $920,506,346; in Missouri, $392,908,047; in Wisconsin, $300,414,064; and in Michigan, $398,240,578; what is the total value of the farms in these States? *Ans* $6,275,201,174.

CONTRACTIONS IN ADDITION.

60. Contractions in Addition are abbreviated methods of adding.

61. There are several methods of abbreviating the process of addition, a few of which will be stated:

1. *In adding omit naming the numbers added; merely name the results.*

2. *When two or more terms of a column can be easily grouped together, use their sum instead of adding each separately; combining with especial reference to* TENS.

3. *When a term is repeated several times in a column, multiply it by the number of times it is repeated, and use the result.*

1. What is the sum of 8752+3687+2573+8576+2857+6872?

1st Method.—We add thus: 2, 9, 15, 18, 25, 27; write the 7 and carry the 2; 2, 9, 14, 21, 28, 36, 41; write the 1 and carry the 4; etc.

2d Method.—Thus: 9, 15, 25, 27, in which the 3 and 7 are grouped and used as 10. We may also group 2 and 6, 7 and 3, and then 7 and 2; thus, 8, 18, 27, etc.

3d Method.—In the second column, we may take the *three* 7's, then the *two* 5's, and then the 8; thus, 21, 31, 39, etc.

OPERATION.

8752
3687
2573
8576
2857
6872
33317
442

WRITTEN EXERCISES.

(2)	(3)	(4)	(5)	(6)
67854	41235	32106	$875.67	$365.75
61387	47368	21876	321.08	406.26
57854	47376	52382	345.67	57.38
87543	51847	36875	891.01	693.84
78587	63529	42356	121.31	746.38
18234	17152	52873	415.16	25.06
34387	50381	61874	178.20	8.72
21857	27596	18027	222.32	45.50
31242	38273	25718	425.26	482.38
82356	54368	43281	272.82	29.45
14181	70305	25728	930.35	8.65
82815	75056	71890	363.73	682.50
71281	18293	12134	839.46	14.75
85427	52738	51617	876.54	937.18
18315	25382	82365	324.36	8.28
61038	32728	57634	484.95	473.54

INTRODUCTION TO SUBTRACTION.

MENTAL EXERCISES.

1. If I have 12 books and sell 5 of them, how many books shall I have remaining?

SOLUTION.—If I have 12 books and sell 5 of them, I have remaining 12 books minus 5 books, which are 7 books.

2. James had 14 cents and spent 9 of them; how many cents had he remaining?

3. Susan has 25 plums and Jane 13; how many more has Susan than Jane?

4. In a school numbering 45 pupils, 15 are absent; how many pupils are present?

5. A watch was bought for 28 dollars and sold for 21 dollars; what was the loss?

6. A cow was bought for 18 dollars and sold for 28 dollars; what was the gain?

7. Begin at 2 and count by 2's to 40; begin at 40 and count by 2's backward to 2.

8. Begin at 45 and count by 3's backward to 3; begin at 44 and count by 3's backward to 2.

9. Count by 4's from 48 back to 4; from 55 back to 3; from 54 back to 2; from 53 back to 1.

10. Count by 5's from 60 back to 5; from 64 back to 4; from 63 back to 3; from 62 back to 2; from 61 back to 1.

11. In a similar manner begin at different numbers, and count backward by 6's, 7's, 8's and 9's.

12. Take the number 3, add 5, subtract 6, add 7, subtract 5, add 8, subtract 7, add 9, subtract 4, and name the result.

13. Take the number 11, add 4, subtract 5, add 5, subtract 4, add 6, subtract 5, add 7, subtract 6, add 8, subtract 7, and name the result.

14. How many are 3 plus 5 minus 7? 4 plus 7 minus 8? 5 plus 6 minus 4? 8 plus 5 minus 9? 9 plus 10 minus 12? 12 plus 13 minus 15?

15. How many are 8 plus 12 minus 13? 9 plus 16 minus 14? 10 plus 15 minus 16? 16 plus 17 minus 18? 18 plus 19 minus 20?

The process of finding the *difference* between two numbers is called *subtraction*. The sign of subtraction is —, and is read *minus*.

16. Required the value of the following:

5+3—2	2+5—4	9+8—6	10+ 7—12	13+14—15
6+4—3	3+6—5	8+6—7	11+ 8—13	14+15—17
7+2—4	4+7—8	7+5—4	12+ 7— 9	15+18—19
8+5—6	6+8—7	6+8—3	8+14—10	16+17—12
9+3—5	8+7—9	5+9—4	9+15—16	17+19—20

SUBTRACTION.

62. Subtraction is the process of finding the *difference* between two numbers.

63. The **Difference** between two numbers is a number which added to the less, equals the greater.

64. The **Minuend** is the number from which we subtract. The **Subtrahend** is the number to be subtracted.

65. The **Sign of Subtraction** is —, and is read *minus*. It denotes that the number immediately following it is to be subtracted from the number preceding it.

NOTES 1. The *Sign of Subtraction* is a short line in the line of writing.

2. The symbol — was introduced by *Stifelius*, a German mathematician, in a work published in 1544.

PRINCIPLES.

1. *The minuend and subtrahend must be similar numbers.*

2. *The difference is a number similar to the minuend and subtrahend.*

CASE I.

66. ***To subtract when no term of the subtrahend is greater than the corresponding term of the minuend.***

1. What is the difference between 486 and 243?

OPERATION.

$$\begin{array}{r} 486 \\ \underline{243} \\ 243 \end{array}$$

SOLUTION.—We write the subtrahend under the minuend, placing terms of the same order in the same column, draw a line beneath, and begin at the right to subtract. 3 units from 6 units leave 3 units, which we write under the units; 4 tens from 8 tens leave 4 tens, which we write under the tens; 2 hundreds from 4 hundreds leave 2 hundreds, which we write under the hundreds. Therefore, the difference between 486 and 243 is 243.

Rule.—I. *Write the subtrahend under the minuend, placing terms of the same order in the same column, and draw a line beneath.*

II. *Begin at the right and subtract each term of the subtrahend from the corresponding term of the minuend, writing the remainder beneath.*

WRITTEN EXERCISES.

	(2)	(3)	(4)	(5)
From	364	487	524	876
Subtract	123	235	321	235

	(6)	(7)	(8)	(9)
From	387	369	879	872
Subtract	215	123	354	351

	(10)	(11)	(12)	(13)
From	895	718	\$38.57	\$51.83
Subtract	713	315	12.35	21.62

	(14)	(15)	(16)	(17)
From	8917	8975	\$37.59	\$75.68
Subtract	5302	3623	25.36	34.15

	(18)	(19)	(20)	(21)
From	6753	8917	7369	\$98.76
Subtract	5241	7214	2134	54.32

22. Subtract 34512 from 67856. *Ans.* 33344.
23. Subtract 41231 from 81275. *Ans.* 40044.
24. Subtract 32125 from 96576. *Ans.* 64451.
25. Subtract 14114 from 85237. *Ans.* 71123.
26. Subtract 23121 from 64875. *Ans.* 41754.
27. Subtract 23254 from 48796. *Ans.* 25542.
28. Subtract \$123.42 from \$9876.54. *Ans.* \$9753.12.
29. Subtract \$2413.25 from \$7654.87. *Ans.* \$5241.62.

CASE II.

67. ***To subtract when one or more terms of the subtrahend are greater than the corresponding terms of the minuend.***

68. There are two methods of explaining this case, called the *Method of Borrowing* and the *Method of Adding Ten.* We will solve the same problem by both methods

1. From 836 subtract 472.

OPERATION.
836
472
364

SOLUTION BY BORROWING.—We write the subtrahend under the minuend, and begin at the right to subtract. 2 units from 6 units leave 4 units, which we write under the units; we cannot take 7 tens from 3 tens, we will therefore take 1 hundred from the 8 hundreds, and add it to the 3 tens; 1 hundred equals 10 tens, which, added to 3 tens, equals 13 tens; 7 tens from 13 tens leave 6 tens, which we write in tens place; 4 hundreds from 7 hundreds (the number remaining after taking away 1 hundred) leave 3 hundreds, which we write in *the hundreds place.*

SOLUTION BY ADDING TEN.—2 units from 6 units leave 4 units; we cannot take 7 tens from 3 tens, we will therefore add 10 tens to the 3 tens, making 13 tens; 7 tens from 13 tens leave 6 tens; now, since we have added 10 tens, or 1 hundred, to the minuend, our remainder will be 1 hundred too large; hence to obtain the correct remainder we must add 1 hundred to the subtrahend; 1 hundred and 4 hundreds are 5 hundreds; 5 hundreds from 8 hundreds leave 3 hundreds.

Rule.—I. *Write the subtrahend under the minuend, placing terms of the same order in the same column, and draw a line beneath.*

II. *Begin at the right and subtract each term of the subtrahend from the corresponding term of the minuend, writing the remainder beneath.*

III. *If any term of the subtrahend is greater than the corresponding term of the minuend, add* 10 *to the latter, and then subtract.*

IV. *Add* 1 *to the next term of the subtrahend (or subtract* 1 *from the next term of the minuend), and proceed as before.*

Proof.—Add the difference to the subtrahend; and if the work is correct the sum will equal the minuend.

SECOND METHOD.—Subtract the difference from the minuend, and if the work is correct, the result will equal the subtrahend.

NOTE.—The taking one from a term of the minuend is called *borrowing* and the adding one to the next term of the subtrahend is called *carrying*.

WRITTEN EXERCISES.

	(2)	(3)	(4)	(5)
From	386	462	$5.23	$6.15
Take	157	175	1.46	1.47
	(6)	(7)	(8)	(9)
From	3123	6357	7518	4712
Take	1415	2829	2036	3508
	(10)	(11)	(12)	(13)
From	4075	3987	3013	6514
Take	2867	1989	2187	2823
	(14)	(15)	(16)	(17)
From	$571.23	$605.32	$657.00	$451.00
Take	278.51	518.18	472.83	132.18

18. From 94278 subtract 62573. *Ans.* 31705.
19. From 70532 subtract 25824. *Ans.* 44708.
20. From 53715 subtract 15186. *Ans.* 38529.
21. From 80526 subtract 25367. *Ans.* 55159.
22. From 75138 subtract 61859. *Ans.* 13279.
23. From $200.87 subtract $123.48. *Ans.* $77.39

What is the value

24. Of 12857+3659—2768? *Ans.* 13748.
25. Of 34821+3127—5879? *Ans.* 32069.
26. Of 8000732—5001916? *Ans.* 2998816.
27. Of 1000000+10000—1? *Ans.* 1009999.
28. Of 8032—6257+18765? *Ans.* 20540.
29. Of 2875—2382+67876? *Ans.* 68369.
30. Of $2.85—$1.86+$3.92—$2.56? *Ans.* $2.35.

31. From ten thousand take nine hundred and ninety-nine. *Ans.* 9001.

32. From one million and one, take three thousand and three. *Ans.* 996998.

33. From four hundred thousand take four hundred and forty-four. *Ans.* 399556.

34. From seventy-seven thousand take seven thousand and seventy-seven. *Ans.* 69923.

35. From one hundred million and one take one million and five. *Ans.* 98999996.

36. From 6 billion, 6 million, 6 thousand and 6, take 80 million, 80 thousand and 8. *Ans.* 5925925998.

MENTAL EXERCISES.

1. Mary bought 20 apples and gave Ella 6 of them; how many did Mary then have?

SOLUTION.—If Mary bought 20 apples and gave Ella 6 of them, Mary then had the difference between 20 apples and 6 apples, which is 14 apples

2. A boy found 16 eggs in the barn and broke 8 of them; how many remained unbroken?

3. How many birds will escape if I shoot 6 birds out of a flock of 40 birds?

4. How many are 17 less 8? 14 less 9? 21 less 7? 23 less 6? 25 less 8? 31 less 4?

5. How many are 32 less 3? 41 less 2? 50 less 6? 49 less 7? 78 less 8? 100 less 9? 124 less 12?

6. Required the value of 112 minus 8; 120 minus 7; 125 minus 8; 132 minus 5; 140 minus 4; 150 minus 5; 150 minus 10.

7. A boy having 12 apples, bought 6 more, and then sold 8; how many had he then?

8. Mary having 14 cents, spent 8 cents, after which her father gave her 12 cents; how many cents had she then?

9. Edwin had $24 and earned $6, and then spent $12; how much did he then have?

10. A merchant gave $65 for goods, and paid $6 for transportation; at what price must he sell the goods to gain $12?

11. Subtract 36 from 84.

SOLUTION.—30 from 84 leaves 54, and 6 from 54 leaves 48.

12. Subtract 26 from 42; 35 from 52; 38 from 60; 42 from 75; 45 from 72; 48 from 84; 52 from 86; 56 from 82.

13. Subtract 123 from 147; 135 from 248; 152 from 285; 206 from 456; 307 from 444; 328 from 565.

14. In a school numbering 75 pupils 28 are absent; how many pupils are present?

15. A lady bought a watch for $250 and sold it at a loss of $35: what did she receive for it?

16. A man bought a horse for $175, paid $15 for his keeping, and sold him for $240; what was the gain?

17. A lady bought some needles for 8 cents, some thread for 15 cents, and some tape for 18 cents, and hands the clerk a 50-cent note: what is her change?

18. A man went into a clothing store and bought a vest for $6, a coat for $25, and a pair of pants for $12; he hands the clerk 5 ten dollar bills; what change will he receive?

19. Count by 6's from 6 to 72 and back; from 5 to 77 and back from 4 to 76 and back; from 3 to 75 and back; etc.

20. Count by 7's from 7 to 84 and back; from 6 to 90 and back from 5 to 89 and back; from 4 to 88 and back; etc.

21. Count by 8's from 0 to 96 and back; from 1 to 97 and back, from 2 to 98 and back; from 3 to 99 and back; etc.

22. Count by 9's from 0 to 108 and back; from 1 to 109 and back from 2 to 110 and back; from 4 to 112 and back; etc.

23. Take the number 12, add 3, subtract 2, add 4, subtract 3, add 5, subtract 4, add 6, subtract 5, add 7, subtract 6, and name the result.

24. How many are 12 plus 8, minus 7, plus 10, minus 6, plus 12, minus 9, plus 14, minus 18, plus 25, minus 16?

25. How many are 25 plus 20, minus 16, plus 36, minus 40, plus 56, minus 25, plus 72, minus 84, plus 120, minus 140?

WRITTEN EXERCISES.

1. A man had 725 bushels of wheat and sold 367 bushels; how many bushels remained?

SOLUTION.—If a man had 725 bushels of wheat and sold 367 bushels, there remained the difference between 725 bushels and 367 bushels, which is 358 bushels.

OPERATION.
725
367
358

2. A farmer raised 627 bushels of potatoes, and sold 386 bushels; how many bushels remained? *Ans.* 241 bushels.

3. From a farm containing 3075 acres, 2528 acres were sold; how many acres remained? *Ans.* 547 acres.

4. A man borrowed $3052 and afterward returned $2527; how much does he still owe? *Ans.* $525.

5. A sold his house for $2572, which was $285 more than it cost; how much did it cost? *Ans.* $2287.

6. Two men together have $5682; how much money has the first if the second has $3826? *Ans.* $1856.

7. A man bought a farm for $10,852, and sold it at a loss of $987; what did he receive for it? *Ans.* $9865.

8. A and B together own 10,840 acres of land; how much does B own if A owns 5284 acres? *Ans.* 5556 acres.

9. Gen. Zachary Taylor was born in 1790 and died in 1850; how old was he when he died? *Ans.* 60 years.

10. Robert Southey was born at Bristol in 1774 and died in 1843; how old was he at his death? *Ans.* 69 years.

11. Socrates was born at Athens 468 years before Christ, and died 398 years before Christ; what was his age at death? *Ans.* 70 years.

12. Walter Scott was born in the year 1771, and died in 1832; how old was he at his death? *Ans.* 61 years.

13. How many years from the birth of Milton in 1608 to the birth of George Washington in 1732? *Ans.* 124 years.

14. A farmer having 478 cows, sold 198 of them, and then bought 226; how many had he then? *Ans.* 506.

15. How many years from the birth of William Penn in 1644 to the birth of Benjamin Franklin in 1706? *Ans.* 62 years.

16. A merchant having 336 pounds of rice, bought 248 pounds, and then sold 467 pounds; how many pounds remained? *Ans.* 117 pounds.

17. Lanterns were invented by King Alfred in 890; how many years from that time to 1876? *Ans.* 986 years.

18. Spectacles were invented by Spina in 1299; how many years from their invention to 1878? *Ans.* 579 years.

19. Watches were invented at Nuremberg in 1477; how many years from that time to 1876? *Ans.* 399 years.

20. The circulation of the blood was discovered by Harvey in 1619; how long since that time?

21. The telescope was invented by Galileo in 1610; how many years since it was invented?

22. Printing is supposed to have been invented in 1441; how many years has the art been used?

23. The first steamship crossed the ocean in 1839; how long is it since that time?

24. The first line of telegraph was established in the United States in 1844; how long has the invention been in use?

25. The distance from Philadelphia to Pittsburgh is 353 miles, and to Harrisburg is 105 miles; what is the distance from Harrisburg to Pittsburgh? *Ans.* 248 miles.

26. A merchant had in bank $4500, and "checked out" $2345; how much will remain in bank? *Ans.* $2155.

27. An architect received $54,750 for building a market-house, and he expended upon it $51,784; what was his profit? *Ans.* $2966.

28. Mr. Newton bought a house for $10,575, and paid for it in yearly installments of $1057.50; how many years did it require to pay for the house? *Ans.* 10 years.

29. At an election the successful candidate received 40,172 votes, and the unsuccessful candidate 34,789 votes; what was the successful candidate's majority?

Ans. 5383 votes.

CONTRACTIONS IN SUBTRACTION.

69. Contractions in Subtraction are abbreviated methods of subtracting.

70. When two or more numbers are to be successively subtracted from another, the operation can be abbreviated by the following rule:

Rule.—*Add the terms of the numbers to be subtracted, and write for a remainder a term which, added to this sum, will give a number having for its unit term the corresponding term of the minuend.*

1. Subtract 324 and 549 from 1024.

SOLUTION.—4 and 9 are 13, and since 1 more will make 14, we write 1 as the first term of the remainder; 4 and 1 are 5 and 2 are 7, and since 5 more make 12, we write 5 in the remainder; 5 and 1 are 6 and 3 are 9, and since 1 more make 10, we write 1 in the remainder.

OPERATION.

1024
324
549
151

WRITTEN EXERCISES.

(2)	(3)	(4)	(5)
5765	85741	69547	1000000
1334	13333	14937	676543
1562	14792	18356	135792
	10234	3344	187665

6. A raised 785 bushels of grain, and having sold B 250 bushels, C 320 bushels, and D 169 bushels, he retained the remainder; how many bushels did he retain? *Ans.* 46 bu.

7. From a tract of land in a newly settled country containing 4850 acres, there were sold at one time 1560 acres; at another, 2490 acres; and at another 500 acres; how much of the tract then remained? *Ans.* 300 acres.

8. In five granaries there are 7874 bushels of wheat; in the first there are 1160 bushels; in the second 1729 bushels; in the third 1957 bushels; in the fourth 1433 bushels; how many bushels are contained in the fifth? *Ans.* 1595 bu.

9. An estate of $10,000 was divided as follows: the widow received $4397; the youngest son, $2498; the daughter, $3104; and the eldest son the remainder. What was the share of the latter?

WRITTEN EXERCISES.

IN ADDITION AND SUBTRACTION.

1. An estate worth $35,000 has two mortgages upon it, one for $15,000, and the other for $8000; what is it worth above the incumbrances? *Ans.* $12,000.

2. A merchant's gross profits last year were $9000; he pays for rent $1200, insurance $150, salaries $3500, and incidental expenses, $225; what is his net profit? *Ans.* $3925.

3. A clerk receives $800 a year; he pays $150 a year for house rent, his butcher's bill is $177.50 and grocer's bill $145.45, and he has expended for clothes and other expenses $250; what does he save in the year? *Ans.* $77.05.

4. Mr. Wainwright bought a farm which cost him $25,000; but afterwards getting into difficulties, he burdened the property with a mortgage for $7500; for what must he sell the farm to gain $3500? *Ans.* $21,000.

5. Mr. Hatton receives for goods sold a check for $15,495.25 on the Girard National Bank; he wishes to deposit $13,500 in the same bank; what will the bank return him when he presents the check? *Ans.* $1995.25.

6. I had in bank on Monday morning, Dec. 4, $11,275, I checked out Monday afternoon $745; Tuesday, I deposited $1500; Wednesday, I deposited $475; Thursday, I checked out $873.50; Friday, I again checked out $563.75; Saturday, I deposited $560; what is the amount of my deposit on Monday morning, Dec. 11? *Ans.* $11,627.75.

7. The day before Christmas I went into a fancy store and bought for Christmas presents, a writing-desk at $4.25, a work-box at $2.75, a glove-box at $1.50, a handkerchief-box at $1.75, a case of perfumery at $2.25, and a gentleman's dressing-case at $5.75; I handed the clerk a twenty dollar bill; what change should I receive? *Ans.* $1.75.

8. A farmer took to the store 2 dozen eggs, $0.70; 10 pounds of butter, $5; 2 barrels of apples, $11.50; he bought 25 pounds crushed sugar, $3.50; 2 gallons molasses, $0.60; 10 yards red flannel, $3.75, and one piece unbleached sheeting, $10: how much cash must he pay? *Ans.* $0.65.

INTRODUCTION TO MULTIPLICATION.

MENTAL EXERCISES.

1. Repeat the table of 2 times; of 3 times; of 4 times; of 5 times of 6 times; of 7 times; of 8 times; etc.

2. If one barrel of flour costs 6 dollars, what will 5 barrels cost at the same rate?

SOLUTION.—If one barrel of flour costs 6 dollars, 5 barrels will cost 5 times 6 dollars, which are 30 dollars.

3. What will 7 gold pens cost at 3 dollars apiece?

4. At 20 cents apiece what will 6 spelling-books cost?

5. What is the cost of 11 yards of calico at 12 cents a yard?

6. George has 15 dollars and James has 6 times as much; how many dollars has James?

7. A farmer bought 6 cows at 25 dollars apiece; how much did he pay for them?

8. If 6 men can do a piece of work in 20 days, how long will it take 1 man to do it?

9. If 8 men mow a field of grass in 15 days, how long will it take 1 man to mow it?

10. In an orchard there are 16 rows of trees and 10 trees in each row; how many trees are there in the orchard?

11. John borrowed 12 cents from a friend, and then earned 6 times as much as he borrowed; how many cents had he then?

12. Arnold earned 18 dollars a week, and paid 9 dollars a week for his board; how much could he save in 8 weeks?

13. How many are 3 times 5, plus 6? 5 times 6, plus 7? 6 times 7, plus 8? 7 times 8, plus 9? 9 times 10, plus 11? 12 times 10, plus 13?

14. How many are 4 times 6, minus 8? 6 times 7, minus 5? 8 times 9, minus 10? 7 times 11, minus 9? 9 times 12, minus 13?

15. How many are 6 and 6 times 7? 7 and 7 times 9? 8 and 8 times 10? 9 and 9 times 8? 10 and 10 times 11? 11 and 11 times 12?

16. The process of taking one number as many times as there are units in another is called *multiplication*. The number multiplied is the *multiplicand*; the number by which we multiply is the *multiplier*; the result is the *product*. The sign of multiplication is ×, and is read "multiplied by."

17. Read the result of the following:

$5\times4+3$	$7\times3+2$	$9\times7+12$	$12\times9-18$	$12\times12-24$
$6\times5+4$	$6\times4-5$	$8\times6-11$	$10\times8+16$	$13\times6-18$
$7\times6+5$	$5\times5+7$	$7\times8-10$	$11\times4-15$	$14\times7-27$
$8\times7+6$	$4\times6-8$	$8\times4+12$	$13\times3-9$	$16\times5-30$
$9\times8+7$	$3\times7+5$	$9\times3-10$	$14\times4-12$	$17\times3-21$

MULTIPLICATION.

71. **Multiplication** is the process of finding the *product* of two numbers.

72. The **Product** of two numbers is the result obtained by taking one number as many times as there are units in another.

73. The **Multiplicand** is the number to be multiplied.

74. The **Multiplier** is the number by which we multiply.

75. The **Sign of Multiplication** is ×, and is read *multiplied by*, *times*, or *into*. When placed between two numbers it denotes that one is to be multiplied by the other.

NOTES.—1. The *Sign of Multiplication* consists of two short lines of equal length, bisecting each other at an angle of 45 degrees with the line of writing.

2. The symbol × was introduced by *Wm. Oughtred*, an English mathematician, born in 1574.

PRINCIPLES.

1. *The multiplier is always an abstract number.*
2. *The multiplicand may be abstract or concrete.*
3. *The product is always similar to the multiplicand.*

CASE I.

76. ***When the multiplier is not greater than twelve.***

1. Multiply 353 by 8.

OPERATION.

Multiplicand,	354
Multiplier,	8
Product,	2832

SOLUTION.—We write the multiplier under the multiplicand, draw a line beneath, and begin at the right to multiply. 8 times 4 units are 32 units, or 2 units and 3 tens; we write the units in units place, and reserve the tens to add to the next product: 8 times 5 tens are 40 tens, plus the 3 tens equal 43 tens, or 3 tens and 4 hundreds; we write the 3 tens in tens place, and reserve the 4 hundreds to add to the next product: 8 times 3 hundreds are 24 hundreds, plus the 4 hundreds equal 28 hundreds, or 8 hundreds and 2 thousands, which we write in their proper places, and we have 2832. Hence the following

Rule.—I. *Write the multiplier under the multiplicand, and draw a line beneath.*

II. *Begin at the right, and multiply each term of the multiplicand by the multiplier, carrying as in addition.*

WRITTEN EXERCISES.

(2)	(3)	(4)	(5)
567	682	795	821
3	4	2	7
1701	2728	1590	5747

(6)	(7)	(8)	(9)
$3.07	$5.86	$12.58	$25.08
8	6	5	9
$24.56	$35.16	$62.90	$225.72

Multiply

10. 5708 by 8.	*Ans.* 45664.	**16.** 32187 by 3.	*Ans.* 96561.
11. 3218 by 7.	*Ans.* 22526.	**17.** 58799 by 6.	*Ans.* 352794.
12. 2107 by 6.	*Ans.* 12642.	**18.** 25767 by 5.	*Ans.* 128835.
13. 5189 by 7.	*Ans.* 36323.	**19.** 41937 by 4.	*Ans.* 167748.
14. 12076 by 4.	*Ans.* 48304.	**20.** 82709 by 8.	*Ans.* 661672.
15. 21876 by 5.	*Ans.* 109380.	**21.** 70095 by 9.	*Ans.* 630855.

22. Multiply 97548 by 9. *Ans.* 877932.

23. Multiply 847946 by 9. *Ans.* 7631514.

24. If one acre of land is worth 275 dollars, what are 7 acres worth at the same rate?

SOLUTION.—If one acre of land is worth 275 dollars, 7 acres are worth 7 times 275 dollars, which we find by multiplying is 1925 dollars. Therefore, etc.

OPERATION.
275
7
1925 *Ans.*

25. Sound moves 1120 feet in a second; how far will it move in six seconds? *Ans.* 6720 feet.

26. If a horse cost $150, what will 8 horses cost at the same rate? *Ans.* $1200.

27. If there are 1375 letters on one page of a book, how many letters on 7 pages at the same rate? *Ans.* 9625.

28. If there are 231 cubic inches in one gallon, how many cubic inches in five gallons? *Ans.* 1155.

29. Light moves about 192,500 miles a second; how far will it move in four seconds? *Ans.* 770,000.

30. The distance of the moon from the earth is about 240,000 miles; what is nine times the distance?

Ans. 2,160,000.

CASE II.

77. ***When the multiplier is greater than twelve.***

1. Multiply 465 by 37.

SOLUTION.—We write the multiplier under the multiplicand, units under units, tens under tens, etc. Since 37 equals 7 units and 3 tens, it is evident that 37 times a number equals 7 times the number plus 3 tens times the number. Seven times 465 equals 3255; 3 times 465 equals 1395, hence 3 tens times 465 equals 1395 tens. Taking the sum of the partial products we have 17205. Hence, etc.

OPERATION.

```
  465
   37
 ----
 3255
1395
-----
17205
```

Rule.—I. *Write the multiplier under the multiplicand, placing terms of the same order in the same column, and draw a line beneath.*

II. *Begin at the right, and multiply the multiplicand by each term of the multiplier, writing the first term of each product under the term of the multiplier which produces it.*

III. *Add the partial products, and their sum will be the entire product.*

Proof.—Multiply the multiplier by the multiplicand; if the work is correct this product will equal the first product.

NOTES.—1. When there are ciphers between the significant terms of the multiplier, pass over them and multiply by the significant terms alone.

2. We begin at the right to multiply, so that when any product exceeds nine we may add the number expressed by the left hand figure to the next product.

2. Multiply 247 by 24; also by 204.

OPERATION.

```
 247
  24
 ---
 988
494
----
5928
```

OPERATION.

```
  247
  204
  ---
  988
494
-----
50388
```

WRITTEN EXERCISES.

(3)	(4)	(5)	(6)
354	234	314	321
14	15	16	18
4956	3510	5024	5778

(7)	(8)	(9)	(10)
$4.16	$4.17	$2.45	$2.18
21	23	24	25
$87.36	$95.91	$58.80	$54.50

11. Multiply 372 by 26.	*Ans.* 9672.
12. Multiply 274 by 29.	*Ans.* 7946.
13. Multiply 418 by 27.	*Ans.* 11286.
14. Multiply 625 by 28	*Ans.* 17500.
15. Multiply 408 by 34.	*Ans.* 13872
16. Multiply 316 by 35	*Ans.* 11060.
17. Multiply 683 by 36.	*Ans.* 24588
18. Multiply 672 by 38.	*Ans.* 25536
19. Multiply 811 by 39.	*Ans.* 31629.
20. Multiply 725 by 41.	*Ans.* 29725.
21. Multiply 437 by 43.	*Ans.* 18791.
22. Multiply 1621 by 45	*Ans.* 72945.
23. Multiply 1763 by 46.	*Ans.* 81098.
24. Multiply 1234 by 49.	*Ans.* 60466.
25. Multiply 6171 by 54.	*Ans.* 333234.
26. Multiply 1876 by 56.	*Ans.* 105056.
27. Multiply 2181 by 58.	*Ans.* 126498.
28. Multiply 2931 by 59.	*Ans.* 172929
29. Multiply 3082 by 68.	*Ans.* 209576.
30. Multiply 4107 by 93.	*Ans.* 381951.
31. Multiply 2573 by 76.	*Ans.* 195548.
32. Multiply 3871 by 87.	*Ans.* 336777.
33. Multiply 5482 by 135.	*Ans.* 740070.
34. Multiply 3257 by 246.	*Ans.* 801222
35. Multiply 4185 by 368.	*Ans.* 1540080.
36. Multiply 5682 by 543.	*Ans.* 3085326.
37. Multiply 7328 by 1021.	*Ans.* 7481888.
38. Multiply 2567 by 2036.	*Ans.* 5226412.
39. Multiply 2135 by 3007.	*Ans.* 6419945.
40. Multiply 6328 by 1276.	*Ans.* 8074528.
41. Multiply 5682 by 1083.	*Ans.* 6153606.
42. Multiply 4185 by 1472.	*Ans.* 6160320.
43. Multiply 3257 by 1576.	*Ans.* 5133032.
44. Multiply 7328 by 2042.	*Ans.* 14963776.
45. Multiply 2567 by 4072.	*Ans.* 10452824
46. Multiply 2135 by 6014.	*Ans.* 12839890.
47. Multiply 71873 by 3018.	*Ans.* 216912714

48. Multiply 6328 by 2552. *Ans.* 16149056.
49. Multiply 71873 by 6036. *Ans.* 433825428.
50. Multiply 11364 by 1629. *Ans.* 18511956
51. Multiply 21984 by 1021. *Ans.* 22445664.
52. Multiply 12835 by 2036. *Ans.* 26132060.
53. Multiply 10675 by 3007. *Ans.* 32099725.
54. Multiply 25312 by 1276 *Ans.* 32298112.
55. Multiply 25312 by 2552. *Ans.* 64596224.
56. Multiply 57873 by 4812. *Ans.* 278484876.
57. Multiply 73218 by 4167. *Ans.* 305099406.
58. Multiply 35821 by 3782. *Ans.* 135475022.
59. Multiply 85765 by 81072. *Ans.* 6953140080.
60. Multiply 180071 by 23456. *Ans.* 4223745376.
61. Multiply 871352 by 40709. *Ans.* 35471868568.
62. Multiply 826107 by 57018. *Ans.* 47102968926.
63. Multiply 360073 by 360073. *Ans.* 129652565329.
64. Multiply 473009 by 874030. *Ans.* 413424056270.

MENTAL EXERCISES.

1. What will 6 yards of muslin cost at 10 cents a yard?

SOLUTION.—If 1 yard of muslin costs 10 cents, 6 yards of muslin will cost 6 times 10 cents, which are 60 cents.

2. What will 8 sheep cost at 12 dollars apiece?

3. What will 12 turkeys cost at 60 cents apiece?

4. At 80 cents each, how much will 9 arithmetics cost?

5. If 5 men can do a piece of work in 6 days, how long will it take 1 man to do it?

6. If 9 men dig a trench in 14 days, how long would it take one man to do it?

7. How many trees in an orchard in which there are 12 rows with 20 trees in each row?

8. A grocer had 24 melons, sold 10 of them, and then bought 5 times as many as he sold; how many had he then?

9. Mason earned $15 a week and paid $7 for his board; how much could he save in 8 weeks?

10. Thompson earned $50 a month and paid $5 a week for his board, and $4 for other expenses; how much could he save in a year?

11. Multiply 76 by 4.

SOLUTION.—4 times 6 units are 24 units, which equals 2 *tens* and 4 *units*, *4 times 7 tens* are 28 *tens*, which increased by the 2 tens equals 30 *tens*; hence *the product is 304.*

12. Multiply 26 by 3; 28 by 4: 37 by 5; 46 by 6; 53 by 7; 75 by 8· 65 by 12; 88 by 11; 125 by 10.

13. Multiply 142 by 4; 325 by 3; 463 by 5; 372 by 6; 408 by 7; 1234 by 6; 2007 by 8; 1306 by 5; 508 by 11; 407 by 12.

14. If a railroad train runs 36 miles an hour, how far will it run at that rate in 6 hours?

15. If a newsboy clears 56 cents a day for 8 days, how much will he clear in all?

16. A jeweler bought 8 watches for $125 apiece, and sold them at a gain of $25 each; what did he receive for them?

17. At $6 a barrel, how much are 3 loads of flour worth, each load containing 10 barrels?

18. A man bought 5 books at 40 cents each, and 2 pairs of shoes at $1.50 a pair; what change will be left out of a $5 bill?

19. A farmer's wife took to a store 6 dozen eggs at 20 cents a dozen, and bought 8 yards of muslin at 12 cents a yard; what balance should she receive?

20. A lady went "a shopping" and bought 6 yards of silk at $3 a yard, 2 yards of velvet at $5 a yard, and 4 pairs of hose at 75 cents a pair; she handed the clerk two $20 bills; what was her change?

WRITTEN EXERCISES.

1. If a steamer sails 235 miles in 1 day, how far will she sail in 37 days?

OPERATION.

$$\begin{array}{r} 235 \\ 37 \\ \hline 1645 \\ 705 \\ \hline 8695 \end{array}$$

SOLUTION.—If a steamer sails 235 miles in 1 day, in 37 days she will sail 37 times 235 miles, or 8695 miles.

2. How many bushels of potatoes can be raised on 56 acres of land, if each acre produces 196 bushels? *Ans.* 10,976.

3. How many oranges in 59 boxes, if each box contains 275 oranges? *Ans.* 16,225 oranges.

4. If a clerk deposits $575 annually in a savings bank, how much will he deposit in 25 years? *Ans.* $14,375.

5. Mr. Gibbs bought a farm of 95 acres at $187 an acre; what was the cost of the farm? *Ans.* $17,765.

6. What will a hogshead of wine, containing 63 gallons, cost at the rate of $4.25 a gallon? *Ans.* $267.75.

7. I bought 356 sacks of Liverpool salt at $2.50 per sack; what did I pay for the whole amount? *Ans.* $890.

8. Sold 265 barrels of mess pork at $16.50 a barrel; what did I receive for the whole? *Ans.* $4372.50.

9. If a locomotive-wheel revolves 478 times in going 1 mile, how often will it revolve in 248 miles? *Ans.* 118,544.

10. When the market-price of butter is 18 cents a pound, what must I pay for 256 pounds? *Ans.* $46.08.

NOTE.—In this and similar problems we multiply by the smaller number, using both numbers abstractly, though in the explanation we should use the proper number as the multiplier.

11. If a barrel of flour is worth $8, how much are 596 barrels worth at the same rate? *Ans.* $4768.

12. A dealer bought 2875 bushels of potatoes at 37 cents a bushel; what was the cost? *Ans.* $1063.75.

13. A Western farmer sold 9876 bushels of wheat at 85 cents a bushel; what did he receive? *Ans.* $8394.60.

14. Henry earned $28 a week and paid $5 for his board; how much could he save in 52 weeks? *Ans.* $1196.

15. A travels 25 miles a day, and B 36 miles a day; how far will both travel in 286 days? *Ans.* 17,446 miles.

16. The President's cabinet consists of 7 members, who receive a salary of $8000 each; what is the amount of their salaries? *Ans.* $56,000.

17. A dealer shipped 2396 bushels of apples at 87 cents a bushel, and the same quantity of potatoes at 56 cents a bushel; what was the value of both? *Ans.* $3426.28.

18. A bought 2960 acres of prairie land at $38 an acre, and the same number of acres of woodland at $45 an acre; what did he pay for all? *Ans.* $245,680.

19. In a row of houses there are 46 rooms, in each room 4 windows, and in each window 12 panes of glass; how many panes in all the houses? *Ans.* 2208 panes.

20. If Thomas sells 48 papers a day, and Henry 72 papers, how many more papers does Henry sell in 94 days than Thomas? *Ans.* 2256 papers.

21. In 1852, 157,548 Irish immigrants arrived in the United States; at this rate what would be the number of immigrants from the beginning of 1852 to the end of 1875? *Ans.* 3,781,152.

22. If two persons should start from the same place and travel in opposite directions, one at the rate of 45 miles a day and the other at the rate of 67 miles a day, how far will they be apart in 67 days? *Ans.* 7504 miles.

23. A planter put up his cotton into 275 bales, averaging 576 pounds each, and sold it at 15 cents per pound; what did he receive for the whole? *Ans.* $23,760.

24. Mr. Angell's annual income is $9750, and his average daily expenditure is $9.75; what can he save in a year of 365 days? *Ans.* $6191.25.

25. A bank teller receives $2500 salary, and spends $520 for board, $475 for clothing, $112.65 for books, and $117.25 for other expenses annually; how much will he save in 5 years? *Ans.* $6375.50.

26. How many miles will a railroad conductor travel in a year if he makes a trip from Harrisburg to Philadelphia (106 miles) and back, every day except Sundays, reckoning 52 Sundays to a year? *Ans.* 66,356 miles.

27. A shoe-dealer bought 35 cases of French calf boots, each case containing 12 pairs, at $6.50 a pair; what did the bill amount to? *Ans.* $2730.

28. A division consisted of 4 regiments, each regiment of 9 companies, and each company of 98 men; how many men in the division? *Ans.* 3528 men.

29. In a block of houses there are 45 buildings, each building containing 32 windows, and each window 12 panes of glass; how many panes of glass in the whole block of buildings? *Ans.* 17,280 panes.

30. If it requires 124 tons of iron worth $72 a ton to build one mile of a railroad, what will be the cost of iron enough to build a railroad 236 miles long? *Ans.* $2,107,008.

31. A field contains 5076 rows of corn, each row containing 4005 hills, and each hill 4 stalks; how many stalks of corn in the field? *Ans.* 81,317,520 stalks.

CONTRACTIONS IN MULTIPLICATION.

78. Contractions in Multiplication are abbreviated methods of multiplying.

79. A Composite Number is the product of two or more numbers, each greater than a unit, called *factors*. Thus, 24 is a composite number, whose factors are 4 and 6, or 3 and 8, or 2, 3, and 4.

CASE I.

80. ***When the multiplier is a composite number.***

1. Multiply 53 by 24.

SOLUTION.—24 equals 4 times 6, hence 24 times 53 equals 4 times 6 times 53. 6 times 53 equals 318, and 4 times 6 times 53 equals 4 times 318, which equals 1272; therefore, 53 multiplied by 24 equals 1272. Hence the

OPERATION.

$$\begin{array}{r} 53 \\ 6 \\ \hline 318 \\ 4 \\ \hline 1272 \end{array}$$

Rule.—*Multiply the multiplicand by one factor, this product by another factor, and thus continue until all the factors have been used; the last product will be the result required.*

Multiply

2. 85 by 35.	*Ans.* 2975.	**8.** 362 by 45.	*Ans.* 16290.
3. 98 by 16.	*Ans.* 1568.	**9.** 893 by 49.	*Ans.* 43757.
4. 75 by 72.	*Ans.* 5400.	**10.** 3572 by 35.	*Ans.* 125020
5. 87 by 56.	*Ans.* 4872.	**11.** 4087 by 27.	*Ans.* 110349
6. 123 by 63.	*Ans.* 7749.	**12.** 8937 by 42.	*Ans.* 375354.
7. 248 by 54.	*Ans.* 13392.	**13.** 40729 by 72.	*Ans.* 2932488

14. What cost 35 cows at 28 dollars apiece?

SOLUTION.—35 equals 5 times 7. If one cow costs 28 dollars, 7 cows will cost 7 times 28 dollars, which are 196 dollars; and 35 cows, which are 5 times 7 cows, will cost 5 times 196 dollars, which are 980 dollars. Therefore, etc.

OPERATION.

$$\begin{array}{r} 28 \\ 7 \\ \hline 196 \\ 5 \\ \hline 980 \end{array}$$

15. What cost 21 yards of muslin, at 14 cents a yard?
Ans. $2.94.

16. What cost 15 barrels of fish, at $13 a barrel?
Ans. $195.

17. What cost 24 horses, at the rate of $165 each?
Ans. $3960

18. What will 42 cows cost at the rate of $27 apiece?
Ans. $1134.

19. What must I pay for 84 books, at the rate of $3.25 each? *Ans.* $273.

20. If one bushel of rye is worth $0.84, how much are 64 bushels worth? *Ans.* $53.76.

21. If a yoke of oxen cost $137, what will 56 yoke cost at the same rate? *Ans.* $7672.

22. What cost 64 acres of land, at $256 an acre; and what cost 81 acres at the same rate? *Ans.* 2d, $20,736

CASE II.

81. ***When there are ciphers at the right of one or both factors.***

Principle.—*Annexing one cipher to a number multiplies it by* 10; *annexing two ciphers multiplies it by* 100; *annexing three, multiplies it by* 1000, *etc.*

For, adding one cipher removes each term one place to the left, and thus makes it denote 10 times as many units as before, hence the entire number is ten times as great as before.

1. Multiply 26 by 140, also 2600 by 140.

OPERATION.	OPERATION.
26	2600
140	140
104	104
26	26
3640	364000

SOLUTION 1ST.—26 multiplied by 14 equals 364; hence 26 multiplied by 140, which is 10 times 14, equals 10 times 364, which, by annexing one cipher, equals 3640.

SOLUTION 2D.—14 times 26 equals 364, hence 14 times 26 *hundred* equals 100 times as much, which by annexing two ciphers, is 36400; and 140 times 2600 equals 10 times 36400 which, by annexing one cipher, is 364000. Hence the following rule:

Rule.—*Take the product of the numbers denoted by the significant figures, and annex as many ciphers to the result as are found at the right of both factors.*

WRITTEN EXERCISES.

2. Multiply 725 by 60. *Ans.* 43500.

3. Multiply 927 by 80. *Ans.* 74160

4. Multiply 2187 by 300. *Ans.* 656100.

5. Multiply 4109 by 500. *Ans.* 2054500.
6. Multiply 5090 by 700. *Ans.* 3563000.
7. Multiply 7355 by 6000. *Ans.* 44130000.
8. Multiply 2170 by 9000. *Ans.* 19530000.
9. Multiply 8542 by 2500. *Ans.* 21355000.
10. Multiply 7681 by 7300. *Ans.* 56071300.
11. Multiply 6500 by 2500. *Ans.* 16250000.
12. Multiply 7800 by 6700. *Ans.* 52260000.
13. Multiply 35600 by 12500. *Ans.* 445000000.
14. Multiply 87200 by 23000. *Ans.* 2005600000.
15. Multiply 90700 by 50700. *Ans.* 4598490000.
16. Multiply 876000 by 25600. *Ans.* 22425600000.
17. Multiply 807000 by 29000. *Ans.* 23403000000.
18. Multiply 908000 by 327000. *Ans.* 296916000000.
19. Multiply 2870000 by 108000. *Ans.* 309960000000.
20. Multiply 607000 by 9080000. *Ans.* 5511560000000.

WRITTEN EXERCISES.

IN ADDITION, SUBTRACTION, AND MULTIPLICATION.

1. I bought 3 spades at $1.50 apiece, 5 shovels at $1.25 apiece, and 4 rakes at $0.75 apiece; what was the entire bill? *Ans.* $13.75.

2. A's income is $5875 a year; he pays $975 for house-rent, and 3 times as much for other expenses; how much can he save annually? *Ans.* $1975.

3. An army of 24,500 men lost in a battle 246 killed, 4 times as many, plus 145, wounded, and 1687 captured; how many remained in the army? *Ans.* 21,438 men.

4. Mary and Martha tried which could count the greater number in 25 minutes; Mary averaged 76 a minute, and Martha 81 a minute; how many did Martha count more than Mary? *Ans.* 125.

5. A poor boy attempted to walk from Philadelphia to Pittsburgh, a distance of 353 miles; after walking 18 miles a day for 15 days, how far was he from Pittsburgh? *Ans.* 83 miles.

6. The light from a certain star is 7 years in coming to the earth; what is its distance, supposing there are 86,400 sec-

onds in one day, and 365 days in a year, light moving at the rate of 192,500 miles a second? *Ans.* 42494760 millions.

7. A wholesale druggist bought as follows: 500 lb. opium at $7.20 per lb.; 375 lb. morphine at $4.75 per lb.; 240 lb. quinine, at $2.70 per lb.; 200 lb. camphor, at $0.29 per lb.; what was his whole bill? *Ans.* $6087.25.

8. A clerk's salary is $1500 a year; he pays $5 a week for his board and washing, $4 a month for car fare, and his other expenses amount to about $2 a day; how much can he save in a year? *Ans.* $462.

9. Mrs. Wilson bought at a furniture store an oiled walnut chamber set for $150, two cottage sets for $55 each, a dozen dining-room chairs at $3.50 each, and 3 camp chairs at $4.25 each; what was her bill? *Ans.* $314.75.

10. A shipping firm bought 150 lb. Manila bolt rope yarns at 16 cents per lb.; 450 lb. Manila whale lines at 17 cents a pound; 260 lb. Russia hemp tarred cordage, at 13 cents per lb.; and 850 lb. Sisal rope at 11 cents per lb.; what was the whole cost? *Ans.* $227.80.

11. A dry goods merchant sold to a Western customer as follows; 125 yards Cocheco fancy prints, at 7 cents a yard; 147 yards Cocheco robes, 8 cents; 156 yards Merrimac robes, 8 cents; 122 yards Merrimac checks and stripes, 7 cents; 125 yards French cashmeres, $1.15; what was the amount of the bill? *Ans.* $185.28.

12. James Chauncey & Co. sold to Barton Brothers, Dec. 23, 1875, the following: 450 bushels White Western oats, at 38 cents a bushel; 375 bushels New York No. 2 White oats, 43 cents; 240 bushels choice new Southern Black-eye peas, $1.45. Jan. 10, 1876, J. C. & Co. bought of Barton Brothers, 460 bushels Pennsylvania rye, 91 cents; 175 bushels No. 2 Canada barley, $1.10; 720 bushels common Winter Red Western wheat, $1.25; required the balance of the account. *Ans.* $830.85 favor of Chauncey & Co

INTRODUCTION TO DIVISION.

MENTAL EXERCISES.

1. At 3 cents each, how many melons can I buy for 15 cents?

SOLUTION.—If 1 melon costs 3 cents, for 15 cents I can buy as many melons as 3 is contained times in 15, which are 5.

2. How many yards of ribbon, at 8 cents a yard, can be bought for 56 cents?

3. A man gave 60 dollars for sheep, at the rate of 5 dollars a head; how many did he buy?

4. How many kegs, of 9 gallons each, can be filled from a hogshead containing 90 gallons of vinegar?

5. How many days must a man work, at the rate of $3 a day, to earn $45?

6. How many lemons at 6 cents apiece may be bought for 84 cents?

7. How many are 15 plus 5, divided by 5? 18 plus 6, divided by 6? 40 plus 8, divided by 8? 35 plus 7, divided by 7?

8. How many are 3 times 8 divided by 4? 5 times 9 divided by 3? 6 times 10 divided by 12? 8 times 7 divided by 4?

9. How many are 3 times 33 divided by 11? 4 times 21 divided by 7? 3 times 25 divided by 5? 3 times 24 divided by 8?

10. How many pencils, worth 5 cents each, may be bought for 4 erasers worth 15 cents each?

11. How many barrels of flour, at $9 a barrel, can be bought for 15 barrels of apples at $3 a barrel?

12. A woman carried to the store 6 dozen eggs at 25 cents a dozen; how many yards of calico, at 10 cents a yard, will pay her?

13. If mackerel is worth $12 a barrel, how many barrels can be bought for $6 in money and 6 barrels of pork at $15 a barrel?

14. The process of finding how often one number is contained in another is called *division*. The result is called the *quotient*.

15. The number divided is the *dividend*; the number we divide by is the *divisor*. The sign of division is ÷, and is read, "divided by."

16. The sign ( ) signifies that the quantities included are to be subjected to the same operation.

17. Find the result of the following:

$(13+5)\div3$	$(25-5)\div4$	$(6\times6)\div4$	$32\div(2\times4)$
$(14+6)\div4$	$(26-2)\div3$	$(8\times9)\div12$	$36\div(3\times4)$
$(15+7)\div2$	$(28-3)\div5$	$(5\times12)\div10$	$42\div(2\times7)$
$(17+4)\div7$	$(32-2)\div6$	$(4\times12)\div8$	$48\div(3\times2)$
$(18+6)\div8$	$(34-6)\div7$	$(7\times9)\div3$	$54\div(3\times3)$

DIVISION.

82. Division is the process of finding the *quotient* of two numbers.

83. The **Quotient** of two numbers is a number which expresses how often one number is contained in another.

84. The **Dividend** is the number to be divided.

85. The **Divisor** is the number by which we divide.

86. The **Remainder** is the number which is sometimes left after dividing.

87. The **Sign of Division** is ÷, and is read *divided by.* It denotes that the number preceding it is to be divided by the number following it.

NOTES.—1. The sign of division is a short line, in the line of writing, with one dot above and another below the middle of it.

2. The symbol ÷ was introduced by Dr. John Pell, an English mathematician, born in 1610.

3. Division is also indicated by writing the divisor beneath the dividend, with a straight line between them; or by writing the divisor at the left of the dividend, with a curved line between them; thus $\frac{27}{9}$, and 9)27, mean 27 divided by 9.

PRINCIPLES.

1. *The divisor and dividend are similar numbers.*

2. *The quotient is an abstract number; the remainder is similar to the dividend.*

3. *In dividing a number into equal parts, the dividend and quotient are similar, and the divisor is abstract.*

METHODS OF DIVISION.

88. There are **Two Methods** of performing division, called *Short Division* and *Long Division.*

SHORT DIVISION.

89. Short Division is that method in which only the dividend, divisor, and quotient are written, the operation being performed mentally.

90. Short division is generally employed when the divisor does not exceed *twelve*, the largest multiplier in the *multiplication table*.

1. How many times is 2 contained in 358?

OPERATION.

$$\begin{array}{r} 2)\underline{358} \\ 179 \end{array}$$

SOLUTION.—We write the divisor at the left of the dividend, with a curved line between them, draw a line beneath the dividend, and begin at the left to divide. 2 is contained in 3 hundreds 1 hundred times, and 1 hundred remaining; 1 hundred equals 10 tens, which with 5 tens are 15 tens: 2 is contained in 15 tens 7 tens times, with a remainder of 1 ten; 1 ten equals 10 units, which with 8 units equals 18 units: 2 is contained in 18 units 9 units times, and we have for the quotient, 179. Hence we have the following

Rule.—I. *Write the divisor at the left of the dividend, with a curved line between them and a line beneath the dividend.*

II. *Begin at the left, divide each term of the dividend by the divisor, and write the quotient beneath.*

III. *If there is a remainder after any division, regard it as prefixed to the next term of the dividend, and divide as before.*

IV. *If any partial dividend is less than the divisor, write a cipher in the quotient and prefix the dividend to the next term.*

V. *When there is a final remainder, annex it, with the divisor written beneath, to the integral part of the quotient.*

Proof.—Multiply the quotient by the divisor, and add the remainder, if any, to the product; if the work is correct, the result will equal the dividend.

NOTE.—In practice we need not name the denomination of the different partial dividends. Thus, in the above solution we say 2 is contained in 3 once and one remaining; 2 is contained in 15, 7 times, etc.

WRITTEN EXERCISES.

(2)	(3)	(4)	(5)	(6)
2)372	2)456	3)618	3)852	3)783
186	228	206	284	261

(7)	(8)	(9)	(10)	(11)
4)968	4)556	4)988	5)850	5)960
242	139	247	170	192

12. Divide 1395 by 5. *Ans.* 279.

13. Divide 6906 by 6. *Ans.* 1151.

14. Divide 3248 by 7. *Ans.* 464.

15. Divide 7185 by 7. *Ans.* $1026\frac{3}{7}$.

16. Divide 8097 by 7. *Ans.* $1156\frac{5}{7}$.

17. Divide 9136 by 8. *Ans.* 1142.

18. Divide 72352 by 8. *Ans.* 9044.

19. Divide 91672 by 8. *Ans.* 11459.

20. Divide 23769 by 9. *Ans.* 2641.

21. Divide 30564 by 9. *Ans.* 3396.

22. Divide 9876504 by 7. *Ans.* $1410929\frac{1}{7}$.

23. Divide 3201567 by 6. *Ans.* $533594\frac{3}{6}$.

24. Divide 4187002 by 8. *Ans.* $523375\frac{2}{8}$.

25. Divide 83702507 by 9. *Ans.* $9300278\frac{5}{9}$.

26. If a book cost 4 dollars, how many books, at the same rate, can you buy for 252 dollars?

OPERATION.

$$4\overline{)252}$$
$$63$$

SOLUTION.—If 1 book cost 4 dollars, for 252 dollars we can buy as many books as 4 dollars are contained times in 252 dollars, which are 63. Therefore, etc.

27. There are 3 feet in one yard; how many yards in 291 feet? *Ans.* 97 yards.

28. There are 8 quarts in one peck; how many pecks are there in 1728 quarts? *Ans.* 216 pecks.

29. There are 7 days in one week; how many weeks in 364 days? *Ans.* 52 weeks.

30. A man gave 324 dollars to some boys, giving 6 dollars to each; how many boys were there? *Ans.* 54.

31. If one sheep cost 9 dollars, how many sheep can you buy for 1935 dollars? *Ans.* 215 sheep.

32. If there are 12 pence in 1 shilling, how many shillings are there in 571,836 pence? *Ans.* 47,653 shillings.

33. If it require one sheet of paper to print 12 pages of a book, how many sheets will be required for a book of 504 pages? *Ans.* 42 sheets.

34. How long will it take two boys, starting at the same place, and traveling in opposite directions, to be 29,076 rods apart, if one goes 5 and the other 7 rods in a minute?

Ans. 2423 minutes.

LONG DIVISION.

91. Long Division is the method of dividing when the work is written out in full. It is generally used when the divisor exceeds 12.

1. Divide 5848 by 23.

OPERATION.

```
23)5848(254
   46
   ---
   124
   115
   ---
     98
     92
     --
      6
```

SOLUTION.—23 is not contained in 5 thousands any thousands times, hence there are no thousands in the quotient. 5 thousands and 8 hundreds are 58 hundreds; 23 is contained in 58 hundreds 2 hundreds times: 2 hundreds times 23 are 46 hundreds, which subtracted from 58 hundreds leave 12 hundreds: 12 hundreds and 4 tens are 124 tens; 23 is contained in 124 tens 5 tens times: 5 tens times 23 are 115 tens, which subtracted from 124 tens leave 9 tens: 9 tens and 8 units are 98 units; 23 is contained in 98 units 4 times; 4 times 23 equals 92: subtracting there is a remainder of 6, which will not contain 23; hence the quotient is 2 hundreds, 5 tens, and 4 units, or 254, with a remainder of 6.

Rule.—I. *Draw curved lines at both sides of the dividend, and place the divisor at the left.*

II. *Divide the number expressed by the fewest terms at the left that will contain the divisor, and place the quotient at the right.*

III. *Multiply the divisor by this quotient, write the product under the partial dividend, subtract, and to the remainder annex the next term of the dividend.*

IV. *Divide as before, and thus continue until all the terms of the dividend have been used.*

V. *If any partial dividend will not contain the divisor, place a cipher in the quotient, annex the next term of the dividend, and proceed as before.*

VI. *When there is a final remainder, annex it, with the divisor written beneath, to the integral part of the quotient.*

Proof.—Multiply the integral part of the quotient by the divisor, and add the remainder, if any, to the product; if the work is correct the result will be equal to the dividend.

NOTES.—I. The pupils will notice that there are *five* operations: 1st. *Write the number*; 2d. *Divide*; 3d. *Multiply*; 4th. *Subtract*; 5th. *Bring down*.

II Pupils often have difficulty in finding the correct quotient figure: this difficulty can be greatly diminished by attention to the following *suggestions*.

1st. Notice how often the left-hand term of the divisor is contained in the term or terms of the partial dividend, as far from the right hand term as the left hand term in the divisor is from the right hand term.

2d. If, when we multiply, the product is greater than the partial dividend, the quotient term must be diminished.

3d. If, when we subtract, the remainder is greater than the divisor, the quotient term must be increased.

III. We commence at the left to divide, so that the remainder can be united to the number of units of the next lower order, giving a new partial dividend. The sign + is used to denote a remainder.

WRITTEN EXERCISES.

1. Divide 840 by 24.	*Ans.* 35.
2. Divide 903 by 21	*Ans.* 43.
3. Divide 455 by 13.	*Ans.* 35.
4. Divide 4956 by 14.	*Ans.* 354.
5. Divide 3510 by 15.	*Ans.* 234.
6. Divide 5024 by 16.	*Ans.* 314.
7. Divide 4720 by 20.	*Ans.* 236
8. Divide 5778 by 18.	*Ans.* 321
9. Divide 8736 by 21.	*Ans.* 416
10. Divide 9591 by 23.	*Ans.* 417.
11. Divide 4056 by 24.	*Ans.* 169
12. Divide 5450 by 25.	*Ans.* 218.
13. Divide 9672 by 26.	*Ans.* 372.
14. Divide 7946 by 29.	*Ans.* 274.
15. Divide 9840 by 30.	*Ans.* 328.
16. Divide 11286 by 27.	*Ans.* 418.
17. Divide 17500 by 28.	*Ans.* 625.
18. Divide 13872 by 34.	*Ans.* 408
19. Divide 24588 by 36	*Ans.* 683.
20. Divide 25536 by 38.	*Ans.* 672.
21. Divide 31629 by 39.	*Ans.* 811
22. Divide 29725 by 41.	*Ans.* 725.
23. Divide 28896 by 43.	*Ans.* 672.
24. Divide 72945 by 45.	*Ans.* 1621
25. Divide 81098 by 46.	*Ans.* 1763
26. Divide 60466 by 49.	*Ans.* 1234
27. Divide 141050 by 50.	*Ans.* 2821
28. Divide 316160 by 52.	*Ans.* 6080
29. Divide 333234 by 54.	Ans 6171.

30. Divide 105056 by 56. *Ans.* 1876.
31. Divide 126498 by 58. *Ans.* 2181.
32. Divide 172929 by 59. *Ans.* 2931.
33. Divide 173911 by 61. *Ans.* 2851.
34. Divide 178857 by 63. *Ans.* 2839.
35. Divide 1568580 by 65 *Ans.* 21132.
36. Divide 1380536 by 68. *Ans.* 20302.
37. Divide 692208 by 69. *Ans.* 10032.
38. Divide 2434380 by 78. *Ans.* 31210.
39. Divide 1031475 by 85. *Ans.* 12135.
40 Divide 1680137 by 97. *Ans.* 17321.
41. Divide 2138654 by 98. *Ans.* 21823.
42. Divide 317646 by 126. *Ans.* 2521.
43. Divide 238788 by 134. *Ans.* 1782.
44. Divide 456104 by 146. *Ans.* 3124.
45. Divide 603264 by 192. *Ans.* 3142.
46. Divide 711287 by 321. *Rem.* 272.
47. Divide 811332 by 372. *Ans.* 2181.
48. Divide 1646301 by 381. *Ans.* 4321.
49. Divide 1985175 by 425. *Ans.* 4671.
50. Divide 1957413 by 453 *Ans.* 4321.
51. Divide 1787160 by 562. *Ans.* 3180.
52. Divide 2100315 by 581. *Ans.* 3615.
53. Divide 3207594 by 767. *Ans.* 4182.
54. Divide 1019806 by 893. *Ans.* 1142.
55. Divide 7481888 by 1021. *Ans.* 7328.
56. Divide 5226412 by 2036. *Ans.* 2567.
57. Divide 6419945 by 2135 *Ans.* 3007.
58. Divide 13824902 by 3367. *Ans.* 4106.
59. Divide 8074528 by 6328. *Ans.* 1276.
60. Divide 97547337 by 3891. *Rem.* 3858.
61. Divide 4223745376 by 23456. *Ans.* 180071.
62. Divide 170627676887 by 413071. *Rem.* 25846.
63. Divide 129652565329 by 360073. *Ans.* 360073.
64. Multiply 37602 by 608, and divide the product by 304. *Ans.* 75204.

PROBLEMS IN DIVISION.

92. In **Division** there are two classes of practical problems:

1st. To find the number of equal parts of a number.

2d To divide a number into equal parts.

CASE I.

93. *To find the number of equal parts of a number.*

MENTAL EXERCISES.

1. At 8 cents a yard how many yards of ribbon can I buy for 72 cents?

SOLUTION.—If 1 yard of ribbon costs 8 cents, for 72 cents I can buy as many yards as 8 is contained times in 72, which are 9.

2. How many oranges can I buy for 48 cents, at the rate of 8 cents apiece?

3. A farmer expended $96 in buying sheep at $8 apiece; how many sheep did he buy?

4. How many melons at 7 cents apiece may be bought for 84 cents?

5. How many oranges at 5 cents apiece may be bought for 7 lemons worth 10 cents apiece?

6. How many yards of lace at 8 cents a yard may be bought for 6 yards of muslin at 12 cents a yard?

7. How much wheat worth 6 dimes a bushel may be bought for 12 bushels of corn at 5 dimes a bushel?

8. If flour is worth $8 a barrel, how many barrels could be exchanged for $12 in money and 4 barrels of fish worth $12 a barrel?

9. A man gave 7 pencils, worth 5 cents each, for 4 packages of envelopes, worth 10 cents each; what was the gain?

10. Divide 96 by 4.

SOLUTION.—4 is contained in 9 tens 2 tens times and 1 ten remaining; 1 ten and 6 units equal 16 units; 4 is contained in 16 units 4 units times; hence the quotient is 2 tens and 4 units, or 24.

11. Divide 68 by 2; 69 by 3; 56 by 4; 72 by 3; 75 by 5; 76 by 4; 126 by 6; 147 by 7; 176 by 8.

12. Divide 720 by 8; 350 by 7; 720 by 9; 156 by 12; 231 by 11; 1224 by 12; 2408 by 8; 2718 by 9; 1728 by 12.

13. If you earn $15 a week and pay $5 for board and $4 for other expenses, in what time could you save $72?

14. Two men start from places 180 miles apart, and travel towards each other, one going 4 and the other 5 miles an hour; in how many hours will they meet?

WRITTEN EXERCISES.

1. At 25 dollars each, how many cows can be bought for 575 dollars?

SOLUTION.—If 25 dollars will buy one cow, 575 dollars will buy as many cows as 25 dollars are contained times in 575 dollars, which are 23.

OPERATION.
25)575(23

2. In one hogshead there are 63 gallons; how many hogsheads in 15,435 gallons? *Ans.* 245 hhd.

3. How many horses can you get for 1824 dollars, at the rate of 152 dollars each? *Ans.* 12 horses.

4. If a boat sails 25 miles an hour, how long will it be in sailing 1800 miles? *Ans.* 72 hours.

5. How many years must a person labor to earn $13,140, at the rate of $730 a year? *Ans.* 18 years.

6. How many acres of land can you purchase for $11,696, at the rate of $86 an acre? *Ans.* 136.

7. How many cows, at $37 each, can be bought for 74 horses, at $150 each? *Ans.* 300 cows.

8. How many oxen, at $54 each, can be bought for 108 mules, at $94 each? *Ans.* 188 oxen.

9. A labors 72 weeks, at $14 a week; how much wheat at 42 cents a bushel will pay him? *Ans* 2400 bushels.

10. The circumference of the earth is 25,000 miles; how long would it take a vessel to sail around it, going at the rate of 125 miles per day? *Ans.* 200 days.

11. The distance from the earth to the sun is 93,000,000 miles; how long will it take light to reach us from the sun, moving 11,520,000 miles a minute? *Ans.* 8 min.+.

12. The moon is 240,000 miles from the earth; how long would it take a balloon to reach it, moving at the rate of 75 miles an hour? *Ans.* 3200 hours.

13. A builder received Western lands at $25 per acre and a balance in cash of $7300, in trade for a row of 15 houses at $2575 each; how many acres did he receive? *Ans.* 1253.

14. A person wishes to trade land worth 150 dollars an acre, for 125 acres, at 75 dollars an acre, and gain 675 dollars by the bargain; how many acres will be exchanged?

Ans. 58.

CASE II.

94. ***To divide a number into equal parts.***

MENTAL EXERCISES.

1 What is one of the 4 equal parts of 24?

SOLUTION.—One of the 4 equal parts of 24 is 6, since 6 taken 4 times is 24.

2. Find one of the 4 equal parts of 12; of 20; of 36; of 48; of 32; of 40; of 44.

3. Find one of the 5 equal parts of 15; of 25; of 30; one of the 6 equal parts of 42; of 48; of 60; of 72.

4. Find one of the 7 equal parts of 35; of 56; of 63; one of the 8 equal parts of 40; of 72; of 88; of 96.

5. What is one-third of anything? *Ans.* One third of anything is one of the 3 equal parts into which it can be divided.

6. What is one-half of anything? one-fourth? one-fifth? one-sixth? one-seventh? one-eighth? one-ninth? one-tenth?

7. What is one-fourth of 24 dollars?

SOLUTION.—One-fourth of $24 is $6, since $6 taken 4 times equals $24.

8. What is one-fourth of 20? of 36? of 48? of 56? of 72?

9. What is one-third of 21? of 27? of 30? of 36? of 48?

10. What is one-fifth of 15? of 30? of 40? of 45? of 60?

11. What is one-sixth of 48? of 72? of 84? of 90? of 96?

12. If 6 slates cost $1.20, how much is that apiece?

SOLUTION.—If 6 slates cost $1.20, one slate cost one-sixth of $1.20, which is 20 cents.

13. If 8 pairs of gloves cost $2.40, how much is that a pair?

14. If 6 cords of wood cost $48, what is the price per cord?

15. What do I pay a bushel for apples if 12 bushels cost me $4.80?

16. What does a lady pay a yard for lace if 12 yards cost her $7.20?

WRITTEN EXERCISES.

1. Divide 235 into 5 equal parts.

SOLUTION 1ST.—If we divide 235 into 5 equal parts, each part is $\frac{1}{5}$ of 235: $\frac{1}{5}$ of 23 tens is 4 tens and 3 tens remaining; 3 tens and 5 units equal 35; $\frac{1}{5}$ of 35 is 7; hence $\frac{1}{5}$ of 235 is 47, or 47 is one of the 5 equal parts of 235.

OPERATION.

5)235
47

SOLUTION 2D.—One of the five equal parts of 5 is *one*, hence one of the 5 equal parts of 235 is as many times *one* as 5 is contained times in 235, which are 47. Therefore, etc.

2. Divide 212 into 4 equal parts. *Ans.* 53.

3. Divide 222 into 6 equal parts. Ans. 37.

4. Divide 455 into 7 equal parts. *Ans.* 65

5. Divide 592 into 8 equal parts. *Ans.* 74.

6. Divide 425 into 17 equal parts. *Ans.* 25

7. Divide 608 into 19 equal parts. *Ans.* 32.

8. Divide 1107 into 9 equal parts. *Ans.* 123.

9. Divide 2574 into 11 equal parts. *Ans.* 234.

10. Divide 3780 into 12 equal parts. *Ans.* 315.

11. Divide 3168 into 24 equal parts. *Ans.* 132.

12. Divide 10725 into 33 equal parts. *Ans.* 325.

13. Divide 205896 into 46 equal parts. *Ans.* 4476.

14. A man divides $318 among 6 boys; how much will each one receive?

SOLUTION 1ST.—It will require $6 to give each boy $1; and in giving $318, each boy will receive as many dollars as $6 are contained times in $318, which are 53. Therefore, etc.

OPERATION.

6)318
53

SOLUTION 2D.—If 6 boys receive $318, one boy will receive one-sixth of $318, which, by division, we find is $53. Therefore, etc.

15. If 12 men earn $384 in a week, how much does one man earn? *Ans.* $32.

16. A boat goes 1584 miles in 24 hours; how far will it go in 1 hour? *Ans.* 66 miles.

17. There are 1575 gallons in 25 hogsheads; how many gallons are there in 1 hogshead? *Ans.* 63 gallons.

18. There are 8316 cubic inches in 36 gallons of wine; how many cubic inches are there in 1 gallon? *Ans.* 231.

19. There are 7614 cubic inches in 27 gallons of beer; how many cubic inches in one gallon? *Ans.* 282.

20. There are 221,760 feet in 42 miles; how many feet are there in one mile? *Ans.* 5280.

21. Sound moves 61,545 feet in 55 seconds; how far does it move in one second? *Ans.* 1119 feet.

22. If a turnpike 132 miles long cost $339,240, how much did it cost per mile? *Ans.* $2570.

23. The salary of the President of the United States is $50,000 a year; what is it a day? *Ans.* $137 nearly.

24. A man having $20,000 buys 150 acres of land, at $75 an acre; how much land can he buy with what remains, at $125 an acre? *Ans.* 70 acres.

CONTRACTIONS IN DIVISION.

95. Contractions in Division are abbreviated forms of dividing.

CASE I.

96. ***When the divisor is a composite number.***

1. Divide 2952 by 24, using the factors 4 and 6.

SOLUTION 1ST.—To *multiply* by 24 we may multiply by 6, and then multiply that product by 4; hence, to *divide* by 24 we may divide by 4, and then divide that quotient by 6. Dividing by 4 we have 738, and dividing 738 by 6 we have 123; hence, etc.

OPERATION.

$$\begin{array}{r} 4)\underline{2952} \\ 6)\underline{738} \\ 123 \end{array}$$

SOLUTION 2D.—Since 24 times a number equals 6 times 4 times the number, $\frac{1}{24}$ of the number equals $\frac{1}{6}$ of $\frac{1}{4}$ of the number; $\frac{1}{4}$ of 2952 is 738, and $\frac{1}{6}$ of 738 is 123; hence, etc.

Rule.—*Divide the dividend by one factor of the divisor, the quotient by another factor, and thus continue for all the factors used; the last quotient will be the quotient required.*

WRITTEN EXERCISES.

Divide the following, using the factors:

2. 570 by 15.	(15=3×5)	*Ans.* 38.
3. 492 by 12.	(12=4×3)	*Ans.* 41.
4. 594 by 18.	(18=3×6)	*Ans.* 33.
5. 2480 by 20.	(20=4×5)	*Ans.* 124.
6. 4494 by 14.	(14=2×7)	*Ans.* 321.
7. 10950 by 30.	(30=5×6)	*Ans.* 365.
8. 7875 by 35.		*Ans.* 225.
9. 12560 by 40.		*Ans.* 314.
10. 22824 by 72.		*Ans.* 317.
11. 47412 by 108.		*Ans.* 439.
12. 64440 by 120.		*Ans.* 537.
13. 54576 by 144.		*Ans.* 379.

TO FIND THE TRUE REMAINDER.

97. The **True Remainder** in successive division, it is evident, is not the last remainder, nor the sum of all the remainders; it is necessary, therefore, to explain the method of finding the true remainder.

1. Divide 791 by 24, using the factors 2, 3, and 4.

SOLUTION.—Dividing by 2 we find that 791 equals 395 *twos* and 1 remaining; dividing 395 *twos* by 3, we find 395 *twos* equals 131 *sixes* and 2 *twos*, or 4, remaining; dividing by 4, we find that 131 *sixes* consists of 32 *twenty-fours* and 3 *sixes*, or 18, remaining. Hence the true remainder is 18+4+1, which is 23. Hence, to find the correct remainder we have the following

OPERATION.

2)791
3)395 1
4)131, 2 *twos* = 4
32, 3 *sixes* =18
True remainder, 23.

Rule.—*Multiply each remainder by all the divisors preceding the one which obtained it, and take the sum of the products and the remainder arising from the first division*

Divide the following and find the true remainder:

2.	582 by 15.		*Rem.* 12.
3.	503 by 12.		*Rem.* 11.
4.	2497 by 20.		*Rem.* 17.
5.	4507 by 14.		*Rem.* 13.
6.	3717 by 30.	2, 3, 5.	*Rem.* 27.
7.	13853 by 105.	3, 5, 7.	*Rem.* 98.
8.	41837 by 180.	4, 5, 9.	*Rem.* 77.
9.	47117 by 308.	4, 7, 11.	*Rem.* 301.
10.	96711 by 310.	2, 5, 31.	*Rem.* 301.
11.	87831 by 720.	2, 3, 4, 5, 6.	*Rem.* 711.

CASE II.

98. ***When there are ciphers at the right of the divisor.***

1. Divide 8254 by 600.

SOLUTION.—6 *hundreds* are contained in 82 *hundreds* 13 times, and 400 remaining; 600 is not contained in 54, hence the entire remainder is 400+54, or 454. From this solution we may derive the following

OPERATION.

6|00)82|54
13–454

Rule.—I. *Cut off the ciphers at the right of the divisor, and as many terms at the right of the dividend.*

II. *Divide the remaining part of the dividend by the remaining part of the divisor.*

III. *Prefix the remainder to the part of the dividend cut off, and the result will be the true remainder.*

NOTES.—1. When the divisor is a unit of any order with ciphers, the remainder will be the figures cut off at the right, and the quotient the figures at the left.

3. When the part of the divisor at the left of the naughts is greater than 12, divide by long division.

2. Divide 876 by 50. *Ans.* 17; *Rem.* 26.
3. Divide 953 by 400. *Ans.* 2; *Rem.* 153.
4. Divide 1733 by 500. *Ans.* 3; *Rem.* 233.
5. Divide 2765 by 700. *Ans.* 3; *Rem.* 665.
6. Divide 7859 by 800. *Ans.* 9; *Rem.* 659.
7. Divide 9763 by 900. *Ans.* 10; *Rem.* 763.
8. Divide 14873 by 1900. *Ans.* 7; *Rem.* 1573.
9. Divide 25075 by 2300. *Ans.* 10; *Rem.* 2075.
10. Divide 187654 by 14700. *Rem.* 11254.
11. Divide 269856 by 237000. *Rem.* 32856.
12. Divide 5767220 by 4730000. *Rem.* 1037220.
13. Divide 9235700 by 7340000. *Rem.* 1895700.

EXERCISE UPON THE PARENTHESIS.

99. The **Parenthesis** ( ), denotes that the quantities included are to be subjected to the same operation; thus, $(8+6-4)\times3$ denotes that the value of $8+6-4$, which is 10, is to be multiplied by 3.

1. What is the value of $(12+9-7)\times5$?

SOLUTION.—12+9 equals 21, and 21 minus 7 equals 14, and 14 multiplied by 5 equals 70. Therefore, etc.

Required the value

2. Of $(25+23-18)\times7$. *Ans.* 210.
3. Of $(46+97-82)\times9$. *Ans.* 549.
4. Of $(98-75+87)\times14$. *Ans.* 1540.
5. Of $(89+96-47)\div6$. *Ans.* 23.
6. Of $(145-110+117)\div8$. *Ans.* 19.
7. Of $(396-128+483)\times32$. *Ans.* 24032.
8. Of $(860+980-1120)\div45$. *Ans.* 16.
9. Of $(320-98)\times(860-145)$. *Ans* 158730.
10. Of $(689-327+986-397)\times428$. *Ans.* 407028.
11. Of $(729+487-244)\div(247-210+71)$. *Ans.* 9.
12. Of $(3014-2601)\times(2477-1325)\div(295\div5)$. *Ans.* 8064.
13. Of $(2247+349-480)\div(3411-2882)+227\times4$. *Ans.* 912.

WRITTEN EXERCISES.

ON THE FOUR FUNDAMENTAL RULES.

1. The minuend is 4160, and the subtrahend is 3425; what is the remainder? *Ans.* 735.

2. The minuend is 9164 and the remainder is 3426; what is the subtrahend? *Ans.* 5738.

3. The subtrahend is 3872 and the remainder 4648; what is the minuend? *Ans.* 8520.

4. The multiplicand is 745 and the multiplier 456; what is the product? *Ans.* 339720.

5. The multiplicand is 2463 and the product 854661; what is the multiplier? *Ans.* 347.

6. The product is 881919 and the multiplier 981; what is the multiplicand? *Ans.* 899.

7. The dividend is 518077 and the divisor 763; what is the quotient? *Ans.* 679.

8. The dividend is 801222 and the quotient 3257; what is the divisor? *Ans.* 246.

9. The divisor is 587 and the quotient 8723; what is the dividend? *Ans.* 5120401.

10. The dividend is 72987 and divisor 45; required the quotient and remainder. *Ans* 1621; 42.

11. The dividend is 7972, the quotient is 274, and remainder 26; what is the divisor? *Ans.* 29.

12. The divisor is 26, the quotient 372, and remainder 23; what is the dividend? *Ans.* 9695.

13. Thomas read 789 pages of history in a week, which lacks 324 of being as many as Walton read; how many did both read? *Ans.* 1902 pages.

14. A freight car ran 365 miles one week, and 3 times as far, lacking 246 miles, the next week; how far did it run the second week? *Ans.* 849 miles.

15. A sold 8318 bushels of wheat, then bought 2514 bushels, and then had 3146 bushels; how many bushels had he at first? *Ans.* 8950 bushels.

16. My barn cost $3156; my house cost 3 times as much as my barn, and my farm cost as much as both; what was the cost of all? *Ans.* $25,248.

17. The value of 5 horses and 7 mules is $2436; now if the value of each mule is $208, what is the value of each horse? *Ans.* $196.

18. A man left $2535 each to his four children, but one of them dying, the three remaining children divided the money; how much did each receive? *Ans.* $3380.

19. Mr. Smith left $6264 to each of three sons and $7240 to each of two daughters, but one daughter dying, her share was equally divided among the remaining children; what did each receive? *Ans.* Son, $8074; daughter, $9050.

20. The income of a man who "struck oil" was $480 a day; how many teachers would this employ at a salary of $438 a year? *Ans.* 400.

21. A stock dealer bought 325 cows at $28 each, and sold 124 of them at cost; how much must he receive a head for the remainder to gain $804? *Ans.* $32.

22. Mr. Galton buys a farm of 110 acres at $75 an acre, $2200 to be paid down and the remainder in five yearly installments; what must he pay each year? *Ans.* $1210.

23. A farmer raised 765 bushels of oats, of which he kept 65 bushels for seed, and after retaining enough for the use of his horses till next harvest, allowing 60 bushels to each horse, sold the balance at 85 cents a bushel, and received $442; how many horses had he? *Ans.* 3 horses.

24. Mr. Milman bequeathed $6500 to each of two sons, to a third son $1000, $5000 to each of 3 daughters, and the balance of his estate, amounting to $25,000, to several benevolent institutions; the will, however, being set aside, the property was divided equally among his children; what was the share of each? *Ans.* $9000.

25. If a soldier enlisting in the late war for 3 years, received a bounty of $930; then served one year as a private, at $13 a month, 6 months as a corporal, at $14 a month, and 18 months as a sergeant at $17 a month; what was the whole amount of his pay and his average pay per month?

Ans. $41 a month.

GENERAL PRINCIPLES

OF THE FUNDAMENTAL RULES.

PRINCIPLES OF ADDITION.

1. The sum of all the parts equals the whole.

2. The whole diminished by one or more parts equals the sum of the other parts.

PRINCIPLES OF SUBTRACTION.

1. The Remainder equals the Minuend minus the Subtrahend.
2. The Minuend equals the Subtrahend plus the Remainder.
3. The Subtrahend equals the Minuend minus the Remainder.

PRINCIPLES OF MULTIPLICATION.

1. The Product equals the Multiplicand into the Multiplier.
2. The Multiplicand equals the Product divided by the Multiplier.
3. The Multiplier equals the Product divided by the Multiplicand.

PRINCIPLES OF DIVISION.

1. The Quotient equals the Dividend divided by the Divisor.
2. The Dividend equals the Divisor multiplied by the Quotient.
3. The Divisor equals the Dividend divided by the Quotient.
4. The Dividend equals the Divisor multiplied by the Quotient plus the Remainder.
5. The Divisor equals the Dividend minus the Remainder, divided by the Quotient.

OTHER PRINCIPLES OF DIVISION.

1. Multiplying the Dividend or dividing the Divisor by any number, multiplies the Quotient by that number.
2. Dividing the Dividend or multiplying the Divisor by any number, divides the Quotient by that number.
3. Multiplying or dividing both Dividend and Divisor by the same number, does not change the Quotient.
4. *A General Law.*—A change in the Dividend by multiplication or division produces a *similar* change in the Quotient; but such a change in the Divisor produces an *opposite* change in the Quotient.

NOTE TO TEACHER.—Let the pupils be required to show the reason for the above principles, and give illustrations of them. No demonstrations are given, since it is better for the pupil to learn to depend somewhat upon himself, that he may become, not a mere imitator, but an *original thinker*.

INTRODUCTION TO SECONDARY OPERATIONS.

MENTAL EXERCISES.

1. What numbers multiplied together will produce 4? 6? 8? 10? 12? 14? 16? 18? 20? 24? 26? 28? 30?

2. What numbers can be produced out of the numbers 2 and 3? 3 and 5? 2, 3, and 5? 3, 4, and 5? 2, 3, 4, and 5?

3. Will the product of any two numbers, each greater than a unit produce 1, 2, 3, 5, 7, 11, 13, 17, 19, 23, 29, 31, and 37?

4. What may we call a number which is *composed* by multiplying several numbers together? *Ans.* A *Composite Number.*

5. What shall we call numbers that cannot be produced by multiplying several numbers together? *Ans. Prime Numbers.*

6. Which are prime and which composite numbers in the following list: 1, 2, 3, 4, 5, 6, 7, 8, 9, 10, 11, 12, 13, 14, 15?

7. What may we call the numbers whose product makes a composite number? *Ans. Makers* of the numbers.

8. If the word *Factor* means the same as *maker*, what may we call the *makers* of a composite number? *Ans. Factors.*

9. Form composite numbers out of the factors 3 and 4; 3, 4, and 6; 4, 5, and 6. What are the factors of 12? 15? 18? 20? 21? 24?

10. Form a composite number by using 2 twice as a factor; 3 twice as a factor; 2 three times as a factor; 3 four times as a factor.

11. Required *one* of the *two equal* factors of 9; of 16; of 25; of 36 *one* of the *three equal* factors of 8; of 27; of 64; of 125.

12. A number composed of two equal factors is called the *second power* of that factor; of three equal factors, the *third power*, etc.

13. Required the second power of 3; of 4; of 6; of 7; of 8: the third power of 2; of 3; of 4; of 6.

14. *One* of the *two equal* factors of a number is the *second root* of a number; *one* of the *three equal* factors is the *third root*, etc.

15. What is the second root of 16? of 25? of 36? of 49? What is the third root of 8? of 27? of 64? of 125?

16. What would it seem natural to call the process of making composite numbers? *Ans. Composition.*

17. What would it seem natural to call the process of finding the factors of a number? *Ans. Factoring.*

18. What are the first four operations of arithmetic called? *Ans* The *Fundamental* or *Primary Operations* of arithmetic.

19. What would it be natural to call these operations which are derived from the fundamental operations? *Ans* The *Derivative* or *Secondary Operations.*

SECTION III.

SECONDARY OPERATIONS.

100. The **Primary Operations** of arithmetic are those of synthesis and analysis, including the four fundamental rules.

101. The **Secondary, or Derivative Operations**, are those which arise from or grow out of the primary operations of synthesis and analysis.

102. The **Secondary Operations** are *Composition*, *Factoring*, *Greatest Common Divisor*, *Least Common Multiple*, *Involution*, and *Evolution*.

COMPOSITION.

103. **Composition** is the process of forming composite numbers when their factors are given.

104. A **Composite Number** is a number which can be produced by multiplying together two or more numbers, each greater than a unit; as 8, 12, 15, etc.

105. The **Factors** of a composite number are the numbers, which, when multiplied together, will produce it; thus 4 and 2 are the factors of 8.

106. A **Prime Number** is one that cannot be produced by multiplying together two or more numbers, each greater than a unit; as, 2, 5, 7, 11, etc.

107. A composite number consisting of two equal factors is said to be the 2d *power* of that factor; of three equal factors, the 3d *power*, etc.; thus, 9 is the 2d power of 3, and 64 is the 3d power of 4.

NOTE.—An even number is one that is exactly divisible by 2; an odd number is one that is not exactly divisible by 2.

MENTAL AND WRITTEN EXERCISES.

1. Tell which of the following numbers are prime or composite: 2, 3, 4, 5, 6, 7, 8, 9, 10, 11, 12, 13, 14, 15, 16.

2. Name the prime numbers from 1 to 53. Name the prime numbers from 53 to 101.

3. Write the numbers from 1 to 100, and cut out all the composite numbers, leaving the primes.

4. What is the second power of 1, 2, 3, 4, 5, 6, 7, 8, 9, 10, 11, 12, 13, 14, 15, 16, 17, 18, 19, 20?

5. What is the 3d power of each of the above? The 4th power? The 5th power? The 6th power?

PRINCIPLES.

1. *Every composite number is equal to the product of its factors.*

2. *A factor of a number is a factor of any number of times that number.*

108. ***To form composite numbers out of any factors.***

1. Form a composite number out of 4, 9, and 5.

SOLUTION.—A composite number formed out of the factors, 4, 9, and 5, is equal to their product, which is 180.

OPERATION.
$4 \times 9 \times 5 = 180$

WRITTEN EXERCISES.

Form composite numbers out

2. Of 5, 6, 7, and 8. *Ans.* 1680.

3. Of two 2's, 3, and 7. *Ans.* 84.

4. Of three 3's, four 2's, and two 5's. *Ans.* 10800.

5. Find a number consisting of four 5's. *Ans.* 625

6. Find the fifth power of 3, of 4, of 7.
Ans. 243; 1024; 16807.

7. Form a composite number out of the first four prime numbers after unity. *Ans.* 210.

8. Form a composite number out of all the prime numbers between 11 and 29. *Ans.* 96577.

9. Form all the composite numbers you can out of 2, 3, 5, and 7. *Ans.* 6; 10; 14; 15; 21; etc.

10. Form all the composite numbers you can out of 2, 3, 5, 7, and 11. *Ans.* 6; 10; 14; 22; 15; 21; etc.

11. Find a composite number consisting of three factors, the first being 2, the second being twice as great, and the third three times as great. *Ans.* 48.

4

DIVISIBILITY OF COMPOSITE NUMBERS.

109. Composite Numbers can be divided by the factors which produce them.

110. The **Factors** of many composite numbers may be seen by inspection from the following principles:

PRINCIPLES.

1. *A number is divisible by 2 when the right hand term is zero or an even digit.*

For, the number is evidently an even number, and all even numbers are divisible by 2.

2. *A number is divisible by 3 when the sum of the digits is divisible by 3.*

This may be shown by trying several numbers, and, seeing that it is true with these, we infer that it is true with all. A rigid demonstration is too difficult for this place.

3. *A number is divisible by 4 when the two right hand terms are ciphers, or when the number they express is divisible by 4.*

If the two right hand terms are ciphers, the number equals a number of hundreds, and since 100 is divisible by 4, any number of hundreds is divisible by 4.

If the number expressed by the two right hand digits is divisible by 4, the number will consist of a number of *hundreds* plus the number expressed by the two right hand digits (thus 1232=1200+32); and since both of these are divisible by 4, their sum, which is the number itself, is divisible by 4.

4. *A number is divisible by 5 when its right hand term is 0 or 5.*

When the unit figure is 0, the last partial dividend must be 0, 10, 20, 30, or 40, each of which is divisible by 5. When the unit figure is 5, the last partial dividend must be 15, 25, 35, or 45, each of which is divisible by 5. Therefore, etc.

5. *A number is divisible by 6 when it is even, and the sum of the digits is divisible by 3.*

Since the number is even it is divisible by 2, and since the sum of the digits is divisible by 3 the number is divisible by 3, and since it contains both 2 and 3, it will contain their product, 3×2, or 6.

6. *A number is divisible by 8 when the three right hand terms are ciphers, or when the number expressed by them is divisible by 8.*

If the three right hand terms are ciphers, the number equals a number of *thousands*, and since 1000 is divisible by 8, any number of thousands is divisible by 8.

If the number expressed by the three right hand digits is divisible by 8, the entire number will consist of a number of *thousands* plus the number expressed by the three right hand digits (thus 17368=17000+368) and since both of these parts are divisible by 8, their sum, which is the number itself, is divisible by 8.

7. *A number is divisible by* 9 *when the sum of the digits is divisible by* 9.

This may be shown by trying several numbers, and, seeing that it is true with these, we can infer that it is true with all. It may also be rigidly demonstrated.

8. *A number is divisible by* 10 *when the unit figure is* 0.

For, such a number equals a number of *tens*, and any number of tens is divisible by 10, hence the number is divisible by 10.

NOTE.—1. *A number is divisible by* 7 *when the sum of the odd numerical periods, minus the sum of the even numerical periods, is divisible by* 7.

2. *A number is divisible by* 11 *when the difference between the sums of the digits in the odd places and in the even places is divisible by* 11, *or when this difference is* 0.

3 These two principles are rather curious than useful. For their demonstration see *Higher Arithmetic.*

INTRODUCTION TO FACTORING.

MENTAL EXERCISES.

1. Name the prime numbers from 1 to 50.
2. Name the composite numbers from 0 to 50.
3. Name some of the factors or makers of 12, 15, 21, 28, 36, 54, 72.
4. Name the prime numbers which are factors of 12, 18, 20, 24, 36, 54, and 72.
5. What shall we call the factors of numbers when they are prime numbers? *Ans.* The *Prime Factors.*
6. Name the prime factors of 12, 16, 18, 20, 24, 30, 32, 36, 40, 45, 50, 60, and 80.
7. Illustrate the principle that the factors of a number are *divisors* of the number.
8. How then can we find the factors of a number? *Ans.* By finding the divisors of a number.
9. How can we find the divisors of a number? *Ans.* By trial, aided by the principles of Art. 110.
10. What do we call that subject of arithmetic which treats of finding the factors of numbers? *Ans. Factoring.*
11. How then may we define the subject of Factoring?

FACTORING.

111. Factoring is the process of finding the factors of composite numbers. Unity and the number itself are not regarded as facters.

112. The **Factors** of a composite number are the numbers which multiplied together will produce it; thus, 3 and 4 are factors of 12.

113. The **Prime Factors** of a composite number are the prime numbers which multiplied together will produce it; thus, 2, 2, and 3 are the prime factors of 12.

114. One of the two equal factors of a number is called its 2d, or *square root;* one of the 3 equal factors, its 3d, or *cube root,* etc.; thus, 3 is the 2d root of 9, 2 the 3d root of 8.

PRINCIPLES.

1. *A divisor of a number, excepting unity and itself, is a factor of that number.*

2. *A divisor of a factor of a number, excepting unity, is a factor of the number.*

3. *A number is divisible by its prime factors or by any product of them.*

4. *A number is divisible only by its prime factors or by some product of them, or by unity.*

CASE I.

115. ***To resolve a number into its prime factors.***

1. Find the prime factors of 105.

SOLUTION.—Dividing by 3, we find that 3 is a factor of 105 (Prin. 1). Dividing the quotient by 5, we find that 5 and 7 are factors of 35 (Prin. 2), and since these numbers 3, 5, and 7, are prime numbers, they are the prime factors of 105.

OPERATION.

$$\begin{array}{r} 3\overline{)105} \\ 5\overline{)35} \\ \overline{7} \end{array}$$

Rule.—I. *Divide the given number by any prime number greater than 1, that will exactly divide it.*

II. *Divide the quotient, if composite, in the same manner, and thus continue until the quotient is prime.*

III. *The divisors and last quotient will be the prime factors required.*

What are the prime factors

2. Of 28? *Ans.* 2, 2, 7.

3. Of 84? *Ans.* 2, 2, 3, 7.

4. Of 125? *Ans.* 5, 5, 5

5. Of 325? *Ans.* 5, 5, 13.

6. Of 210? *Ans.* 2, 3, 5, 7.

7. Of 114? *Ans.* 2, 3, 19.

8. Of 432? *Ans.* 2, 2, 2, etc.

9. Of 426? *Ans.* 2, 3, 71.

10. Of 1872? *Ans.* 2, 2, 2, 2, 3, 3, 13.

11. Of 7644? *Ans.* 2, 2, 3, 7, 7, 13.

12. Of 1184? *Ans.* 2, 2, 2, 2, 2, 37.

13. Of 1140? *Ans.* 2, 2, 3, 5, 19.

CASE II.

116. ***To resolve a number into equal factors.***

1. Find the two equal factors of 1225.

OPERATION.

```
5)1225
5)245
7)49
  7
5×7=35
```

SOLUTION.—We first resolve the number into its prime factors. Now since there are *two* 5's, we take *one* 5 for each factor; and since there are *two* 7's, we take *one* 7 for each factor; hence each of the two equal factors is 5×7, or 35. Therefore 35 is one of the two equal factors of 1225.

Rule.—I. *Resolve the number into its prime factors.*

II. *Take the continued product, of one of each of the two equal factors, when we wish the two equal factors, one of each of the three, for the three equal factors, etc.*

2. Find the two equal factors of 16, 36, 100, 196, 256, 324, 900, 1296, 2025. *Ans.* 4, 4; 6, 6; 10, 10; 14, 14, etc.

3. Find the square root of 225, 576, 1764, 3136, 3969, 5184
Ans. 15; 24; 42; 56; 63; 72.

4. Find the cube root of 27, 64, 125, 216, 512, 1000.
Ans. 3; 4; 5; 6; 8; 10.

5. Find the fourth root of 16, 81, 256, 1296, 4096, 20736
Ans. 2; 3; 4; 6; 8; 12.

6. Find the fifth root of 32; 243; 1024; 3125.
Ans. 2; 3; 4; 5.

INTRODUCTION TO COMMON DIVISOR.

MENTAL EXERCISES.

1. Name an exact divisor of 6; of 9; of 12; of 15; of 18; of 20; of 24; of 36.

2. What exact divisors are common to 8 and 12? to 10 and 15? to 12 and 18? to 24 and 36? to 32 and 48? to 48 and 72?

3. What may a divisor *common* to two or more numbers be called? *Ans.* Their *common divisor.*

4. What is a common divisor of 16 and 24? of 15 and 20? of 18 and 30? of 48 and 54?

5. What is the greatest divisor common to 24 and 32? to 32 and 56? to 48 and 72? to 72 and 96?

6. What may the greatest divisor common to two or more numbers be called? *Ans.* Their *greatest common divisor.*

7. What is the greatest common divisor of 24 and 30? of 45 and 50? of 54 and 60? of 64 and 72?

8. What prime factors are common to 18 and 24? 27 and 30? 30 and 35? 36 and 40?

9. The product of what two prime factors of 12 and 18 will divide both? of 20 and 30?

GREATEST COMMON DIVISOR.

117. A **Divisor** of a number is a number that exactly divides it. Thus 4 is a divisor of 20.

118. A **Common Divisor** of two or more numbers is a number that exactly divides each of them. Thus 4 is a common divisor of 16 and 20.

119. The **Greatest Common Divisor** of two or more numbers is the greatest number that exactly divides each of them. Thus 8 is the greatest common divisor of 16 and 24.

NOTE.—The greatest common divisor may be represented by the initials G. C. D.

PRINCIPLES.

1. *A common factor of two or more numbers is a factor of their greatest common divisor.*

2. *The product of all the common prime factors of two or more numbers is their greatest common divisor.*

3. *A common divisor of two numbers is a divisor of their sum and also of their difference.*

DEM.—Take any two numbers, as 12 and 20, of which 4 is a common divisor. Now, 12 equals *three* times 4, and 20 equals *five* times 4, and their sum equals *three* times 4 plus *five* times 4, or *eight* times 4, which contains 4, or is divisible by 4. Their difference is *five* times 4 minus *three* times 4, or *two* times 4, which is also divisible by 4.

CASE I.

120. ***When the numbers are small and can be readily factored.***

FIRST METHOD.

121. This method consists in finding the common factors, and taking their product.

1. Find the greatest common divisor of 42, 84, and 126.

OPERATION.

2	42	84	126
3	21	42	63
7	7	14	21
	1	2	3

$2 \times 3 \times 7 = 42$

SOLUTION.—We write the numbers one beside another, as in the margin. Dividing by 2, we see that 2 is a factor of each number; it is therefore a factor of the G. C. D. (Prin. 1). Dividing the quotients by 3, we see that 3 is a factor of each number and therefore a factor of the G. C. D.; and in the same way we see that 7 is a factor of the G. C. D. Now since the quotients 1, 2, and 3 are prime to each other, 2, 3, and 7 are all the common factors; hence their product, which is 42, is the G. C. D. (Prin. 2).

Rule.—I. *Write the numbers one beside another, with a vertical line at the left, and divide by any common factor of all the numbers.*

II. *Divide the quotients in the same manner, and thus continue until the quotients have no common factor.*

III. *Take the product of all the divisors; the result will be the greatest common divisor.*

What is the greatest common divisor of

2. 10, 20, and 30? *Ans.* 10.
3. 36, 48, and 54? *Ans.* 6.
4. 18, 36, and 72? *Ans.* 18.
5. 48, 72, and 96? *Ans.* 24.
6. 120, 210, and 360? *Ans.* 30.
7. 210, 315, and 420? *Ans.* 105.
8. 252, 336, and 420? *Ans.* 84.
9. 330, 495, and 660? *Ans.* 165.
10. 468, 780, and 1092? *Ans.* 156.

SECOND METHOD.

122. This method consists in resolving the numbers into their prime factors, and taking the product of the common factors.

1. Find the greatest common divisor of 42, 84, and 126.

SOLUTION.—The factors of 42 are 2, 3, and 7; the factors of 84 are 2, 2, 3, and 7; the factors of 126 are 2, 3, 3, and 7. We see that 2, 3, and 7 are all the prime factors common to the three numbers; hence their product, which is 42, is the greatest common divisor of the numbers (Prin. 2). Hence the following

OPERATION.

$$42=2\times3\times7$$
$$84=2\times2\times3\times7$$
$$126=2\times3\times3\times7$$
$$2\times3\times7=42$$

Rule.—*Resolve the numbers into their prime factors, and take the product of all the common factors.*

Find the greatest common divisor of

2. 270, 315, and 405. *Ans.* 45.
3. 168, 192, and 216. *Ans.* 24.
4. 252, 308, and 364. *Ans* 28.
5. 504, 546, and 588. *Ans* 42.
6. 392, 448, and 504. *Ans.* 56.
7. 432, 504, and 648. *Ans.* 72.
8. 792, 864, 936, and 1008. *Ans.* 72.
9. 384, 576, 768, and 960. *Ans.* 192.

CASE II.

123. ***When the numbers are large and cannot be readily factored.***

1. Find the greatest common divisor of 32 and 56.

SOLUTION.—We divide 56 by 32, the divisor 32 by 24, and the divisor 24 by the remainder 8, and have no remainder; then will 8 be the greatest common divisor of 32 and 56. For

1st. *The last remainder*, 8, *is a* NUMBER OF TIMES *the G. C. D.* Since 32 and 56 are each a number of times the G. C. D., their difference, 24, is a number of times the G. C. D., by Prin. 3; and since 24 and 32 are each a number of times the G. C. D., their difference, 8, is also a number of times the G. C. D.

OPERATION.

```
32)56(1
   32
   --
   24)32(1
      24
      --
       8)24(3
         24
         --
```

2d. *The last divisor*, 8, *is* ONCE *the G. C. D.* Since 8 divides 24 it will divide 24+8, or 32, by Prin. 3; and since it divides 32 and 24, it will divide 32+24, or 56; and now since 8 divides 32 and 56, and is a *number of times the G. C. D.*, it must be *once* the G. C. D.

OPERATION.

32	116	3
	96	
20	20	1
12	12	1
8	8	1
4	8	2

Another Form.—In the margin on the right is another form of writing the division, which in practice we prefer to the above. The problem is to find the greatest common divisor of 32 and 116. The method will be clear from a slight inspection of the work. Let the pupils adopt it after they are familiar with the common form.

Rule.—*Divide the greater number by the less, the divisor by the remainder, and thus continue to divide the last divisor by the last remainder until there is no remainder; the last divisor will be the greatest common divisor.*

Note.—To find the greatest common divisor of more than two numbers, we first find the greatest common divisor of two of them, then of that divisor and one of the other numbers, etc.

WRITTEN EXERCISES.

Find the greatest common divisor of

2. 115 and 161. *Ans.* 23.

3. 91 and 143. *Ans.* 13.

4. 333 and 592. *Ans.* 37.

5. 697 and 820. *Ans.* 41.

6. 1220 and 2013. *Ans.* 61.

7. 730, 1241, and 1460. *Ans.* 73.

8. 72491 and 103121. *Ans.* 1021.

9. 347387 and 561851. *Ans.* 1117.

10. A farmer has two heaps of apples, one containing 364, and the other 585, which he wishes to divide into smaller heaps, each containing the same number; what is the largest number that the heaps may contain? *Ans.* 13.

11. A benevolent society distributed $678, $906, and $1146 in equal sums to the poor of three wards of a city, the sums being as large as possible. Required the amount of the equal sums and the number of persons receiving relief in each ward. *Ans.* $6; 113; 151; 191.

12. A Western landholder has three tracts, the first containing 533 acres, the second 574 acres, and the third 861 acres, which he wishes to divide into fields of equal size, having the least number possible. Required the number of fields, and the number of acres in each.

Ans. 48 fields; 41 acres.

INTRODUCTION TO COMMON MULTIPLE.

MENTAL EXERCISES.

1. What number is three times 5? four times 6? five times 6? six times 8?

2. A number which is one or more times another number is called a *multiple* of that number.

3. What is the multiple of 4? of 5? of 6? of 7? of 8? of 9? of 10? of 11? of 12?

4. What multiple is common to 2 and 3? to 3 and 4? to 4 and 6? to 6 and 8? to 6 and 9?

5. What may we call a multiple common to two or more numbers? *Ans.* A *common multiple.*

6. What is a common multiple of 4 and 5? 8 and 9? 6 and 7? 4 and 6? 5 and 8?

7. What is the least multiple common to 2 and 4? 4 and 6? 4 and 8? 6 and 8? 8 and 12?

8. What shall we call the least multiple common to two or more numbers? *Ans.* Their *least common multiple.*

9. What is the least common multiple of 4 and 6? 9 and 12? 10 and 16? 20 and 24? 25 and 30?

LEAST COMMON MULTIPLE.

124. A **Multiple** of a number is one or more times the number; thus, 4 times 5, or 20, is a multiple of 5.

125. A **Common Multiple** of two or more numbers is a number which is a multiple of each of them; thus, 24 is a common multiple of 2, 3, and 4.

126. The **Least Common Multiple** of two or more numbers is the least number which is a multiple of each of them; thus, 12 is the least common multiple of 2, 3, and 4.

NOTE.—The least common multiple may be represented by the initials L. C. M.

PRINCIPLES.

1. *A multiple of a number is exactly divisible by that number.*

2. *A multiple of a number must contain all the prime factors of that number.*

3. *A common multiple of two or more numbers must contain all the prime factors of each of those numbers.*

4. *The least common multiple of two or more numbers must contain all the prime factors of each number, and no other factors.*

CASE I.

127. ***When the numbers are small and easily factored.***

FIRST METHOD.

128. This method consists in resolving the numbers into their prime factors, and taking the product of all the different factors.

1. Find the least common multiple of 12, 30, and 70.

SOLUTION.—We first resolve the numbers into their prime factors. A multiple of 12 must contain the factors of 12, 2, 2, 3; a multiple of 30 must contain the factors of 30, 2, 3, 5; a multiple of 70 must contain the factors of 70, 2, 5, 7; hence the common multiple of 12, 30, and 70 must contain all these different factors and no others; therefore $2\times2\times5\times3\times7$, or 420, is the L. C. M. of 12, 30, and 70 (Prin. 4).

OPERATION.

$12=2\times2\times3$
$30=2\times3\times5$
$70=2\times5\times7$
$2\times2\times3\times5\times7=420$

Rule.—I. *Resolve the numbers into their prime factors.*

II. *Take the product of all the different factors, using each factor the greatest number of times it occurs in either number.*

NOTE.—Any numbers which are divisors of the others may be omitted, since the multiple of the other numbers will be a multiple of these.

Find the least common multiple of

2. 24, 30, and 36. *Ans.* 360.
3. 16, 24, and 56. *Ans.* 336.
4. 28, 36, and 60. *Ans.* 1260.
5. 36, 48, and 84. *Ans.* 1008.
6. 63, 72, and 108. *Ans.* 1512.
7. 15, 30, 42, and 72. *Ans.* 2520.
8. 22, 55, 77, and 110. *Ans.* 770.
9. 33, 99, 36, 108, and 135. *Ans.* 5940.

10. A has $14, B $15, C $36, and D as many as the least common multiple of the amounts of the others; how many has D? *Ans.* $1260.

SECOND METHOD.

129. This method consists in taking out the prime factors of the least common multiple and finding their product.

1. Find the least common multiple of 12, 30, and 70.

OPERATION.

2	12	30	70
3	6	15	35
5	2	5	35
	2	1	7

$2 \times 3 \times 5 \times 2 \times 7 = 420$

SOLUTION.—Placing the numbers one beside another and dividing by 2, we see that 2 is a factor of each of them, it is therefore a factor of the L. C. M. (Prin. 3); dividing the quotients that will contain it by 3, we see that 3 is a factor of some of the numbers, it is therefore a factor of the L. C. M. Dividing the next quotients by 5, we see that 5 is a factor of some of them, hence 5 is a factor of the L. C. M.; and the quotients having no other common factor, we see that the factors of the given numbers are 2, 3, 5, 2, and 7, hence their product, which is 420, is the L. C. M. Hence the following

Rule.—I. *Write the numbers one beside another, divide by any prime number that will exactly divide two or more, and write the quotients and undivided numbers beneath.*

II. *Divide the quotients in the same manner, and thus continue until no two numbers in the lowest line have a common factor.*

III. *Take the product of the divisors and final quotients; the result will be the least common multiple required.*

Find the least common multiple of

2. 16, 20, and 30. *Ans.* 240.

3. 28, 56, and 84. *Ans.* 168.

4. 48, 60, and 30. *Ans.* 240.

5. 150, 200, and 250. *Ans.* 3000.

6. 40, 96, 100, and 120. *Ans.* 2400.

7. 120, 180, 200, and 240. *Ans.* 3600.

8. 140, 280, 160, and 320. *Ans.* 2240.

CASE II.

130. *When the numbers are large and cannot be readily factored.*

1. Find the least common multiple of 28 and 63.

OPERATION.

$28 = 4 \times 7;\ 63 = 9 \times 7$

$\text{L. C. M.} = 4 \times 7 \times 9$

$= 28 \times \frac{63}{7}$

SOLUTION.—The greatest common divisor of these numbers is 7; 28 equals 4 times 7, and 63 equals 9 times 7; hence the L. C. M., as found in the first method, is $4 \times 7 \times 9$, which equals 28 multiplied by 63 divided by 7; *or the first number multiplied by the second divided by their greatest common divisor.*

Rule.—I. *Find the greatest common divisor of two numbers, divide one number by it, and multiply the other number by the quotient.*

II. *When there are more than two numbers, find the least common multiple of two of the numbers, and then of this number and the third number*, etc.

WRITTEN EXERCISES.

Find the least common multiple of

2. 671 and 793. *Ans.* 8723.
3. 3503 and 4859. *Ans.* 150629.
4. 6527 and 7597. *Ans.* 463417.
5. 7205 and 9432. *Ans.* 518760.
6. 11183 and 15403. *Ans.* 816359.
7. 357, 612, and 663. *Ans.* 55692.
8. 4141, 6161, and 7171. *Ans.* 17934671.

9. What is the smallest sum of money for which I could hire workmen for one month, paying either $16, $20, $28, or $35 a month? *Ans.* $560.

10. A can dig 14 rods of ditch in a week, B 18 rods, C 22 rods, and D 24 rods; what is the least number of rods that would afford an exact number of weeks' work for each one of them? *Ans.* 5544 rods.

11. What is the smallest number of bushels of corn that will fill a number of barrels containing 3 bushels each, a number of sacks containing 5 bushels each, a number of casks containing 14 bushels each, or a number of bins containing 48 bushels each? *Ans.* 1680.

12. Four men start at the same place to walk around a garden; A can go around in 9 minutes, B in 10 minutes, C in 12 minutes, and D in 15 minutes; in what time will they all meet at the starting point? *Ans.* 180 minutes.

13. A, B, C, and D start from the same point, A traveling a mile in 18 minutes, B in 24 minutes, C in 30 minutes, and D in 35 minutes; what is the least whole number of miles each may travel that they may return to the starting point at the same moment? A, 140; B, 105; C, 84; D, 72

INTRODUCTION TO CANCELLATION.

MENTAL EXERCISES.

1. If we omit the factor 2 from 12 and 6, what factors will remain?

2. Divide 24 by 6. Divide 24 by $\frac{1}{2}$ of 6. Divide $\frac{1}{2}$ of 24 by 6.

3. Divide $\frac{1}{2}$ of 24 by $\frac{1}{2}$ of 6. Divide 36 by 18, first taking out the common factor 6.

4. Is there any difference in the quotient of 48 divided by 12, and $\frac{1}{4}$ of 48 divided by $\frac{1}{4}$ of 12?

5. Divide 72 by 48, first omitting common factors. Divide 90 by 60 in the same way; 144 by 96.

6. Divide $2\times2\times2$ by 2×2; $3\times3\times4$ by 2×3; $3\times4\times5$ by 3×5.

7. Divide $2\times3\times7$ by 2×7; $2\times3\times4$ by 2×3; $3\times5\times8$ by 3×8; $6\times5\times3$ by 3×6; $2\times7\times9\times10$ by 9×2.

CANCELLATION.

131. Cancellation is a process of abbreviating arithmetical operations by rejecting common factors in both dividend and divisor.

PRINCIPLES.

1. *The cancelling of a factor from any number divides the number by that factor.*

DEM.—Thus if we take the factor 3 out of 24 we shall divide 24 by 3.

2. *The cancelling of a factor in both dividend and divisor will not change the quotient.*

DEM.—Cancelling a factor in both dividend and divisor is the same as dividing them both by the same number, which, by the principles of division, does not change the quotient.

1. Divide 84×60 by 24×63.

SOLUTION.—We cancel the common factor 12 from 60 and 24, writing 5, the other factor of 60, above 60, and 2, the other factor of 24, below 24; we then cancel the common factor 21 from 84 and 63, writing 4, the other factor of 84, above 84, and 3, the other factor of 63, below 63; we then cancel 2 from 2 and 4, writing 2 above the 4; the product of the remaining factors of the dividend is 10, the product of the remaining factors of the divisor is 3; hence the quotient is 10 divided by 3, or $3\frac{1}{3}$.

OPERATION.

$$\frac{\overset{\overset{2}{\not4}}{\not8\not4}\times\overset{5}{\not6\not0}}{\underset{\not2}{\not2\not4}\times\underset{3}{\not6\not3}}=\frac{10}{3}=3\frac{1}{3}$$

Rule.—I. *Cancel the common factors from the dividend and divisor.*

II. *Then divide the product of the remaining factors of the dividend by the product of the remaining factors of the divisor.*

NOTES.—1. The unit 1 takes the place of a cancelled factor, but need not be written, except in the dividend of the quotient, when there are no other factors of the dividend remaining.

2. A factor in one term will cancel two or more factors in the other term, when their product is equal to the former.

3. Some prefer to place the dividend upon the right and the divisor upon the left, of a vertical line.

2. Divide $12\times 14\times 16$ by $6\times 7\times 8$. *Ans.* 8.

3. Divide $20\times 32\times 35$ by $4\times 5\times 16$. *Ans.* 70.

4. Divide 125×250 by $15\times 50\times 75$. *Ans.* $\frac{5}{9}$.

5. Divide 180×270 by 45×108. *Ans.* 10.

6. Divide 120×140 by 60×350. *Ans.* $\frac{4}{5}$.

7. Divide $45\times 49\times 81$ by $35\times 84\times 63$. *Ans.* $\frac{27}{28}$

8. Divide $60\times 77\times 320$ by $25\times 42\times 33$. *Ans.* $42\frac{2}{3}$.

9. Divide $75\times 42\times 99$ by $125\times 63\times 33$. *Ans.* $1\frac{1}{5}$.

WRITTEN EXERCISES.

1. How many yards of muslin, worth 12 cents a yard, may be bought for 16 pounds of butter, worth 15 cents a pound?

SOLUTION.—If one pound of butter is worth 15 cents, 16 pounds are worth 16×15 cents; for 16×15 cents, at 12 cents a yard, we can get as many yards of muslin as 12 is contained times in 15×16, which we find, by cancellation, to be 20.

OPERATION.

$$\frac{\overset{5}{\cancel{15}}\times\overset{4}{\cancel{16}}}{\underset{4}{\cancel{12}}}=20$$

2. How many bushels of corn, worth 45 cents a bushel, must be exchanged for 125 pounds of butter, at 18 cents a pound? *Ans.* 50.

3. A exchanged rye, worth 84 cents per bushel, for 78 bushels of wheat, worth 98 cents per bushel; required the number of bushels of rye. *Ans.* 91.

4. How many bushels of corn, at 42 cents a bushel, must be given in exchange for 7 pieces of cloth, each containing 40 yards, at 36 cents a yard? *Ans.* 240.

5. How many boxes of tea, each containing 24 pounds at 90 cents a pound, must be given for 27 firkins of butter of 56 pounds each, at 20 cents a pound? *Ans.* 14.

6. A farmer sold a grocer 9 loads of apples, each load containing 18 bags, and each bag 2 bushels, at 35 cents a bushel, and received in payment 12 boxes of sugar, each containing 135 pounds; what was the sugar worth a pound?

Ans. 7 cents

PRIME NUMBERS.

132. No general method of determining prime numbers, beyond a certain limit, has yet been discovered, although much time has been spent in the investigation.

133. We give the following practical method, which consists in writing a series of numbers, and sifting out those which are composite.

METHOD.—Since the even numbers after 2 are composite, we write the series of odd numbers; thus,

1, 3, 5, 7, $\overset{3}{9}$, 11, 13, $\overset{3,5}{15}$, 17, 19, $\overset{3,7}{21}$, 23, $\overset{5}{25}$, $\overset{3}{27}$, 29, 31, $\overset{3}{33}$, $\overset{5,7}{35}$, 37, $\overset{3}{39}$, 41,
43, $\overset{3,5}{45}$, 47, $\overset{7}{49}$, $\overset{3}{51}$, 53, $\overset{5}{55}$, $\overset{3}{57}$, 59, 61, $\overset{3,7}{63}$, $\overset{5}{65}$, 67, $\overset{3}{69}$, 71, 73, $\overset{3,5}{75}$, $\overset{7}{77}$, 79, $\overset{3}{81}$,
83, $\overset{5}{85}$, $\overset{3}{87}$, 89, $\overset{7}{91}$, $\overset{3}{93}$, $\overset{5}{95}$, 97, $\overset{3}{99}$.

Now, commencing at 3, since every third term is divisible by 3, every third number is composite, which we indicate by putting the figure 3 over it.

Commencing at 5, every fifth number is divisible by 5, and is therefore composite, hence we place a figure 5 over every fifth number. Proceed in the same manner with 7, and the numbers unmarked will be the prime numbers up to 100.

This method was discovered by Eratosthenes, a Greek mathematician. He inscribed the series of odd numbers on parchment, and then cutting out the composite numbers, his parchment with its holes resembled a sieve; hence the method has been called *Eratosthenes' Sieve.*

TABLE OF PRIME NUMBERS.

1	47	113	197	281	379	463	571	659	761	863
2	53	127	199	283	383	467	577	661	769	877
3	59	131	211	293	389	479	587	673	773	881
5	61	137	223	307	397	487	593	677	787	883
7	67	139	227	311	401	491	599	683	797	887
11	71	149	229	313	409	499	601	691	809	907
13	73	151	233	317	419	503	607	701	811	911
17	79	157	239	331	421	509	613	709	821	919
19	83	163	241	337	431	521	617	719	823	927
23	89	167	251	347	433	523	619	727	827	937
29	97	173	257	349	439	541	631	733	829	941
31	101	179	263	353	443	547	641	739	839	947
37	103	181	269	359	449	557	643	743	853	953
41	107	191	271	367	457	563	647	751	857	967
43	109	193	277	373	461	569	653	757	859	971

INTRODUCTION TO FRACTIONS.

MENTAL EXERCISES.

1. If an apple is divided into two equal parts, what is one of these parts called?

2. What are two of these parts called? How many halves in anything?

3. What is $\frac{1}{2}$ of 6? of 4? of 12? of 16? of 10? of 18? of 20? of 24? of 28? of 36?

4. If I divide an apple into 3 equal parts, what is one of these parts called?

5. What are 2 and 3 of these parts called? How many thirds in anything?

6. The number of equal parts into which a unit may be divided is represented by a figure below the line; thus $_2$ represents *halves*; $_3$, *thirds*; $_4$, *fourths*, etc.

7. The number of fractional parts taken may be represented by a figure above the line; thus, $\frac{2}{3}$ represents 2 *thirds*; $\frac{3}{4}$, 3 *fourths*; $\frac{5}{6}$, 5 *sixths*, etc.

8. What is $\frac{1}{3}$ of 6? of 9? of 12? of 18? of 15? of 21? of 27? What are $\frac{2}{3}$ of 12? of 15? of 21? of 18? of 24? of 33?

9. If I divide an apple into 4 equal parts, what is one of these parts called? If I divide in 5, or 6, etc., equal parts?

10. How many fourths make a whole? How many fifths? Sixths? Sevenths? Eighths? Ninths? Tenths?

11. What is $\frac{1}{4}$ of 12? $\frac{1}{5}$ of 20? $\frac{1}{6}$ of 24? $\frac{3}{4}$ of 16? $\frac{2}{5}$ of 30? $\frac{4}{7}$ of 28? $\frac{5}{8}$ of 40? $\frac{4}{7}$ of 35?

12. What is $\frac{3}{4}$ of 20? $\frac{2}{5}$ of 15? $\frac{5}{6}$ of 12? $\frac{7}{8}$ of 24? $\frac{5}{9}$ of 27? $\frac{4}{5}$ of 30? $\frac{5}{11}$ of 22? $\frac{7}{8}$ of 64?

13. If a yard of muslin cost 24 cents, what will $\frac{1}{4}$ of a yard cost? What will $\frac{3}{4}$ of a yard cost?

14. Henry's age is 36 years, and his wife's age is $\frac{5}{6}$ as much; what is his wife's age?

15. If 5 melons cost 60 cents, what will 7 melons cost at the same rate?

16. What will 4 yards of satin cost at the rate of $6 for $\frac{2}{3}$ of a yard?

17. What must I pay for three-fourths of a ton of hay if five-sixths of a ton cost $20?

18. What will $\frac{7}{8}$ of a ton of coal cost at the rate of $4.50 for $\frac{3}{4}$ of a ton?

SECTION IV.

COMMON FRACTIONS.

134. A **Fraction** is a number of the equal parts of a unit.

135. **Fractions** are divided into two classes; *common fractions* and *decimal fractions.*

136. A **Common Fraction** is one in which the unit is divided into *any number* of equal parts.

137. A **Decimal Fraction** is one in which the unit is divided into *tenths, hundredths,* etc.

138. A **Common Fraction** is expressed by two numbers, one written above the other, with a short line between them; thus, $\frac{3}{4}$ expresses 3 *fourths.*

139. The **Denominator** of a fraction denotes the number of equal parts into which the unit is divided.

140. The **Numerator** of a fraction denotes the number of equal parts which are taken

141. The **Numerator** and **Denominator** are called the *Terms* of the fraction. The numerator is written above the line, and the denominator below it.

142. **Common Fractions** consist of three principal classes; namely, *Simple, Compound,* and *Complex.*

143. A **Simple Fraction** is a fraction having a single integral numerator and denominator; as $\frac{2}{3}$, $\frac{4}{5}$, etc.

144. A **Proper Fraction** is a simple fraction whose value is less than a unit; as $\frac{2}{3}$, $\frac{3}{4}$.

145. An **Improper Fraction** is a simple fraction whose value is equal to or greater than a unit; as $\frac{5}{5}$, $\frac{7}{6}$, $\frac{12}{8}$, etc.

146. A **Compound Fraction** is a fraction of a fraction; as $\frac{2}{3}$ of $\frac{3}{4}$, $\frac{4}{7}$ of $\frac{5}{6}$ of $\frac{7}{8}$, etc.

147. A **Complex Fraction** is one whose numerator, or denominator, or both, are fractional; as $\frac{5}{\frac{4}{5}}$, $\frac{\frac{2}{3} \text{ of } \frac{4}{5}}{\frac{3}{4} \text{ of } \frac{5}{6}}$.

148. A **Mixed Number** consists of an integer and a fraction; as, $4\frac{2}{3}$, $5\frac{3}{4}$, etc.

149. The **Reciprocal** of a number is a unit divided by that number; thus, the reciprocal of 3 is $\frac{1}{3}$.

150. The **Number of Cases** of common fractions is eight. They are as follows:

1. Reduction,
2. Addition,
3. Subtraction,
4. Multiplication,
5. Division,
6. Relation of Fractions,
7. Greatest Common Divisor,
8. Least Common Multiple.

NOTES.—1. Each fractional part is used as a single thing and is therefore a unit; hence, we have *Units* and *fractional units*.

2. The primary conception of a fraction is that it is a number of equal parts of a unit. It may, however, be regarded as a *number* of parts of *one* thing, or as *one* part of a *number* of things. Thus, $\frac{3}{4}$ may be regarded as $\frac{3}{4}$ of *one* or $\frac{1}{4}$ of *three*.

NUMERATION AND NOTATION.

151. Numeration of Fractions is the art of reading a fraction when expressed by figures.

Rule.—*Read the number of fractional units expressed by the numerator, and give them the name indicated by the denominator.*

Name the kind and read the following:—

1. $\frac{2}{3}$.
2. $\frac{4}{5}$.
3. $\frac{7}{4}$.
4. $\frac{9}{10}$.
5. $\frac{8}{8}$.
6. $\frac{14}{17}$.
7. $5\frac{3}{7}$.
8. $\frac{16}{16}$.
9. $\frac{4}{5}$ of $\frac{3}{7}$.

152. Notation of Fractions is the art of expressing fractions by means of figures.

Rule.—*Write the number of fractional units, draw a line beneath, under which write the number which indicates the kind of fractional units.*

Write the following fractions:—

1. Two-thirds.
2. Five-sixths.
3. Six-eighths.
4. Nine-tenths.
5. Seven-elevenths.
6. Eight-tenths.
7. Eleven-fifteenths.
8. Twelve-twentieths.

ANALYSIS OF FRACTIONS.

153. To **Analyze** a fraction is to explain what is expressed by the fractional notation.

1. Analyze the fraction $\frac{4}{5}$.

SOLUTION.—In the fraction $\frac{4}{5}$, the denominator 5 indicates that the unit is divided into 5 equal parts, and the numerator 4 denotes that 4 of these parts are taken.

Analyze the following fractions:

2. $\frac{3}{5}$.	**5.** $\frac{9}{10}$.	**8.** $\frac{14}{11}$.
3. $\frac{5}{9}$.	**6.** $\frac{13}{14}$.	**9.** $\frac{17}{15}$.
4. $\frac{7}{8}$.	**7.** $\frac{10}{11}$.	**10.** $\frac{25}{37}$.

METHOD OF TREATMENT.

154. There are **Two Methods** of treating common fractions, which may be distinguished as the *Inductive* and *Deductive Methods.*

155. By the **Inductive Method** we solve all the different cases by analysis, and derive the *rules* or *methods of operation* from these analyses by *inference* or *induction.*

156. By the **Deductive Method** we first establish a few general principles, and then derive all the *rules* or *methods of operation* from these general principles.

NOTE.—The Inductive Method will be used with the mental exercises; with the written exercises the method which is thought to be the simplest is used.

PRINCIPLES OF FRACTIONS.

1. *Multiplying the numerator of a fraction by any number multiplies the value of the fraction by that number.*

If we multiply the numerator of a fraction by any number, as 5, the resulting fraction will express 5 times as many fractional units, each of the same size as before, hence the value of the fraction is 5 times as great.

2. *Dividing the numerator of a fraction by any number divides the value of the fraction by that number.*

If we divide the numerator of a fraction by any number, as 4, the resulting fraction will express $\frac{1}{4}$ as many fractional units, each of the same size as before, hence the value of the fraction is divided by 4.

3. *Multiplying the denominator of a fraction by any number divides the value of the fraction by that number.*

Since the denominator denotes the number of equal parts into which the unit is divided, if we multiply the denominator of a fraction by any number, as 5, the unit will be divided into 5 times as many equal parts, hence each fractional unit will be $\frac{1}{5}$ as large as before, and the same number of fractional units being taken, the value of the fraction is $\frac{1}{5}$ as great.

4. *Dividing the denominator of a fraction by any number multiplies the value of the fraction by that number.*

Since the denominator denotes the number of equal parts into which the unit is divided, if we divide it by any number, as 4, the unit will be divided into $\frac{1}{4}$ as many equal parts, hence each fractional unit will be 4 times as large as before, and the same number of fractional units being taken, the value of the fraction will be 4 times as great.

5. *Multiplying both numerator and denominator of a fraction by the same number does not change the value of the fraction.*

Since multiplying the numerator multiplies the value of the fraction, and multiplying the denominator divides the value of the fraction, multiplying both numerator and denominator both multiplies and divides the value of the fraction by the same number, and hence does not change its value.

6. *Dividing both numerator and denominator of a fraction by the same number does not change its value.*

Since dividing the numerator divides the value of the fraction, and dividing the denominator multiplies the value, dividing both numerator and denominator both divides and multiplies the value of the fraction, and hence does not change its value.

157. These principles may be embodied in one general law as follows:

General Principle.—*A change in the* NUMERATOR *by multiplication or division produces a* SIMILAR *change in the value of the fraction, but such a change in the* DENOMINATOR *produces an* OPPOSITE *change in the value of the fraction.*

REDUCTION OF FRACTIONS.

158. The **Reduction of Fractions** is the process of changing their form without altering their value.

159. There are **Six Cases** of reduction:

1st Numbers to fractions.	4th. To lower terms.
2d. Fractions to numbers.	5th. Compound to simple
3d. To higher terms.	6th. Complex to simple.

NOTE.—Reducing to a *Common Denominator* is included in these six cases.

CASE I.

160. ***To reduce whole or mixed numbers to improper fractions.***

MENTAL EXERCISES.

1. How many fifths in $4\frac{3}{5}$?

SOLUTION.—In *one* there are 5 *fifths*, and in 4 there are 4 times 5 fifths, or 20 fifths, which added to 3 fifths, equal 23 fifths; therefore $4\frac{3}{5} = \frac{23}{5}$.

2. How many fourths in $7\frac{3}{4}$? in $5\frac{3}{4}$? in $9\frac{1}{4}$? in $12\frac{3}{4}$?

3. How many fifths in $6\frac{3}{5}$? in $7\frac{2}{5}$? in $8\frac{1}{5}$? in $10\frac{4}{5}$?

4. How many sixths in $4\frac{5}{6}$? in $8\frac{5}{6}$? in $9\frac{1}{6}$? in $11\frac{5}{6}$?

5. How many sevenths in $7\frac{4}{7}$? in $9\frac{3}{7}$? in $8\frac{5}{7}$? in $12\frac{6}{7}$?

6. Describe the operation we perform in reducing a mixed number to a fraction.

WRITTEN EXERCISES.

1. Reduce $27\frac{3}{4}$ to fourths.

OPERATION.

$$\begin{array}{r} 27\frac{3}{4} \\ 4 \\ \hline \frac{111}{4} \end{array}$$

SOLUTION.—In *one* there are 4 fourths, and in 27 there are 27 times 4 *fourths*, or $\frac{108}{4}$, which added to the $\frac{3}{4}$, equals $\frac{111}{4}$. Therefore, etc.

Rule.—*Multiply the whole number by the denominator of the fraction, add the numerator to the product, and write the denominator under the sum.*

Reduce to improper fractions,

2. $5\frac{2}{3}$. *Ans.* $\frac{17}{3}$.

3. $12\frac{3}{5}$. *Ans.* $\frac{63}{5}$.

4. $18\frac{4}{5}$. *Ans.* $\frac{94}{5}$.

5. $11\frac{5}{9}$. *Ans.* $\frac{104}{9}$.

6. $27\frac{5}{8}$. *Ans.* $\frac{221}{8}$.

7. $51\frac{5}{6}$. *Ans.* $\frac{311}{6}$.

8. $35\frac{11}{12}$. *Ans.* $\frac{431}{12}$.

9. $24\frac{7}{13}$; $51\frac{7}{13}$. *Ans.* $\frac{319}{13}$; $\frac{670}{13}$.

10. $82\frac{14}{15}$; $100\frac{99}{100}$. *Ans.* $\frac{1244}{15}$; $\frac{10099}{100}$.

11. $49\frac{49}{50}$; $235\frac{42}{65}$. *Ans.* $\frac{2499}{50}$; $\frac{15317}{65}$.

CASE II.

161. ***To reduce improper fractions to whole or mixed numbers.***

MENTAL EXERCISES.

1. How many units in $\frac{23}{4}$?

SOLUTION.—In *one* there are $\frac{4}{4}$, hence in 23 fourths there are as many ones as 4 is contained times in 23, which are $5\frac{3}{4}$. Therefore $\frac{23}{4} = 5\frac{3}{4}$.

2. How many units in $\frac{18}{4}$? in $\frac{26}{5}$? in $\frac{35}{6}$?

3. How many units in $\frac{17}{3}$? in $\frac{21}{4}$? in $\frac{25}{6}$?

4. How many units in $\frac{30}{9}$? in $\frac{43}{11}$? in $\frac{45}{13}$?

5. How many units in $\frac{27}{14}$? in $\frac{38}{16}$? in $\frac{63}{25}$?

6. Describe the process of reducing an improper fraction to a mixed number.

WRITTEN EXERCISES.

1. How many units in $\frac{79}{5}$?

SOLUTION.—Since dividing both terms of a fraction by the same number does not change its value (Prin. 5), by dividing both terms by 5, we have $\frac{15\frac{4}{5}}{1}$, or $15\frac{4}{5}$.

OPERATION.

$\frac{79}{5}=15\frac{4}{5}$.

Rule.—*Divide the numerator by the denominator, and the quotient will be the whole or mixed number.*

Reduce to whole or mixed numbers,

2. $\frac{16}{4}$. *Ans.* 4.

3. $\frac{13}{3}$. *Ans.* $4\frac{1}{3}$.

4. $\frac{25}{6}$. *Ans.* $4\frac{1}{6}$.

5. $\frac{79}{9}$. *Ans.* $8\frac{7}{9}$.

6. $\frac{87}{11}$. *Ans.* $7\frac{10}{11}$.

7. $\frac{144}{16}$. *Ans.* 9.

8. $\frac{296}{16}$. *Ans.* $18\frac{8}{16}$.

9. $\frac{982}{15}$. *Ans.* $65\frac{7}{15}$.

10. $\frac{1208}{148}$. *Ans.* $8\frac{24}{148}$.

11. $\frac{1000}{40}$. *Ans.* 25.

12. $\frac{725}{45}$ ft.; $\frac{329}{25}$ yd. *Ans.* $16\frac{5}{45}$; $13\frac{4}{25}$.

13. $\frac{7284}{128}$; $\frac{6875}{284}$; $\frac{57896}{7864}$. *Ans.* $59\frac{27}{128}$; $29\frac{89}{284}$; $7\frac{2848}{7864}$.

CASE III.

162. *To reduce fractions to higher terms.*

163. **Reducing a Fraction to higher terms** is the process of reducing it to an equivalent fraction, having a greater numerator and denominator.

MENTAL EXERCISES.

1. Change $\frac{2}{3}$ to twelfths.

SOLUTION.—In *one* there are $\frac{12}{12}$, and in $\frac{1}{3}$ there are $\frac{1}{3}$ of $\frac{12}{12}$, which are $\frac{4}{12}$ and in $\frac{2}{3}$ there are 2 times $\frac{4}{12}$, which are $\frac{8}{12}$; therefore $\frac{2}{3}=\frac{8}{12}$.

2. Change $\frac{3}{4}$ and $\frac{2}{5}$ to twentieths; $\frac{2}{3}$ and $\frac{3}{5}$ to fifteenths.

3. Change $\frac{3}{5}$ and $\frac{5}{6}$ to thirtieths; $\frac{3}{4}$ and $\frac{7}{8}$ to sixteenths.

4. Change $\frac{4}{5}$ and $\frac{5}{7}$ to seventieths; $\frac{5}{6}$ and $\frac{7}{9}$ to eighteenths.

5. Change $\frac{7}{8}$ and $\frac{5}{9}$ to seventy-seconds; $\frac{9}{10}$ and $\frac{11}{12}$ to 120ths.

6. Describe the process of reducing a fraction to higher terms.

WRITTEN EXERCISES.

1. How many twentieths in $\frac{3}{5}$?

SOLUTION.—Since multiplying both terms of a fraction by the same number does not change its value (Prin. 5), we multiply both terms by the number which will give the required denominator, which we see is 4; hence, $\frac{3}{5}=\frac{12}{20}$.

OPERATION.

$\frac{3}{5}=\frac{3\times4}{5\times4}=\frac{12}{20}$.

Rule.—*Multiply both numerator and denominator by the number which will give the required denominator.*

2. Reduce $\frac{3}{4}$, $\frac{2}{3}$, and $\frac{5}{6}$ to twelfths. *Ans.* $\frac{9}{12}$, $\frac{8}{12}$, $\frac{10}{12}$.
3. Reduce $\frac{3}{4}$, $\frac{4}{5}$, and $\frac{9}{10}$ to twentieths. *Ans.* $\frac{15}{20}$, $\frac{16}{20}$, $\frac{18}{20}$.
4. Reduce $\frac{1}{2}$, $\frac{3}{4}$, and $\frac{5}{8}$ to sixteenths. *Ans.* $\frac{8}{16}$, $\frac{12}{16}$, $\frac{10}{16}$.
5. Reduce $\frac{4}{5}$, $\frac{5}{6}$, and $\frac{7}{10}$ to thirtieths. *Ans.* $\frac{24}{30}$, $\frac{25}{30}$, $\frac{21}{30}$.
6. Reduce $\frac{6}{8}$, $\frac{4}{6}$, and $\frac{17}{24}$ to forty-eighths. *Ans.* $\frac{36}{48}$, $\frac{32}{48}$, $\frac{34}{48}$.
7. Reduce $\frac{4}{5}$, $\frac{7}{8}$, $\frac{9}{10}$, and $\frac{11}{12}$ to 120ths.
Ans. $\frac{96}{120}$, $\frac{105}{120}$, $\frac{108}{120}$, $\frac{110}{120}$.

CASE IV.

164. *To reduce fractions to lower terms.*

165. Reducing a Fraction to lower terms is the process of reducing it to an equivalent fraction having a smaller numerator and denominator.

Principle.—*A fraction is in its lowest terms when the numerator and denominator are prime to each other.*

MENTAL EXERCISES.

1. Reduce $\frac{9}{12}$ to fourths.

SOLUTION.—*One* equals $\frac{12}{12}$, and $\frac{1}{4}$ equals $\frac{1}{4}$ of $\frac{12}{12}$, or $\frac{3}{12}$; since $\frac{3}{12}$ equals $\frac{1}{4}$, $\frac{9}{12}$ equals as many fourths as 3 is contained times in 9, which is 3; hence $\frac{9}{12}$ equals $\frac{3}{4}$. Therefore, etc.

Reduce

2. $\frac{15}{20}$ to 4ths. *Ans.* $\frac{3}{4}$.
3. $\frac{16}{20}$ to 5ths. *Ans.* $\frac{4}{5}$.
4. $\frac{8}{24}$ to 6ths. *Ans.* $\frac{2}{6}$.
5. $\frac{54}{63}$ to 7ths. *Ans.* $\frac{6}{7}$.
6. $\frac{54}{72}$ to 8ths. *Ans.* $\frac{6}{8}$.
7. $\frac{48}{108}$, $\frac{84}{126}$ to 9ths. *Ans.* $\frac{4}{9}$, $\frac{6}{9}$.
8. Describe the process of reducing a fraction to lower terms.

WRITTEN EXERCISES.

1. Reduce $\frac{24}{30}$ to fifths.

SOLUTION.—Since dividing both terms of a fraction by the same number does not change its value (Prin. 6), we may reduce $\frac{24}{30}$ to lower terms by dividing both numerator and denominator by 6; dividing, we have $\frac{24}{30}$ equal to $\frac{4}{5}$; and since the terms 4 and 5 are prime to each other, the fraction is in its lowest terms. Therefore, etc.

OPERATION.

$6)\frac{24}{30}=\frac{4}{5}$.

Rule I.—*Divide both terms successively by their common factors.*

Rule II.—*Divide both terms by their greatest common divisor.*

Reduce the following fractions to lowest terms:—

2. $\frac{14}{21}$, $\frac{15}{18}$.	*Ans.* $\frac{2}{3}$, $\frac{5}{6}$.	7. $\frac{120}{168}$, $\frac{648}{720}$.	*Ans.* $\frac{5}{7}$, $\frac{9}{10}$.
3. $\frac{16}{24}$, $\frac{18}{30}$.	*Ans.* $\frac{2}{3}$, $\frac{3}{5}$.	8. $\frac{792}{864}$, $\frac{192}{216}$.	*Ans.* $\frac{11}{12}$, $\frac{8}{9}$.
4. $\frac{15}{25}$, $\frac{18}{36}$.	*Ans.* $\frac{3}{5}$, $\frac{1}{2}$.	9. $\frac{840}{4312}$, $\frac{1176}{1512}$.	*Ans.* $\frac{15}{77}$, $\frac{7}{9}$.
5. $\frac{21}{84}$, $\frac{24}{36}$.	*Ans.* $\frac{1}{4}$, $\frac{2}{3}$.	10. $\frac{726}{792}$, $\frac{9540}{13356}$.	*Ans.* $\frac{11}{12}$, $\frac{5}{7}$.
6. $\frac{72}{288}$, $\frac{288}{864}$.	*Ans.* $\frac{1}{4}$, $\frac{1}{3}$.	11. $\frac{6161}{7171}$, $\frac{42521}{118301}$.	*Ans.* $\frac{61}{71}$, $\frac{101}{281}$.

CASE V.

166. ***To reduce compound fractions to simple ones.***

MENTAL EXERCISES.

1. What is $\frac{1}{3}$ of $\frac{1}{4}$?

SOLUTION.—$\frac{1}{3}$ of $\frac{1}{4}$ is *one* of the *three* equal parts into which $\frac{1}{4}$ may be divided; if each fourth is divided into 3 equal parts, 4 fourths, or the unit, will be divided into 4 times 3, or 12 equal parts; hence each part is $\frac{1}{12}$ of a unit. Therefore, $\frac{1}{3}$ of $\frac{1}{4}$ is $\frac{1}{12}$.

2. What is $\frac{1}{3}$ of $\frac{1}{5}$? $\frac{1}{4}$ of $\frac{1}{6}$? $\frac{1}{6}$ of $\frac{1}{7}$?

3. What is $\frac{1}{8}$ of $\frac{1}{9}$? $\frac{1}{9}$ of $\frac{1}{10}$? $\frac{1}{11}$ of $\frac{1}{12}$?

4. What is $\frac{2}{3}$ of $\frac{1}{5}$? $\frac{3}{4}$ of $\frac{2}{5}$? $\frac{2}{3}$ of $\frac{5}{6}$?

5. What is $\frac{3}{5}$ of $\frac{4}{6}$? $\frac{5}{6}$ of $\frac{5}{7}$? $\frac{4}{5}$ of $\frac{5}{8}$?

6. What is $\frac{4}{7}$ of $\frac{3}{8}$? $\frac{5}{8}$ of $\frac{3}{4}$? $\frac{3}{7}$ of $\frac{14}{9}$?

7. In reducing a compound fraction to a simple one, what do we do with the numerators, and what with the denominators?

WRITTEN EXERCISES.

1. What is the value of $\frac{3}{4}$ of $\frac{5}{6}$?

SOLUTION.—$\frac{1}{4}$ of $\frac{5}{6}$ equals $\frac{5}{24}$ (Prin. 3), and since $\frac{1}{4}$ of $\frac{5}{6}$ equals $\frac{5}{24}$, $\frac{3}{4}$ of $\frac{5}{6}$ equals 3 times $\frac{5}{24}$, which by Prin. 1, equals $\frac{15}{24}$ or $\frac{5}{8}$.

OPERATION.

$\frac{3}{4}$ of $\frac{5}{6}$ =

$\frac{3\times 5}{4\times 6}=\frac{5}{8}$, *Ans.*

Rule.—*Multiply the numerators together and the denominators together, cancelling the factors common to both terms.*

NOTE.—Reduce whole or mixed numbers to fractions before commencing the reduction to a simple fraction. To reduce complex fractions to simple ones, see Art. 183.

What is

2. $\frac{4}{5}$ of $\frac{6}{7}$?	*Ans.* $\frac{24}{35}$.	8. $\frac{2}{3}$ of $\frac{3}{4}$ of $\frac{4}{5}$?	*Ans.* $\frac{2}{5}$.
3. $\frac{5}{6}$ of $\frac{3}{4}$?	*Ans.* $\frac{5}{8}$.	9. $\frac{3}{5}$ of $\frac{4}{7}$ of $\frac{10}{12}$?	*Ans.* $\frac{2}{7}$.
4. $\frac{6}{7}$ of $\frac{9}{10}$?	*Ans.* $\frac{27}{35}$.	10. $\frac{5}{6}$ of $\frac{7}{8}$ of $\frac{24}{35}$?	*Ans.* $\frac{1}{2}$.
5. $\frac{7}{8}$ of $\frac{12}{14}$?	*Ans.* $\frac{3}{4}$.	11. $\frac{2}{3}$ of $\frac{4}{5}$ of $\frac{6}{7}$ of $\frac{21}{28}$?	*Ans.* $\frac{12}{35}$.
6. $\frac{11}{12}$ of $\frac{16}{22}$?	*Ans.* $\frac{2}{3}$.	12. $\frac{4}{9}$ of $\frac{12}{15}$ of $\frac{25}{64}$ of $\frac{96}{75}$?	*Ans.* $\frac{8}{45}$.
7. $\frac{15}{19}$ of $\frac{19}{15}$?	*Ans.* 1.	13. $\frac{18}{25}$ of $\frac{35}{27}$ of $\frac{77}{88}$ of $\frac{117}{154}$?	*Ans.* $\frac{273}{440}$

5

MENTAL EXERCISES.

1. A man gave his son $\frac{1}{2}$ of $\frac{1}{3}$ of a dollar; how much did the boy receive?

2. A man owning $\frac{1}{3}$ of the stock of a bank, sold $\frac{1}{3}$ of his stock; how much did he sell?

3. A boy earned $\frac{3}{4}$ of a dollar and gave his mother $\frac{1}{2}$ of it; how much did he give his mother?

4. Having $\frac{3}{4}$ of a bushel of apples I gave $\frac{2}{3}$ of them to my sister; what part of a bushel did I give her?

5. Having lost $\frac{1}{4}$ of my money I found $\frac{1}{2}$ of what I lost, and then had $140; how much had I at first?

6. At $\frac{3}{4}$ of a dollar a gallon, what will $\frac{5}{6}$ of a gallon of syrup cost?

7. At $\frac{4}{5}$ of a dollar a yard, what will $\frac{2}{3}$ of a yard of cloth cost?

8. At $$6\frac{1}{4}$ a barrel, what will $\frac{3}{5}$ of a barrel of flour cost?

9. At $$8\frac{1}{3}$ a case, what will $\frac{4}{5}$ of a case of slates cost?

WRITTEN EXERCISES.

1. A had $\frac{4}{7}$ of a ton of hay, which is $\frac{3}{4}$ as much as B has; how much has B?

SOLUTION.—If $\frac{4}{7}$ of a ton of hay is $\frac{3}{4}$ of what B has, $\frac{1}{4}$ of what B has is $\frac{1}{3}$ of $\frac{4}{7}$, which is $\frac{4}{21}$ of a ton, and $\frac{4}{4}$ of what B has is 4 times $\frac{4}{21}$ of a ton, which is $\frac{16}{21}$ of a ton. Therefore, etc.

2. A has $\frac{6}{7}$ of a certain sum of money, which is $\frac{4}{5}$ of what B has; how much has B? *Ans.* $\frac{15}{14}$.

3. A barrel of flour cost $$8\frac{4}{5}$, and a barrel of fish cost $\frac{3}{4}$ as much; what was the cost of the fish? *Ans.* $$6\frac{3}{5}$.

4. A lady bought $\frac{7}{8}$ of a yard of velvet, at $$15\frac{3}{5}$ a yard: what did it cost? *Ans.* $$13\frac{13}{20}$.

5. Henry having $\frac{12}{15}$ of a quart of nuts, divided them among 8 of his schoolmates; what did each receive? *Ans.* $\frac{1}{10}$.

6. A owns $\frac{4}{7}$ of the stock of a railroad, and $\frac{3}{4}$ of this is $3\frac{3}{7}$ times what B owns; how much does B own? *Ans.* $\frac{1}{8}$.

7. Mary shared $\frac{18}{19}$ of a bushel of berries with 8 of her schoolmates; what did each receive? *Ans.* $\frac{2}{19}$.

8. I drew $580 from bank, which was $\frac{5}{6}$ of what still remained in bank; what was my bank deposit? *Ans.* $1276.

9. William lost $\frac{2}{3}$ of $\frac{5}{6}$ of his money, and found that $132 was $\frac{3}{4}$ of $\frac{8}{9}$ of $\frac{11}{12}$ of what remained; how much had he at first? *Ans.* $486.

COMMON DENOMINATOR.

167. **A Common Denominator** is a denominator common to several fractions, or a denominator to which all may be reduced.

168. **Similar Fractional Units** are those which are of the same kind; as 3 *fifths* and 2 *fifths*.

169. **Dissimilar Fractional Units** are those which are of different kinds; as, 3 *fourths* and 3 *fifths*.

Principle.—*A common denominator of several fractions must be a common multiple of their denominators.*

MENTAL EXERCISES.

1. Reduce $\frac{2}{3}$ and $\frac{3}{4}$ to a common denominator.

SOLUTION.—A common denominator for *thirds* and *fourths* is *twelfths* In *one* there are $\frac{12}{12}$, and in $\frac{1}{3}$ there are $\frac{1}{3}$ of $\frac{12}{12}$, or $\frac{4}{12}$, and in $\frac{2}{3}$, etc.

Reduce to a common denominator,

2. $\frac{2}{3}$ and $\frac{3}{4}$.
3. $\frac{2}{5}$ and $\frac{5}{6}$.
4. $\frac{4}{5}$ and $\frac{3}{8}$.
5. $\frac{3}{4}$ and $\frac{4}{5}$.
6. $\frac{5}{6}$ and $\frac{7}{8}$.
7. $\frac{7}{8}$ and $\frac{9}{10}$.
8. Describe the process of reducing two fractions to a common denominator.

WRITTEN EXERCISES.

1. Reduce $\frac{3}{4}$, $\frac{4}{5}$, and $\frac{5}{7}$ to a common denominator.

SOLUTION.—Since the product of the denominators of the fractions is a common multiple of their denominators, $4\times5\times7$, which equals 140, will be the common denominator. Then multiplying both terms of $\frac{3}{4}$ by 5×7 we have $\frac{3}{4}=\frac{105}{140}$ (Prin. 5). Multiplying both terms of $\frac{4}{5}$ by 4×7, we have $\frac{4}{5}=\frac{112}{140}$, etc. Hence the following

OPERATION.

$\frac{3}{4}$, $\frac{4}{5}$, $\frac{5}{7}$ = $\frac{105}{140}$, $\frac{112}{140}$, $\frac{100}{140}$

Rule.—*Multiply both terms of each fraction by the denominators of the other fractions.*

Reduce to a common denominator,

2. $\frac{3}{4}$, $\frac{5}{6}$, and $\frac{7}{8}$. *Ans.* $\frac{144}{192}$, $\frac{160}{192}$, $\frac{168}{192}$.
3. $\frac{2}{5}$, $\frac{4}{7}$, and $\frac{5}{6}$. *Ans.* $\frac{84}{210}$, $\frac{120}{210}$, $\frac{175}{210}$.
4. $\frac{4}{5}$, $\frac{3}{4}$, and $\frac{6}{11}$. *Ans.* $\frac{176}{220}$, $\frac{165}{220}$, $\frac{120}{220}$.
5. $\frac{7}{8}$, $\frac{8}{9}$, and $\frac{10}{11}$. *Ans.* $\frac{693}{792}$, $\frac{704}{792}$, $\frac{720}{792}$.
6. $\frac{3}{5}$, $\frac{5}{8}$, and $\frac{11}{12}$. *Ans.* $\frac{288}{480}$, $\frac{300}{480}$, $\frac{440}{480}$.
7. $\frac{14}{16}$, $\frac{13}{12}$, and $4\frac{1}{8}$. *Ans.* $\frac{1344}{1536}$, $\frac{1664}{1536}$, $\frac{6336}{1536}$.
8. $\frac{1}{2}$, $\frac{2}{3}$, $\frac{3}{4}$, $\frac{5}{6}$, and $\frac{7}{8}$. *Ans.* $\frac{576}{1152}$, $\frac{768}{1152}$, $\frac{864}{1152}$, $\frac{960}{1152}$, $\frac{1008}{1152}$.
9. Show that the common denominator of several fractions is a common *multiple of the* denominators of those fractions.

LEAST COMMON DENOMINATOR.

170. The **Least Common Denominator** of several fractions is the smallest denominator to which all may be reduced.

Principle.—*The least common denominator of several fractions is the least common multiple of their denominators.*

1. Reduce $\frac{2}{3}$, $\frac{5}{6}$, and $\frac{7}{8}$ to their least common denominator

SOLUTION.—We find the least common multiple of the denominators to be 24, hence 24 is the least common denominator. Dividing 24 by 3, the denominator of $\frac{2}{3}$, we find we must multiply 3 by 8 to produce 24; hence multiplying both terms of $\frac{2}{3}$ by 8, we have $\frac{2}{3}=\frac{16}{24}$ (Prin. 5). Dividing 24 by 6, the denominator of $\frac{5}{6}$, we find we must multiply 6 by 4 to produce 24; hence, multiplying both terms by 4, we have $\frac{5}{6}=\frac{20}{24}$, etc.

OPERATION.

L. C. M.=24

$$\frac{2}{3}=\frac{2\times8}{3\times8}=\frac{16}{24}$$

$$\frac{5}{6}=\frac{5\times4}{6\times4}=\frac{20}{24}$$

$$\frac{7}{8}=\frac{7\times3}{8\times3}=\frac{21}{24}$$

Rule.—I. *Find the least common multiple of the denominators, for the least common denominator.*

II. *Divide the least common denominator by the denominator of each fraction, and multiply both terms by the quotient.*

NOTE.—Reduce compound fractions to simple ones, mixed numbers to improper fractions, and all to their lowest terms, before finding the least common denominator.

To their least common denominator,

2. Reduce $\frac{5}{6}$, $\frac{4}{9}$, $\frac{7}{12}$. *Ans.* $\frac{30}{36}$, $\frac{16}{36}$, $\frac{21}{36}$.

3. Reduce $\frac{3}{4}$, $\frac{7}{10}$, $\frac{5}{6}$. *Ans.* $\frac{45}{60}$, $\frac{42}{60}$, $\frac{50}{60}$.

4. Reduce $\frac{1}{4}$, $\frac{11}{12}$, $\frac{13}{20}$. *Ans.* $\frac{15}{60}$, $\frac{55}{60}$, $\frac{39}{60}$.

5. Reduce $4\frac{1}{2}$, $5\frac{2}{3}$, $7\frac{3}{4}$. *Ans.* $\frac{54}{12}$, $\frac{68}{12}$, $\frac{93}{12}$.

6. Reduce $\frac{3}{5}$, $\frac{17}{20}$, $\frac{19}{24}$. *Ans.* $\frac{72}{120}$, $\frac{102}{120}$, $\frac{95}{120}$.

7. Reduce $\frac{6}{9}$, $\frac{12}{14}$, $\frac{15}{18}$, $\frac{21}{36}$. *Ans.* $\frac{168}{252}$, $\frac{216}{252}$, $\frac{210}{252}$, $\frac{147}{252}$.

8. Reduce $2\frac{3}{5}$, $5\frac{4}{7}$, $\frac{13}{14}$, $\frac{9}{10}$, $\frac{29}{35}$. *Ans.* $\frac{182}{70}$, $\frac{390}{70}$, $\frac{65}{70}$, $\frac{63}{70}$, $\frac{58}{70}$.

9. Reduce $\frac{2}{3}$ of $\frac{7}{8}$, $\frac{4}{5}$ of $3\frac{3}{4}$, $11\frac{13}{34}$. *Ans.* $\frac{119}{204}$, $\frac{612}{204}$, $\frac{2322}{204}$.

10. Reduce $\frac{1}{2}$, $\frac{1}{3}$, $\frac{1}{4}$, $\frac{1}{5}$, $\frac{1}{6}$, $\frac{1}{7}$, $\frac{1}{8}$, $\frac{1}{9}$. *Ans.* $\frac{1260}{2520}$, $\frac{840}{2520}$, etc.

ADDITION OF FRACTIONS.

171. Addition of Fractions is the process of finding the sum of two or more fractions.

PRINCIPLES.

1. *To add two or more fractions, they must express similar fractional units.*

2. *To add two or more fractions they must be reduced to a common denominator.*

MENTAL EXERCISES.

1. How many sixths in $\frac{1}{2}$ and $\frac{2}{3}$?

SOLUTION.—$\frac{1}{2}$ is equal to $\frac{3}{6}$ and $\frac{2}{3}$ is equal to $\frac{4}{6}$; $\frac{3}{6}$ and $\frac{4}{6}$ are $\frac{7}{6}$, which is equal to $1\frac{1}{6}$.

Find the sum

2. Of $\frac{1}{3}$ and $\frac{1}{4}$.
3. Of $\frac{2}{3}$ and $\frac{3}{4}$.
4. Of $\frac{3}{4}$ and $\frac{4}{5}$.
5. Of $\frac{1}{2}$ and $\frac{2}{5}$.
6. Of $\frac{2}{3}$ and $\frac{5}{6}$.
7. Of $\frac{4}{5}$ and $\frac{3}{7}$.
8. Of $\frac{3}{4}$ and $\frac{7}{8}$.
9. Of $\frac{5}{6}$ and $\frac{7}{8}$.

10. How then shall we add two fractions whose denominators are unlike?

WRITTEN EXERCISES.

1. What is the sum of $\frac{3}{4}$, $\frac{5}{6}$, and $\frac{7}{8}$?

SOLUTION.—Reducing the fractions to a common denominator that they may express similar fractional units, we have $\frac{3}{4}=\frac{18}{24}$, $\frac{5}{6}=\frac{20}{24}$, $\frac{7}{8}=\frac{21}{24}$; 18 *twenty-fourths* plus 20 *twenty-fourths* plus 21 *twenty-fourths* equals 59 *twenty-fourths*. Hence the following

OPERATION.

$$\frac{3}{4}+\frac{5}{6}+\frac{7}{8}=$$
$$\frac{18}{24}+\frac{20}{24}+\frac{21}{24}=\frac{59}{24}.$$

Rule.—*Reduce the fractions to a common denominator, then add the numerators and write the sum over the common denominator.*

NOTES.—1. Reduce compound fractions to simple ones, and reduce each fraction and the sum to lowest terms.

2. To add mixed numbers, add the integers and fractions separately, and then unite their sums.

2. Find the sum of $\frac{4}{5}$, $\frac{6}{8}$, $\frac{9}{12}$. *Ans.* $2\frac{3}{10}$.
3. Find the sum of $\frac{4}{6}$, $\frac{7}{9}$, $\frac{5}{8}$. *Ans.* $2\frac{5}{72}$.
4. Find the sum of $\frac{5}{6}$, $\frac{11}{12}$, $\frac{13}{18}$. *Ans.* $2\frac{17}{36}$.
5. Find the sum of $\frac{4}{7}$, $\frac{3}{4}$, $\frac{11}{14}$. *Ans.* $2\frac{3}{28}$.
6. Find the sum of $2\frac{1}{2}$, $4\frac{3}{4}$, $\frac{11}{12}$. *Ans.* $8\frac{1}{6}$.
7. Find the sum of $\frac{3}{4}$, $\frac{5}{6}$, $\frac{7}{8}$, $\frac{11}{12}$. *Ans.* $3\frac{3}{8}$.
8. Find the sum of $\frac{2}{3}$, $\frac{4}{5}$, $\frac{5}{6}$, $\frac{7}{8}$. *Ans.* $3\frac{7}{40}$.
9. Find the sum of $\frac{5}{6}$, $\frac{7}{8}$, $\frac{9}{10}$, $\frac{11}{12}$. *Ans.* $3\frac{21}{40}$.
10. Find the sum of $2\frac{1}{2}$, $4\frac{2}{3}$, $3\frac{1}{4}$, $1\frac{9}{10}$. *Ans.* $12\frac{19}{60}$.
11. Find the sum of $4\frac{3}{4}$, $7\frac{1}{8}$, $9\frac{1}{16}$, $7\frac{1}{50}$. *Ans.* $27\frac{383}{400}$.
12. Find the sum of $21\frac{4}{7}$, $35\frac{1}{11}$, $22\frac{3}{21}$, and $4\frac{1}{2}$. *Ans.* $83\frac{47}{154}$.
13. Find the sum of $17\frac{7}{8}$, $49\frac{8}{15}$, $24\frac{6}{11}$, $18\frac{1}{33}$. *Ans.* $109\frac{433}{440}$.
14. Find the sum of $\frac{3}{4}$ of $\frac{5}{6}$, $\frac{4}{5}$ of $\frac{6}{7}$, $\frac{5}{6}$ of $\frac{7}{8}$. *Ans.* $2\frac{67}{1680}$.
15. Find the sum of $\frac{1}{2}$, $\frac{1}{3}$, $\frac{1}{4}$, $\frac{1}{5}$, $\frac{1}{6}$, $\frac{1}{7}$, $\frac{1}{8}$, $\frac{1}{9}$, $\frac{1}{10}$. *Ans.* $1\frac{2341}{2520}$.

SUBTRACTION OF FRACTIONS.

172. Subtraction of Fractions is the process of finding the difference between two fractions.

PRINCIPLES.

1. *To subtract two fractions they must express similar fractional units.*

2. *To subtract two fractions they must be reduced to a common denominator.*

MENTAL EXERCISES.

1. How many 12ths in the difference between $\frac{3}{4}$ and $\frac{2}{3}$?

SOLUTION.—$\frac{3}{4}$ equals $\frac{9}{12}$ and $\frac{2}{3}$ equals $\frac{8}{12}$, and the difference between $\frac{9}{12}$ and $\frac{8}{12}$ is $\frac{1}{12}$.

Subtract

2. $\frac{1}{4}$ from $\frac{1}{3}$.
3. $\frac{2}{5}$ from $\frac{3}{4}$.
4. $\frac{3}{4}$ from $\frac{4}{5}$.
5. $\frac{2}{5}$ from $\frac{1}{2}$.
6. $\frac{2}{3}$ from $\frac{5}{6}$.
7. $\frac{4}{7}$ from $\frac{3}{5}$.
8. $\frac{3}{4}$ from $\frac{7}{8}$.
9. $\frac{5}{6}$ from $\frac{7}{8}$.
10. How then shall we subtract two fractions whose denominators are unlike?

WRITTEN EXERCISES.

1. What is the difference between $\frac{5}{8}$ and $\frac{7}{9}$?

SOLUTION.—Reducing the fractions to a common denominator that they may express similar fractional units, we have $\frac{7}{9}=\frac{56}{72}$ and $\frac{5}{8}=\frac{45}{72}$: 56 *seventy-seconds* minus 45 *seventy-seconds* equals 11 *seventy-seconds*. Hence the following

OPERATION.

$$\frac{7}{9}-\frac{5}{8}=$$
$$\frac{56}{72}-\frac{45}{72}=\frac{11}{72}$$

Rule.—*Reduce the fractions to a common denominator, take the difference of the numerators, and write it over the common denominator.*

NOTE.—Reduce compound fractions to simple ones, and reduce each fraction and the difference to its lowest terms.

Subtract

2. $\frac{3}{5}$ from $\frac{5}{7}$. *Ans.* $\frac{4}{35}$.
3. $\frac{2}{7}$ from $\frac{4}{9}$. *Ans.* $\frac{10}{63}$.
4. $\frac{3}{4}$ from $\frac{7}{9}$. *Ans.* $\frac{1}{36}$.
5. $\frac{8}{9}$ from $\frac{9}{10}$. *Ans.* $\frac{1}{90}$.
6. $\frac{10}{11}$ from $\frac{11}{12}$. *Ans.* $\frac{1}{132}$.
7. $\frac{8}{15}$ from $\frac{13}{14}$. *Ans.* $\frac{83}{210}$.
8. $\frac{13}{17}$ from $\frac{17}{18}$. *Ans.* $\frac{1}{68}$.
9. $\frac{13}{15}$ from $\frac{18}{19}$. *Ans.* $\frac{23}{285}$.
10. $\frac{14}{26}$ from $\frac{28}{36}$. *Ans.* $\frac{28}{117}$.
11. $\frac{7}{12}$ from $\frac{3}{4}$ of $\frac{9}{10}$. *Ans.* $\frac{11}{120}$.
12. $1\frac{1}{2}$ from $\frac{4}{7}$ of $3\frac{1}{3}$. *Ans.* $\frac{17}{42}$.
13. $\frac{3}{7}$ from $\frac{8}{9}$ of $\frac{18}{36}$. *Ans.* $\frac{1}{63}$.
14. $\frac{2}{5}$ of $\frac{15}{18}$ from $\frac{7}{8}$ of $\frac{24}{28}$. *Ans.* $\frac{5}{12}$.
15. $\frac{3}{4}$ of $\frac{7}{8}$ from $\frac{4}{5}$ of $\frac{15}{16}$. *Ans.* $\frac{3}{32}$.

16. Subtract $8\frac{5}{7}$ from $12\frac{3}{7}$.

Solution.—We cannot subtract $\frac{5}{7}$ from $\frac{3}{7}$, so we take 1 from 12, which added to $\frac{3}{7}$ equals $1\frac{3}{7}$ or $\frac{10}{7}$; $\frac{5}{7}$ from $\frac{10}{7}$ leaves $\frac{5}{7}$, and 8 from 11 leaves 3; hence the difference is $3\frac{5}{7}$.

17. Subtract $8\frac{5}{6}$ from $12\frac{1}{6}$. *Ans.* $3\frac{1}{3}$.

18. Subtract $5\frac{7}{8}$ from $10\frac{3}{8}$. *Ans.* $4\frac{1}{2}$.

19. Subtract $10\frac{8}{9}$ from $20\frac{7}{9}$. *Ans.* $9\frac{8}{9}$.

20. Subtract $12\frac{7}{11}$ from $24\frac{3}{11}$. *Ans.* $11\frac{7}{11}$.

21. Subtract $20\frac{11}{12}$ from $30\frac{3}{4}$. *Ans.* $9\frac{5}{6}$.

22. Subtract $40\frac{7}{9}$ from $60\frac{5}{12}$. *Ans.* $19\frac{23}{36}$.

WRITTEN PROBLEMS

IN ADDITION AND SUBTRACTION OF FRACTIONS.

1. A has $\$5\frac{1}{2}$, B has $\$4\frac{1}{4}$, and C has $\$6\frac{3}{4}$; how much money have they all? *Ans.* $\$16\frac{1}{2}$.

2. A miller ground $7\frac{1}{3}$ bushels of corn for A, $9\frac{3}{4}$ for B, $10\frac{1}{5}$ for C; how much did he grind in all? *Ans.* $27\frac{17}{60}$ bu.

3. A lady bought material for a wrapper costing $\$1\frac{11}{40}$, and buttons costing $\$\frac{9}{40}$; what change should she receive from a \$5 bill? *Ans.* $\$3\frac{1}{2}$.

4. A lady went shopping with \$100 and paid $\$12\frac{1}{2}$ for a bonnet, $\$32\frac{3}{4}$ for a dress, and $\$52\frac{5}{8}$ for a cloak; how much money did she bring home? *Ans.* $\$2.12\frac{1}{2}$.

5. A boy gave $12\frac{1}{2}$ cents for a slate, $18\frac{3}{4}$ cents for a knife, $87\frac{1}{2}$ cents for a grammar, and $62\frac{1}{2}$ cents for an arithmetic; what did they all cost? *Ans.* $\$1.31\frac{1}{4}$

6. A merchant bought two pieces of muslin, each containing $41\frac{5}{8}$ yd.; after selling $57\frac{3}{4}$ yd. from them, how many yards remained? *Ans.* $25\frac{1}{2}$ yd.

7. Mr. Weeks finds that his family burned last winter $1\frac{1}{2}$ tons of coal in December, $2\frac{5}{8}$ tons of coal in January, $2\frac{3}{16}$ tons of coal in February, and in March $1\frac{3}{8}$ tons; how much was burned during the four months? *Ans.* $7\frac{11}{16}$ tons.

8. Four loads of hay weighed upon the scales $49\frac{7}{75}$ hundredweight, $43\frac{4}{25}$ hundredweight, $39\frac{1}{20}$ hundredweight, and $45\frac{9}{100}$ hundredweight; what was the weight of the hay, the weight of the wagon being $15\frac{7}{25}$ hundredweight?

Ans. $115\frac{23}{50}$ hundredweight.

MULTIPLICATION OF FRACTIONS.

173. Multiplication of Fractions is the process of finding a product when one or both factors are fractions.

174. There are **Three Cases**: 1st. A fraction by an integer; 2d. An integer by a fraction; 3d. A fraction by a fraction.

CASE I.

175. ***To multiply a fraction by an integer.***

MENTAL EXERCISES.

1. How many are 5 times $\frac{2}{15}$?

SOLUTION.—5 times $\frac{2}{15}$ are $\frac{10}{15}$, which reduced to its lowest terms equals $\frac{2}{3}$. Therefore, etc.

SOLUTION 2d.—5 times $\frac{1}{15}$ equal $\frac{5}{15}$, or $\frac{1}{3}$; if 5 times $\frac{1}{15}$ equal $\frac{1}{3}$, 5 times $\frac{2}{15}$ equals 2 times $\frac{1}{3}$ or $\frac{2}{3}$; hence 5 times $\frac{2}{15}$ equals $\frac{2}{3}$.

How many are

2. 3 times $\frac{5}{9}$?
3. 6 times $\frac{7}{18}$?
4. 7 times $\frac{20}{21}$?
5. 8 times $\frac{19}{24}$?
6. 3 times $5\frac{5}{6}$?
7. 4 times $7\frac{7}{8}$?
8. 8 times $6\frac{3}{4}$?
9. 9 times $4\frac{7}{27}$?

10. Since 5 times $\frac{2}{15}$ equals $\frac{2}{3}$, how may this result be obtained by omitting the analysis?

11. How then may a fraction be multiplied by omitting the analysis?

WRITTEN EXERCISES.

1. Multiply $\frac{11}{12}$ by 4

SOLUTION.—Multiplying the numerator (Prin. 1), we have 4 times $\frac{11}{12}$ equals $\frac{44}{12}$; or, dividing the denominator (Prin. 4), we have 4 times $\frac{11}{12}=\frac{11}{3}$, or $3\frac{2}{3}$.

OPERATION.

$$\frac{11}{12}\times 4=\frac{44}{12}$$
$$\frac{11}{12}\times 4=\frac{11}{3}$$

Rule.—*To multiply a fraction by an integer, multiply the numerator, or divide the denominator.*

Multiply

2. $\frac{11}{12}$ by 6. *Ans.* $5\frac{1}{2}$.
3. $\frac{14}{15}$ by 5. *Ans.* $4\frac{2}{3}$.
4. $\frac{17}{18}$ by 12. *Ans.* $11\frac{1}{3}$.
5. $\frac{19}{36}$ by 18. *Ans.* $9\frac{1}{2}$.
6. $\frac{15}{28}$ by 14. *Ans.* $7\frac{1}{2}$.
7. $\frac{115}{112}$ by 28. *Ans.* $28\frac{3}{4}$.
8. $\frac{105}{224}$ by 32. *Ans.* 15.
9. $\frac{229}{1728}$ by 144. *Ans.* $19\frac{1}{12}$.
10. $7\frac{9}{10}$ by 5. *Ans.* $39\frac{1}{2}$
11. $18\frac{7}{9}$ by 3. *Ans.* $56\frac{1}{3}$
12. $28\frac{15}{20}$ by 5. *Ans.* $143\frac{3}{4}$.
13. $17\frac{18}{21}$ by 7. *Ans.* 125.
14. $24\frac{19}{24}$ by 12. *Ans.* $297\frac{1}{2}$.
15. $46\frac{15}{26}$ by 13. *Ans* $605\frac{1}{2}$.
16. $128\frac{19}{54}$ by 18. *Ans.* $2310\frac{1}{3}$.
17. $428\frac{18}{121}$ by 11. *Ans.* $4709\frac{7}{11}$.

CASE II.

176. *To multiply an integer by a fraction.*

MENTAL EXERCISES.

1. Multiply 12 by $\frac{3}{4}$.

SOLUTION.—12 multiplied by *one* equals 12, hence 12 multiplied by $\frac{1}{4}$ equals $\frac{1}{4}$ of 12, which is 3, and 12 multiplied by $\frac{3}{4}$ equals 3 times 3, which are 9.

2. 16 by $\frac{3}{4}$.
3. 25 by $\frac{4}{5}$.
4. 35 by $\frac{5}{7}$.
5. 26 by $\frac{5}{6}$.
6. 30 by $\frac{7}{8}$.
7. 40 by $2\frac{3}{4}$.
8. 50 by $3\frac{4}{5}$.
9. 60 by $4\frac{5}{6}$.

10. In multiplying 12 by $\frac{3}{4}$, by what do we divide 12, and by what do we multiply the quotient?

11. How then may we multiply a whole number by a fraction?

WRITTEN EXERCISES.

1. Multiply 7 by $\frac{4}{5}$.

SOLUTION.—7 multiplied by *one* equals 7, hence 7 multiplied by $\frac{1}{5}$ equals $\frac{1}{5}$ of 7, which is $\frac{7}{5}$, and 7 multiplied by $\frac{4}{5}$ equals 4 times $\frac{7}{5}$, which equals $\frac{28}{5}$. Therefore, etc.

OPERATION.

$$7\times\frac{4}{5}=\frac{28}{5}$$

2. Multiply 28 by $7\frac{3}{5}$.

SOLUTION.—Multiplying 28 by 3 and dividing by 5, we have $28\times\frac{3}{5}$ equals $16\frac{4}{5}$; and then multiplying 28 by 7, we have 196; adding 196 to $16\frac{4}{5}$, we have $212\frac{4}{5}$. Therefore, etc.

OPERATION.

```
   28
   7 3/5
5)84
  16 4/5
 196
 212 4/5  Ans.
```

NOTE.—This method of multiplying by a mixed number is more convenient than the one usually presented.

Rule.—*Multiply the integer by the numerator of the multiplier, and divide the product by the denominator; or divide first and then multiply.*

Multiply

3. 9 by $\frac{2}{3}$. *Ans.* 6.
4. 10 by $\frac{4}{5}$. *Ans.* 8.
5. 18 by $\frac{8}{9}$. *Ans.* 16.
6. 27 by $\frac{5}{7}$. *Ans.* $19\frac{2}{7}$.
7. 28 by $3\frac{4}{7}$. *Ans.* 100.
8. 180 by $15\frac{4}{5}$. *Ans.* 2844.
9. 144 by $\frac{17}{24}$. *Ans.* 102.
10. 284 by $\frac{35}{36}$. *Ans.* $276\frac{1}{9}$.
11. 256 by $7\frac{15}{16}$. *Ans* 2032.
12. 180 by $25\frac{17}{24}$. *Ans* $4627\frac{1}{2}$.
13. 196 by $\frac{9}{15}$ of $\frac{45}{48}$. *Ans.* $110\frac{1}{4}$
14. 648 by $\frac{40}{65}$ of $\frac{78}{84}$. *Ans.* $370\frac{2}{7}$

CASE III.

177. ***To multiply a fraction by a fraction.***

MENTAL EXERCISES.

1. What is the product of $\frac{6}{7}$ by $\frac{3}{5}$?

SOLUTION.—$\frac{6}{7}$ multiplied by *one* equals $\frac{6}{7}$, hence $\frac{6}{7}$ multiplied by $\frac{1}{5}$ equals $\frac{1}{5}$ of $\frac{6}{7}$, which is $\frac{6}{35}$, and $\frac{6}{7}$ multiplied by $\frac{3}{5}$ equals 3 times $\frac{6}{35}$, which is $\frac{18}{35}$. Therefore, etc.

What is the product

2. Of $\frac{5}{6}$ by $\frac{3}{4}$?
3. Of $\frac{7}{8}$ by $\frac{10}{11}$?
4. Of $\frac{9}{20}$ by $\frac{5}{18}$?
5. Of $\frac{7}{24}$ by $\frac{3}{10}$?
6. Of $2\frac{1}{2}$ by $3\frac{1}{5}$?
7. Of $7\frac{1}{2}$ by $\frac{4}{5}$?
8. Of $1\frac{1}{99}$ by $\frac{11}{20}$?
9. Of $5\frac{1}{3}$ by $4\frac{1}{8}$?

10. In multiplying $\frac{6}{7}$ by $\frac{3}{5}$, what do we multiply together for the numerator of the product and what for the denominator?

11. How then do we multiply a fraction by a fraction?

WRITTEN EXERCISES.

1. What is the product of $\frac{5}{6}$ by $\frac{7}{8}$?

SOLUTION.—$\frac{5}{6}$ multiplied by *one* equals $\frac{5}{6}$, hence $\frac{5}{6}$ multiplied by $\frac{1}{8}$ equals $\frac{1}{8}$ of $\frac{5}{6}$, which is $\frac{5}{48}$, and $\frac{5}{6}$ multiplied by $\frac{7}{8}$ equals 7 times $\frac{5}{48}$, which is $\frac{35}{48}$. Therefore, etc.

OPERATION.

$$\frac{5}{6}\times\frac{7}{8}=\frac{35}{48}$$

SOLUTION 2D.—$\frac{5}{6}$ times $\frac{7}{8}$ equals $\frac{1}{6}$ of 5 times $\frac{7}{8}$; 5 times $\frac{7}{8}$ equals $\frac{35}{8}$ (Prin. 1); and $\frac{1}{6}$ of $\frac{35}{8}$ equals $\frac{35}{48}$ (Prin. 3).

Rule.—*Multiply the numerators together and the denominators together, cancelling common factors.*

NOTE.—Practically this case is the same as finding a fractional part of a fraction, but theoretically the two cases are entirely distinct.

What is the product of

2. $\frac{5}{9}$ by $\frac{3}{7}$? *Ans.* $\frac{5}{21}$.
3. $\frac{10}{12}$ by $\frac{6}{15}$? *Ans.* $\frac{1}{3}$.
4. $\frac{49}{99}$ by $\frac{33}{56}$? *Ans.* $\frac{7}{24}$.
5. $\frac{3}{4}\times\frac{5}{6}\times\frac{6}{7}\times\frac{7}{8}$? *Ans.* $\frac{15}{32}$.
6. $\frac{16}{21}\times\frac{9}{28}\times\frac{35}{54}$? *Ans.* $\frac{10}{63}$.
7. $\frac{9}{10} < \frac{16}{27}\times\frac{35}{48}$? *Ans.* $\frac{7}{18}$.
8. $\frac{18}{24}\times\frac{28}{45}\times\frac{115}{116}$? *Ans.* $\frac{161}{348}$.
9. $\frac{45}{49}\times\frac{56}{65}\times\frac{91}{72}$? *Ans.* 1.
10. $12\frac{1}{2}\times 8\frac{1}{3}$? *Ans.* $104\frac{1}{6}$.
11. $17\frac{2}{5}\times 16\frac{2}{3}$? *Ans.* 290.
12. $27\frac{3}{11}\times 21\frac{5}{6}$? *Ans.* $595\frac{5}{11}$.
13. $39\frac{3}{5}\times 33\frac{1}{3}$. *Ans.* 1320.

14. Multiply $\frac{9}{10}$ of $\frac{18}{35}$ by $\frac{7}{24}$ of $\frac{16}{27}$. *Ans.* $\frac{2}{25}$.
15. Multiply $\frac{16}{28}$ of $\frac{18}{24}$ by $\frac{28}{45}$ of $\frac{115}{116}$. *Ans.* $\frac{23}{87}$.
16. Multiply $\frac{19}{27}$ of $\frac{102}{95}$ by $\frac{144}{189}$ of $\frac{273}{1728}$. *Ans.* $\frac{221}{2430}$.

MENTAL EXERCISES.

1. How much will $4\frac{3}{4}$ yards of muslin cost, at 6 cents a yard?

2. If you give to 8 boys each $2\frac{3}{4}$ apples, how many apples will it take?

3. How many dollars does that man give away who gives to each of 10 beggars $\frac{3}{5}$ of a dollar?

4. How much will 15 inkstands cost at the rate of 2 inkstands for $\frac{3}{4}$ of a dollar?

5. What will 12 Letter-writers cost at the rate of 3 for $\frac{3}{10}$ of a dollar?

6. If a man can dig a rod of ditch in $\frac{7}{8}$ of a day, how many rods can he dig in $4\frac{1}{2}$ days?

7. Multiplying by $\frac{1}{4}$ is the same as taking what part of a number?

8. Show that a fractional part of an integer equals the product of the integer by the fraction.

9. What will be the cost of $2\frac{1}{2}$ yards of cloth at $\$3\frac{1}{4}$ a yard?

10. What will $7\frac{1}{4}$ bushels of shelled corn cost at $25\frac{1}{2}$ cents a bushel?

11. A lady bought $4\frac{3}{4}$ dozen spoons at $\$6\frac{3}{5}$ a dozen; what did they cost her?

12. When flour is selling at $\$7\frac{1}{2}$ a barrel, what must I pay for $8\frac{3}{4}$ barrels?

WRITTEN EXERCISES.

Required the cost of

1. 96 lb. of paper, at $2\frac{5}{6}$ cts. per pound? *Ans.* \$2.72.
2. 475 lb. of beef, at $6\frac{3}{4}$ cts. per pound? *Ans.* $\$32.06\frac{1}{4}$.
3. 130 yd. of muslin, at $18\frac{3}{4}$ cts. a yard? *Ans.* $\$24.37\frac{1}{2}$.
4. 387 sheep, if 1 sheep cost $8\frac{3}{5}$ dollars? *Ans.* $\$3328\frac{1}{5}$.
5. 180 tons of coal, at $\$10.62\frac{1}{2}$ per ton? *Ans.* $\$1912\frac{1}{2}$.
6. $729\frac{2}{3}$ lb. of sugar, at 6 cts. a pound? *Ans.* \$43.78.
7. $8\frac{7}{8}$ barrels of flour, at $\$6\frac{2}{5}$ a barrel? *Ans.* $\$56\frac{4}{5}$.
8. $6\frac{3}{8}$ yards of cloth, at $\$5\frac{1}{4}$ a yard? *Ans.* $\$33\frac{15}{32}$.
9. $3\frac{5}{7}$ bushels of apples, at $\$\frac{5}{8}$ a bushel? *Ans.* $\$2\frac{9}{28}$.
10. 24 handkerchiefs, at $87\frac{1}{2}$ cents apiece? *Ans.* \$21
11. $6\frac{3}{4}$ tons of hay, at $\$6\frac{7}{8}$ a ton? *Ans.* $\$46\frac{13}{32}$.
12. $8\frac{2}{3}$ cords of wood, at $\$4\frac{2}{5}$ a cord? *Ans.* $\$38\frac{2}{15}$.
13. 15 barrels of vinegar, at $\$4\frac{3}{4}$ a barrel? *Ans.* $\$71\frac{1}{4}$.
14. $16\frac{1}{2}$ quarts of beans, at $\$0.18\frac{3}{4}$ a quart? *Ans.* $\$3.09\frac{3}{8}$.
15. What is the value of $\frac{24}{25}\times\frac{75}{81}\times\frac{9}{64}\times\frac{256}{289}$? *Ans.* $\frac{32}{289}$.
16. What is the value of $\frac{49}{64}\times\frac{27}{128}\times\frac{144}{729}\times\frac{48}{343}$? *Ans.* $\frac{1}{224}$.
17. Value of $\frac{729}{961}\times\frac{992}{1089}\times\frac{231}{1024}\times\frac{124}{1001}$? *Ans.* $\frac{243}{12584}$.

18. Value of $\frac{289}{441}\times\frac{231}{340}\times\frac{700}{841}\times\frac{870}{999}\times\frac{1998}{935}$? *Ans.* $\frac{20}{29}$.

19. Value of $(\frac{1}{3}+\frac{8}{15})\times\frac{3}{11}+(\frac{4}{17}+\frac{5}{6})\times 3$? *Ans.* $3\frac{827}{1870}$.

20. Value of $(\frac{5}{6}+\frac{10}{11})\times(\frac{3}{4}+\frac{7}{15})\times\frac{72}{73}\times\frac{363}{230}$? *Ans.* $3\frac{3}{10}$.

21. Value of $7\frac{7}{10}\times 4\frac{6}{11}+(4\frac{1}{2}-3\frac{4}{5})-(5\frac{6}{7}+6\frac{3}{11})$?

Ans. $23\frac{439}{770}$.

22. Find the value of $732\frac{1}{11}-(33\frac{1}{3}+37\frac{1}{2})$ diminished by $(6\frac{1}{4}+6\frac{7}{12})\times 49\frac{15}{16}$. *Ans.* $20\frac{415}{1056}$.

23. Mr. Fish rented a house at $\$42\frac{3}{4}$ a month, taking a lease for 5 years; but disposed of the lease at the end of $3\frac{1}{4}$ years; how much rent did he pay? *Ans.* $\$1667\frac{1}{4}$.

24. A bill of books at retail amounts to $\$376\frac{3}{8}$, but I get a reduction of $\frac{1}{3}$ for wholesale and $\frac{3}{50}$ for cash; what was the exact amount of the bill? *Ans.* $\$235\frac{517}{800}$.

25. Mr. Alden bought $52\frac{1}{2}$ bushels of wheat at $62\frac{1}{2}$ cts per bushel, and sold $\frac{1}{3}$ at $69\frac{1}{2}$ cts., and the remainder at $71\frac{3}{8}$ cts. per bushel; what did he clear by the transaction?

Ans. $\$4.43\frac{1}{8}$.

DIVISION OF FRACTIONS.

178. Division of Fractions is the process of dividing when one or both of the terms are fractional.

179. There are **Three Cases**: 1st. A fraction by an integer; 2d. An integer by a fraction; 3d. A fraction by a fraction.

CASE I.

180. ***To divide a fraction by an integer.***

MENTAL EXERCISES.

1. How many times is 4 contained in $\frac{8}{11}$?

SOLUTION.—*One* is contained in $\frac{8}{11}$, $\frac{8}{11}$ times; and if 1 is contained in $\frac{8}{11}$, $\frac{8}{11}$ times, 4 is contained in $\frac{8}{11}$, $\frac{1}{4}$ of $\frac{8}{11}$ times, or $\frac{2}{11}$ times. Therefore, etc.

How many times is

2. 3 contained in $\frac{7}{8}$?
3. 5 contained in $\frac{15}{16}$?
4. 6 contained in $\frac{8}{9}$?
5. 7 contained in $\frac{14}{15}$?
6. 4 contained in $8\frac{4}{5}$?
7. 6 contained in $12\frac{3}{5}$?
8. 7 contained in $6\frac{3}{5}$?
9. 9 contained in $17\frac{1}{10}$?

10. Since $\frac{8}{11}\div 4=\frac{2}{11}$, how may we divide a fraction by a number that will divide the numerator?

11. Since $\frac{10}{11}\div 4=\frac{10}{44}$, how may we divide a fraction by a number *that will not divide* the numerator?

WRITTEN EXERCISES.

1. Divide $\frac{10}{11}$ by 5, and also by 6.

Solution.—Dividing the numerator of $\frac{10}{11}$ by 5, we have $\frac{10}{11} \div 5$ equals $\frac{2}{11}$ (Prin. 2).

OPERATION.

$\frac{10}{11} \div 5 = \frac{2}{11}$

Solution 2d.—Multiplying the denominator of $\frac{10}{11}$ by 6, we have $\frac{10}{11} \div 6$ equals $\frac{10}{66}$, or $\frac{5}{33}$ (Prin. 3).

OPERATION.

$\frac{10}{11} \div 6 = \frac{10}{66} = \frac{5}{33}$

2. Divide $822\frac{5}{8}$ by 4.

Solution.—Dividing 822 by 4, we have 205 and a remainder of 2; 2 equals $\frac{16}{8}$, which, added to $\frac{5}{8}$, equals $\frac{21}{8}$; $\frac{21}{8}$ divided by 4 equals $\frac{21}{32}$; hence the quotient is $205\frac{21}{32}$.

OPERATION.

$4)822\frac{5}{8}$

$205\frac{21}{32}$

Rule.—*Divide the numerator or multiply the denominator of the dividend by the divisor.*

Note.—Reduce a mixed number to a fraction; or divide the integer, unite the remainder with the fraction, and divide the result.

Divide

3. $\frac{9}{10}$ by 3.	*Ans.* $\frac{3}{10}$.	**10.** $785\frac{3}{5}$ by 7.	*Ans.* $112\frac{8}{35}$.
4. $\frac{16}{15}$ by 4.	*Ans.* $\frac{4}{15}$.	**11.** $287\frac{5}{8}$ by 12.	*Ans.* $23\frac{31}{32}$.
5. $\frac{18}{25}$ by 36.	*Ans.* $\frac{1}{50}$.	**12.** $129\frac{3}{5}$ by 16.	*Ans.* $8\frac{1}{10}$.
6. $\frac{3}{4}$ of $\frac{14}{9}$ by 7.	*Ans.* $\frac{1}{6}$.	**13.** $878\frac{4}{11}$ by 15.	*Ans.* $58\frac{92}{165}$.
7. $\frac{5}{6}$ of $\frac{72}{125}$ by 9.	*Ans.* $\frac{4}{75}$.	**14.** $1284\frac{9}{13}$ by 26.	*Ans.* $49\frac{139}{338}$.
8. $\frac{7}{8}$ of $\frac{72}{91}$ by 18.	*Ans.* $\frac{1}{26}$.	**15.** $1347\frac{1}{17}$ by 30.	*Ans.* $44\frac{46}{51}$.
9. $2\frac{4}{21}$ by $\frac{7}{8}$ of $3\frac{3}{7}$.	*Ans.* $\frac{46}{63}$.	**16.** $2892\frac{7}{20}$ by 36.	*Ans.* $80\frac{247}{720}$

CASE II.

181. ***To divide an integer by a fraction.***

MENTAL EXERCISES.

1. How many times is $\frac{3}{4}$ contained in 3?

Solution.—*One* is contained in 3, 3 times; and if 1 is contained in 3, 3 times, $\frac{1}{4}$ is contained in 3, 4 times 3 times, which are 12 times; and $\frac{3}{4}$ is contained in 3, $\frac{1}{3}$ of 12 times, or 4 times. Therefore, etc.

How many times is

2. $\frac{2}{3}$ contained in 6?

3. $\frac{4}{5}$ contained in 8?

4. $\frac{7}{8}$ contained in 7?

5. $\frac{3}{4}$ contained in 11?

6. $1\frac{1}{2}$ contained in 3?

7. $1\frac{3}{4}$ contained in 7?

8. $2\frac{3}{10}$ contained in 46?

9. $1\frac{3}{7}$ contained in 30?

10. In dividing 12 by $\frac{3}{4}$, by what do we multiply 12, and by what do we divide the product?

11. How then may we divide a whole number by a fraction without the analysis?

WRITTEN EXERCISES.

1. Divide 10 by the fraction $\frac{5}{6}$.

SOLUTION.—10 divided by *one* equals 10, hence 10 divided by $\frac{1}{6}$ equals 6 times 10, and 10 divided by 5 sixths equals $\frac{1}{5}$ of 6 times 10, which is $\frac{6}{5}$ times 10, which equals 12. Hence the

OPERATION

$10 \div \frac{5}{6} =$
$10 \times \frac{6}{5} = 12$ *Ans.*

Rule I.—*Multiply the dividend by the denominator of the fraction, and divide the product by the numerator.*

Rule II.—*Invert the divisor and proceed as in multiplication of fractions.*

Divide

2. 5 by $\frac{3}{5}$.	*Ans.* $8\frac{1}{3}$.	8. 15 by $8\frac{3}{4}$.	*Ans.* $1\frac{5}{7}$.
3. 9 by $\frac{4}{7}$.	*Ans.* $15\frac{3}{4}$.	9. 256 by $3\frac{7}{35}$.	*Ans.* 80.
4. 12 by $\frac{6}{7}$.	*Ans.* 14.	10. 552 by $2\frac{11}{29}$.	*Ans.* 232.
5. 27 by $2\frac{1}{4}$.	*Ans.* 12.	11. 558 by $8\frac{6}{7}$.	*Ans.* 63.
6. 30 by $\frac{5}{8}$.	*Ans.* 48.	12. 638 by $19\frac{1}{3}$.	*Ans.* 33.
7. 144 by $2\frac{2}{3}$.	*Ans.* 54.	13. 729 by $22\frac{1}{2}$.	*Ans.* $32\frac{2}{5}$.

CASE III.

182. ***To divide a fraction by a fraction.***

MENTAL EXERCISES.

1. How many times is $\frac{3}{4}$ contained in $\frac{5}{6}$?

SOLUTION.—*One* is contained in $\frac{5}{6}$, $\frac{5}{6}$ times, and if 1 is contained in $\frac{5}{6}$, $\frac{5}{6}$ times, $\frac{1}{4}$ is contained in $\frac{5}{6}$, 4 times $\frac{5}{6}$ times, which is $\frac{20}{6}$, or $\frac{10}{3}$ times; and $\frac{3}{4}$ is contained in $\frac{5}{6}$, $\frac{1}{3}$ of $\frac{10}{3}$, or $\frac{10}{9}$ times. Therefore, etc.

How many times is

2. $\frac{4}{5}$ contained in $\frac{7}{8}$?
3. $\frac{2}{3}$ contained in $\frac{4}{7}$?
4. $\frac{4}{9}$ contained in $\frac{5}{6}$?
5. $\frac{7}{11}$ contained in $1\frac{2}{3}$?
6. $\frac{3}{4}$ contained in $1\frac{4}{5}$?
7. $\frac{5}{9}$ contained in $2\frac{2}{3}$?
8. $\frac{7}{11}$ contained in $8\frac{3}{11}$?
9. $\frac{4}{9}$ contained in $9\frac{2}{3}$?

10. In dividing $\frac{5}{6}$ by $\frac{3}{4}$ we multiply $\frac{5}{6}$ by what, and divide the product by what?

11. How then may we divide a fraction by a fraction without the analysis?

WRITTEN EXERCISES.

1. Divide $\frac{7}{8}$ by $\frac{5}{6}$.

SOLUTION.—$\frac{7}{8}$ divided by *one* equals $\frac{7}{8}$, hence $\frac{7}{8}$ divided by $\frac{1}{6}$ equals 6 times $\frac{7}{8}$, and $\frac{7}{8}$ divided by 5 sixths equals $\frac{1}{5}$ of 6 times $\frac{7}{8}$, which is $\frac{6}{5}$ times $\frac{7}{8}$, which equals $\frac{42}{40}$, or $\frac{21}{20}$. Therefore we have the *following*

OPERATION.

$\frac{7}{8} \div \frac{5}{6} =$
$\frac{7}{8} \times \frac{6}{5} = \frac{42}{40}$ or
$= \frac{21}{20}$. *Ans.*

Rule I.—*Multiply the dividend by the denominator of the divisor, and divide by the numerator.*

Rule II.—*Invert the divisor and proceed as in multiplication of fractions.*

Note.—Reduce mixed numbers to simple fractions. When the divisor is a compound fraction, invert each term and multiply, cancelling when possible.

Divide

2. $\frac{7}{8}$ by $\frac{3}{4}$. *Ans.* $1\frac{1}{6}$.
3. $\frac{9}{10}$ by $\frac{4}{5}$. *Ans.* $1\frac{1}{8}$.
4. $\frac{27}{52}$ by $\frac{36}{65}$. *Ans.* $\frac{15}{16}$.
5. $\frac{14}{25}$ by $1\frac{13}{15}$. *Ans.* $\frac{3}{10}$.
6. $\frac{48}{35}$ by $\frac{72}{105}$. *Ans.* 2.
7. $\frac{84}{99}$ by $\frac{48}{77}$. *Ans.* $1\frac{13}{36}$.
8. $\frac{45}{112}$ by $\frac{15}{56}$. *Ans.* $1\frac{1}{2}$.
9. $\frac{105}{144}$ by $\frac{245}{432}$. *Ans.* $1\frac{2}{7}$.
10. $\frac{105}{121}$ by $\frac{175}{198}$. *Ans.* $\frac{54}{55}$.
11. $7\frac{5}{24}$ by $12\frac{8}{15}$. *Ans.* $\frac{865}{1504}$.
12. $15\frac{15}{16}$ by $8\frac{5}{11}$. *Ans.* $1\frac{439}{496}$.
13. $17\frac{7}{19}$ by $10\frac{4}{9}$. *Ans.* $1\frac{592}{893}$.
14. $21\frac{3}{7}$ by $12\frac{8}{21}$. *Ans.* $1\frac{19}{26}$.
15. $25\frac{4}{9}$ by $15\frac{4}{11}$. *Ans.* $1\frac{998}{1521}$.
16. $33\frac{1}{3}$ by $12\frac{1}{25}$. *Ans.* $2\frac{694}{903}$.
17. $\frac{9}{16}$ by $\frac{9}{20}$ of $\frac{36}{39}$. *Ans.* $1\frac{17}{48}$.
18. $\frac{28}{42}$ of $7\frac{9}{10}$ by $\frac{72}{84}$. *Ans.* $6\frac{13}{90}$.
19. $\frac{45}{80} \div \frac{90}{112}$ by $\frac{27}{40} \times \frac{112}{135}$. *Ans.* $1\frac{1}{4}$.

MENTAL EXERCISES.

1. If a yard of cloth cost $\frac{3}{5}$ of a dollar, how many yards can you buy for 12 dollars?

2. Mary gave \$$12\frac{1}{2}$ for silk at the rate of \$$2\frac{1}{2}$ a yard; how many yards did she buy?

3. How long will it require for a man to earn \$33, at the rate of \$$2\frac{3}{4}$ a day?

4. If $2\frac{1}{2}$ barrels of apples cost \$15, how many barrels can be bought for 12 dollars?

5. How many building lots of $\frac{3}{4}$ of an acre each, can we make out of 12 acres?

6. A lady distributed \$29 equally among some poor people, giving them \$$5\frac{4}{5}$ apiece; how many did she aid?

7. A school girl bought 6 yards of ribbon worth $5\frac{1}{3}$ cents a yard; how many apples worth $1\frac{1}{3}$ cents each would cost the same?

8. Mrs. Bear exchanged 20 pounds of butter, at 15 cents a pound, for calico, worth $12\frac{1}{2}$ cents a yard; how many yards did she receive?

9. A school-boy shared 14 apples equally with his companions, giving to each $3\frac{1}{2}$ apples; required the number of his companions.

10. A grocer sold 8 bushels of potatoes, worth \$$\frac{5}{8}$ a bushel, and received for them eggs at the rate of \$$\frac{1}{8}$ a dozen; how many eggs did he get?

WRITTEN EXERCISES.

1. If a pound of brown sugar cost $7\frac{7}{10}$ cents, how many pounds can you buy for $25\frac{4}{6}$ cents? *Ans.* $3\frac{1}{3}$.

2. If a quart of vinegar cost $18\frac{3}{4}$ cents, how many quarts can be bought for $52\frac{1}{2}$ cents? *Ans.* $2\frac{4}{5}$.

3. If 7 yards of book muslin cost \$$5\frac{37}{50}$, how much will 5 yards cost? *Ans.* \$4.10.

4. If $15\frac{3}{5}$ yards of ribbon cost 45 cents, what will 6 yards cost? *Ans.* $17\frac{4}{13}$ cts.

5. If $21\frac{3}{4}$ pounds of tea cost \$$18\frac{39}{80}$, what will 1 pound cost? *Ans.* \$0.85.

6. If $22\frac{1}{5}$ barrels of sugar cost \$$308\frac{1}{40}$, what is the price per barrel? *Ans.* \$$13\frac{7}{8}$.

7. If $28\frac{1}{4}$ tons of Lehigh coal cost \$$183\frac{5}{8}$, what is the price per ton? *Ans.* \$$6\frac{1}{2}$.

8. Divide $\frac{4}{5}$ of $\frac{7}{12}$ of $\frac{9}{10}$ by $\frac{5}{6}$ of $\frac{15}{64}$ of $\frac{21}{20}$ of $5\frac{1}{3}$. *Ans.* $\frac{48}{125}$.

9. Find the value of $\frac{39}{72} \times (\frac{218}{425} \div \frac{327}{289}) \times \frac{75}{117}$. *Ans.* $\frac{17}{108}$.

10. Find the value of $\frac{5184}{6561} \times \frac{6480}{9801} \div (\frac{392}{726} \times \frac{144}{475})$. *Ans.* $3\frac{691}{5108}$.

11. Find the value of $(1\frac{1}{2} \times 3\frac{1}{5} + \frac{5}{11} \text{ of } \frac{3}{7}) \times \frac{77}{81} \div (6\frac{2}{3} - 4\frac{3}{4})$. *Ans.* $2\frac{494}{1035}$.

12. Find the value of $(\frac{54}{63} + 3\frac{4}{7}) \div (\overline{6\frac{2}{3} - 4\frac{1}{4}} \times 9\frac{7}{9}) + 76\frac{1}{10}$. *Ans.* $94\frac{109}{8090}$.

13. Find the value of $\overline{77\frac{20}{21} - 44\frac{3}{7}} \times (5\frac{5}{11} + 6\frac{3}{7} - 7\frac{21}{77}) \div \overline{42\frac{3}{4} + 16\frac{3}{8}}$. *Ans.* $2\frac{862}{1617}$.

14. If an errand boy earn \$$7\frac{7}{8}$ in a week, how long will it require him to earn \$$20\frac{1}{4}$? *Ans.* $2\frac{4}{7}$ weeks.

15. A lady distributed \$231 among the poor, giving \$$28\frac{7}{8}$ to each person; how many were there? *Ans.* 8 persons.

16. Mr. B divided \$$416\frac{1}{4}$ equally among five persons; how much did each receive? *Ans.* \$$83\frac{1}{4}$.

17. By his father's will, Henry shared \$9600 equally with his five brothers; how much did he receive? *Ans.* \$1600.

18. The product of two numbers, diminished by $112\frac{3}{4}$, is $127\frac{3}{4}$, and one number is 15; required the other.

Ans. $16\frac{1}{30}$.

19. A boy shared 102 apples with his companions, giving to each $6\frac{4}{5}$ apples; required the number of companions.

Ans. 15.

20. A man divided $30\frac{4}{5}$ pounds of flour among the poor, giving to each $2\frac{4}{5}$ pounds; how many persons were there?

Ans. 11 persons.

21. A seamstress bought a sewing-machine for $85.50, and paid $40 down; how much must she save each month, to pay for it in 7 months? *Ans.* $6 50.

22. A man's wages are $$3\frac{4}{5}$ a day and his daily expenses are $$1\frac{7}{8}$; how many days must he labor to enable him to buy a suit of clothes worth $$46\frac{1}{5}$? *Ans.* 24 days.

23. Mr. Landis gained $228 on the sale of $6\frac{7}{8}$ acres of land; how much more would he have gained on each acre, if he had gained $382 on the whole? *Ans.* $22.40

REDUCTION OF COMPLEX FRACTIONS.

183. The **Reduction of Complex Fractions** is the process of changing them to simple fractions.

NOTE.—A complex fraction is not really a fraction, according to the definition of a fraction. It is rather a complex fractional expression of one fraction divided by another.

1. Reduce $\dfrac{\frac{3}{4}}{\frac{5}{6}}$ to a simple fraction.

SOLUTION.—This expression means that $\frac{3}{4}$ is to be divided by $\frac{5}{6}$, and inverting the divisor and multiplying, we have $\frac{3}{4}\times\frac{6}{5}$, which equals $\frac{9}{10}$.

OPERATION.

$$\frac{\frac{3}{4}}{\frac{5}{6}}=\frac{3}{4}\div\frac{5}{6}=\frac{3}{4}\times\frac{6}{5}=\frac{9}{10}$$

SOLUTION 2D.—Multiplying both terms of the complex fraction by 12, the least common multiple of the denominators of the terms, and reducing the resulting fraction to its lowest terms, we have $\frac{9}{10}$.

OPERATION.

$$\frac{\frac{3}{4}}{\frac{5}{6}}=\frac{\frac{3}{4}\times12}{\frac{5}{6}\times12}=\frac{9}{10}, \textit{ Ans.}$$

Rule I.—*Multiply the numerator of the complex fraction by its denominator inverted.*

Rule II.—*Multiply both terms of the complex fraction by the least common multiple of the denominators.*

WRITTEN EXERCISES.

Reduce to simple fractions

2. $\dfrac{\frac{5}{6}}{\frac{7}{8}}$.	*Ans.* $\frac{20}{21}$.	8. $\dfrac{2\frac{1}{2}}{3\frac{2}{3}}$.	*Ans.* $\frac{15}{22}$.
3. $\dfrac{\frac{7}{8}}{\frac{11}{12}}$	*Ans.* $\frac{21}{22}$.	9. $\dfrac{6\frac{5}{8}}{9\frac{7}{11}}$.	*Ans.* $\frac{11}{16}$.
4. $\dfrac{\frac{14}{15}}{\frac{21}{55}}$.	*Ans.* $2\frac{4}{9}$.	10. $\dfrac{10\frac{10}{13}}{12\frac{1}{4}}$.	*Ans.* $\frac{80}{91}$.
5. $\dfrac{\frac{9}{10}\times\frac{25}{27}}{\frac{15}{16}\times\frac{17}{18}}$.	*Ans.* $\frac{16}{17}$.	11. $\dfrac{\frac{1}{5}+\frac{3}{7}}{\frac{4}{9}+\frac{2}{5}}$.	*Ans.* $\frac{99}{133}$.
6. $\dfrac{\frac{10}{13}\times\frac{39}{45}}{\frac{16}{25}\times\frac{45}{64}}$.	*Ans.* $1\frac{13}{27}$.	12. $\dfrac{3\frac{1}{2}-2\frac{1}{3}}{4\frac{1}{4}-2\frac{1}{5}}$.	*Ans.* $\frac{70}{123}$.
7. $\dfrac{\frac{17}{35}\times\frac{95}{85}}{\frac{19}{28}\times\frac{105}{152}}$.	*Ans.* $1\frac{83}{525}$.	13. $\dfrac{12\frac{1}{2}}{41\frac{2}{3}}\times\dfrac{3\frac{7}{8}}{6\frac{1}{5}}$.	*Ans.* $\frac{3}{16}$.

RELATION OF NUMBERS.

184. The **Relation of Numbers** is their relative value as compared with one another.

NOTE.—This subject is equivalent to Ratio, but is presented here as affording an excellent illustration of the analysis of numbers. The treatment of the subject under Ratio is *demonstrative* rather than *analytic.*

CASE I.

185. ***To find the relation of an integer to an integer.***

1. 27 is how many times 6, or what is the relation of 27 to 6?

SOLUTION.—*One* is $\frac{1}{6}$ of 6, and if 1 is $\frac{1}{6}$ of 6, 27 is 27 times one-sixth of 6, which is $\frac{27}{6}$, or $\frac{9}{2}$, or $4\frac{1}{2}$ times 6. Therefore 27 is $4\frac{1}{2}$ times 6. Hence we have the following

Rule.—*Divide the number which you compare by the number with which it is compared.*

NOTE.—The rule is the same for each case, and need not be repeated.

MENTAL EXERCISES.

1. 9 is how many times 6? 25 is how many times 10? 64 is how many times 36?

2. 16 is what part of 36? 18 is what part of 40? 32 is what part of 72? 90 is what part of 108?

3. A watch cost $40, and a chain cost $12; what part of the cost of the watch equals the cost of the chain?

4. If 5 yards of cloth cost \$21, what will 20 yards cost at the same rate?

5. If 6 tons of coal are bought for \$19$\frac{1}{2}$, what will 30 tons cost at the same rate?

6. A has \$20 and B has $\frac{4}{5}$ as much, plus \$3; what part of A's money equals B's money?

WRITTEN EXERCISES.

What is the relation of

2. 25 to 75?	*Ans.* $\frac{1}{3}$.	6. 105 to 225?	*Ans.* $\frac{7}{15}$.
3. 18 to 54?	*Ans.* $\frac{1}{3}$.	7. 420 to 588?	*Ans.* $\frac{5}{7}$.
4. 76 to 24?	*Ans.* $\frac{19}{6}$.	8. 540 to 432?	*Ans.* $\frac{5}{4}$.
5. 84 to 91?	*Ans.* $\frac{12}{13}$.	9. 1899 to 2743?	*Ans.* $\frac{9}{13}$.

10. If 16 apples cost 13 cents, what will 48 apples cost at the same rate?

SOLUTION.—If 16 apples cost 13 cents, 48 apples, which are $\frac{48}{16}$, or 3 times 16 apples, cost 3 times 13 cents, or 39 cents.

11. If 15 oranges cost 14 cents, what will 60 oranges cost at the same rate? *Ans.* 56 cents.

12. What cost 125 sheep, if 25 sheep cost 137 dollars? *Ans.* \$685.

13. If 56 cows cost \$1000, what will 42 cows cost at the same rate? *Ans.* \$750.

14. If 105 pens cost 77 cents, what will 225 pens cost at the same rate? *Ans.* \$1.65.

15. A has 20 cows, which is $\frac{4}{5}$ of B's number; now if A obtains $\frac{3}{5}$ of B's number, what part of A's number will equal B's? *Ans.* $\frac{5}{7}$ of A's.

CASE II.

186. ***To find the relation of a fraction to a number.***

1. The fraction $\frac{3}{4}$ is what part of 6?

SOLUTION.—*One* is $\frac{1}{6}$ of 6, and if 1 is $\frac{1}{6}$ of 6, $\frac{1}{4}$ is $\frac{1}{4}$ of $\frac{1}{6}$, which is $\frac{1}{24}$ of 6, and $\frac{3}{4}$ is 3 times $\frac{1}{24}$, which is $\frac{1}{8}$ of 6.

MENTAL EXERCISES.

1. What part of 3 is $\frac{2}{5}$? Of 4 is $\frac{2}{5}$? Of 7 is $\frac{5}{8}$? Of 9 is $\frac{4}{7}$?

2. What part of 5 is $\frac{1}{2}$ of $\frac{3}{4}$? Of 7 is $\frac{2}{5}$ of $\frac{3}{2}$? Of 5 is $\frac{4}{5}$ of $\frac{5}{8}$?

3. If $\frac{2}{3}$ of a ton of coal cost \$2$\frac{1}{4}$, what will 6 tons cost at the same rate?

4. A hat was bought for \$2 and sold at a gain of \$$\frac{3}{4}$; what part of the cost equals the gain?

5. Sarah bought 5 yards of silk and used $3\frac{1}{4}$ yards; what part of the silk then remained?

6. A bought 6 oranges, and B bought 4 oranges; what did B pay for his oranges, if A paid 26 cents for his?

WRITTEN EXERCISES.

What is the relation of

2. $\frac{3}{4}$ to 9?	*Ans.* $\frac{1}{12}$.	**6.** $\frac{8}{9}$ to 64?	*Ans.* $\frac{1}{72}$.
3. $\frac{3}{7}$ to 12?	*Ans.* $\frac{1}{28}$.	**7.** $\frac{35}{38}$ to 35?	*Ans.* $\frac{1}{38}$.
4. $\frac{7}{10}$ to 14?	*Ans.* $\frac{1}{20}$.	**8.** $\frac{56}{57}$ to 14?	*Ans.* $\frac{4}{57}$.
5. $\frac{14}{15}$ to 28?	*Ans.* $\frac{1}{30}$.	**9.** $\frac{21}{16}$ to 35?	*Ans.* $\frac{3}{80}$.

10. A having \$25, gave \$5 to B and $\frac{1}{4}$ of the remainder to C: what part of \$25 then remained? *Ans.* $\frac{3}{5}$.

CASE III.

187. ***To find the relation of a number to a fraction.***

1. 6 is how many times $\frac{3}{4}$?

SOLUTION.—$\frac{1}{4}$ is $\frac{1}{3}$ of $\frac{3}{4}$, and since $\frac{1}{4}$ is $\frac{1}{3}$ of $\frac{3}{4}$, $\frac{4}{4}$ or 1 is 4 times $\frac{1}{3}$, or $\frac{4}{3}$ of $\frac{3}{4}$, and 6 is 6 times $\frac{4}{3}$, or $\frac{24}{3}$, which equals 8.

MENTAL EXERCISES.

1. 8 is how many times $\frac{4}{5}$? 14 is how many times $\frac{7}{8}$? 24 is how many times $\frac{6}{7}$?

2. 96 is how many times $\frac{16}{17}$? 115 is how many times $\frac{23}{28}$? 144 is how many times $\frac{12}{11}$?

3. If a yard of cloth cost \$$\frac{3}{4}$, how many yards will \$9 buy?

4. A can walk a certain distance in $2\frac{1}{2}$ hours; how many times that distance can he walk in 6 hours?

5. A book cost \$4 and a slate \$$\frac{3}{4}$; how many times the cost of the slate equals the cost of the book?

WRITTEN EXERCISES.

What is the relation of

2. 288 to $\frac{16}{23}$?	*Ans.* 414.	**5.** 462 to $8\frac{5}{9}$?	*Ans.* 54.
3. 495 to $4\frac{2}{5}$?	*Ans.* $112\frac{1}{2}$.	**6.** 588 to $3\frac{9}{11}$?	*Ans.* 154.
4. 1896 to $\frac{24}{27}$?	*Ans.* 2133.	**7.** 63504 to $50\frac{2}{5}$?	*Ans.* 1260

8. Mary had 12 quarts of nuts and Sarah had $\frac{4}{5}$ as many, minus $\frac{3}{4}$ of a quart; how many times Sarah's number equals *Mary's?* *Ans.* $1\frac{21}{59}$.

CASE IV.

188. ***To find the relation of a fraction to a fraction.***

1. What part of $\frac{14}{15}$ is $\frac{7}{12}$?

SOLUTION.—$\frac{1}{15}$ is $\frac{1}{14}$ of $\frac{14}{15}$, and $\frac{15}{15}$, or *one*, is 15 times $\frac{1}{14}$, which equals $\frac{15}{14}$ of $\frac{14}{15}$. If 1 equals $\frac{15}{14}$ of $\frac{14}{15}$, $\frac{1}{12}$ equals $\frac{1}{12}$ of $\frac{15}{14}$, and $\frac{7}{12}$ equals 7 times $\frac{1}{12}$ of $\frac{15}{14}$, which equals $\frac{7}{12}$ of $\frac{15}{14}$, or $\frac{5}{8}$.

MENTAL EXERCISES.

1. What part of $\frac{3}{4}$ is $\frac{2}{5}$? Of $\frac{5}{6}$ is $\frac{3}{4}$? Of $\frac{6}{7}$ is $\frac{3}{8}$? Of $\frac{4}{5}$ is $\frac{5}{6}$?

2. What part of $\frac{5}{7}$ is $2\frac{1}{2}$? Of $\frac{6}{7}$ is $\frac{3}{5}$ of $\frac{4}{7}$? Of $\frac{7}{8}$ is $\frac{3}{4}$ of $\frac{5}{6}$? Of $\frac{9}{10}$ is $\frac{4}{5}$ of $\frac{7}{8}$?

3. A had $\$\frac{3}{4}$ and spent $\$\frac{2}{5}$; what part of his money did he spend?

4. If $2\frac{1}{2}$ yards of silk cost $\$7\frac{1}{2}$, what will $7\frac{1}{2}$ yards cost at the same rate?

5. A merchant bought $\frac{5}{6}$ of a cargo of oranges and sold $\frac{3}{4}$ of them: what part of the cargo did he sell?

6. Edwin and Emma each had \$2; Edwin spent $\frac{1}{4}$ of his, and Emma spent $\frac{1}{3}$ of hers; what part of Emma's money then equaled Edwin's?

WRITTEN EXERCISES.

What is the relation of

2. $\frac{5}{7}$ to $\frac{10}{7}$? *Ans.* $\frac{1}{2}$.
3. $\frac{19}{20}$ to $\frac{57}{40}$? *Ans.* $\frac{2}{3}$.
4. $\frac{17}{21}$ to $\frac{85}{42}$? *Ans.* $\frac{2}{5}$.
5. $\frac{28}{37}$ to $\frac{54}{74}$? *Ans.* $\frac{28}{27}$.
6. $9\frac{11}{15}$ to $11\frac{2}{3}$? *Ans.* $\frac{146}{175}$.
7. $11\frac{12}{13}$ to $7\frac{6}{7}$? *Ans.* $\frac{217}{143}$.
8. $\frac{2}{3}\times\frac{9}{16}$ to $\frac{8}{7}\times\frac{19}{20}$? *Ans.* $\frac{105}{304}$.
9. $\frac{4}{5}\times\frac{16}{18}$ to $\frac{19}{20}\times\frac{24}{57}$? *Ans.* $\frac{16}{9}$.

10. If a merchant sold $\frac{4}{5}$ of $\frac{5}{6}$ of his stock, how many fifths of $\frac{5}{6}$ of his stock remained? *Ans.* $\frac{2}{5}$.

REMARK.—He sold $\frac{4}{6}$, and hence there remained $\frac{2}{6}$, and $\frac{2}{6}$ equals *two fifths* of $\frac{5}{6}$ of his stock; hence the answer is $\frac{2}{5}$.

11. Harry lost $\frac{2}{7}$ of $\frac{7}{9}$ of a certain sum of money; how many sevenths of $\frac{7}{9}$ of the sum remained? *Ans.* $\frac{5}{7}$.

12. Two boys had each $\frac{9}{10}$ of a bushel of walnuts, when the first sold $\frac{2}{3}$ of his to the second; what part of what the second has equals what the first has? *Ans.* $\frac{1}{5}$.

13. James had $\frac{4}{9}$ of a ton of coal and Thomas had $\frac{5}{9}$ of a ton; if James burns $\frac{1}{5}$ of his coal and Thomas $\frac{1}{4}$ of his, how many ninths of a ton will remain? *Ans.* $6\frac{19}{20}$ ninths.

NOTE.—For *Greatest Common Divisor* and *Least Common Multiple of Fractions see Manual.*

ARITHMETICAL ANALYSIS.

189. Analysis is the process of solving problems by a comparison of their elements. In comparing, we reason to the unit and from the unit; the unit being the basis of the reasoning process.

CASE I.

190. ***To pass from one integer to another integer.***

1. If 5 oranges cost $12\frac{1}{2}$ cents, what will 8 oranges cost at the same rate?

SOLUTION.—If 5 oranges cost $12\frac{1}{2}$ or $\frac{25}{2}$ cents, 1 orange cost $\frac{1}{5}$ of $\frac{25}{2}$, which is $\frac{5}{2}$ cents; and if 1 orange cost $\frac{5}{2}$ cents, 8 oranges will cost 8 times $\frac{5}{2}$ cents, or 20 cents.

OPERATION.

5 cost $\frac{25}{2}$

1 " $\frac{5}{2}$

8 " $\frac{5}{2}\times 8=20$, *Ans.*

MENTAL EXERCISES.

1. If 4 apples cost $5\frac{3}{4}$ cents, what will 5 apples cost at the same rate?

2. If 3 girls earn $\$2\frac{1}{4}$ in a day, how much could 5 girls earn at the same rate?

3. How long will it take a boy to plow 8 furrows, if it takes him $13\frac{3}{4}$ minutes to plow 5 furrows?

4. How much will 9 barrels of flour cost, if 2 barrels cost $\$10\frac{3}{8}$?

5. If 5 hats can be bought for $\$11\frac{1}{4}$, how much will 8 hats cost?

6. What cost 5 pounds of butter at the rate of $\$3\frac{1}{3}$ for 10 pounds?

WRITTEN EXERCISES.

2. If 6 hens cost $3.54, what will 7 hens cost at the same rate? *Ans.* $4.13.

3. If 8 turkeys cost $\$19\frac{1}{5}$, what will 10 turkeys cost at the same rate? *Ans.* $24.

4. How much must I pay for 22 yards of muslin, at the rate of 5 yards for $\$\frac{19}{20}$? *Ans.* $4.18.

5. How much will 21 books cost at the rate of 9 books for $15\frac{3}{7}$? *Ans.* $36.

6. What must I pay for 8 arithmetics, at the rate of 5 arithmetics for $4.75? *Ans.* $7.60.

7. If 3 tons of hay cost $\$47\frac{1}{3}$, how much will 12 tons cost at the same rate? *Ans.* $\$189\frac{1}{3}$.

8. Required the cost of 21 barrels of flour, at the rate of $17.50 for 3 barrels. Ans. $122.50

9. A gave \$43.50 for pigs, at the rate of \$7.25 for 3 pigs; how many did he buy? *Ans.* 18 pigs.

10. B gave \$54 for flour, at the rate of \$6 for $4\frac{1}{3}$ bushels; how many bushels did he buy? *Ans.* 39 bushels.

CASE II.

191. *To pass from a fraction to an integer.*

1. If $\frac{3}{4}$ of an acre of land cost \$108, what will one acre cost?

	OPERATION.
SOLUTION.—If $\frac{3}{4}$ of an acre cost \$108, $\frac{1}{4}$ of an acre cost $\frac{1}{3}$ of \$108, which is \$36; and if $\frac{1}{4}$ of an acre cost \$36, $\frac{4}{4}$ of an acre, or one acre, will cost 4 times \$36, which are \$144.	$\frac{3}{4}$ cost \$108 $\frac{1}{4}$ " \$36 $\frac{4}{4}$ " \$144, *Ans.*

MENTAL EXERCISES.

1. If $\frac{4}{5}$ of a yard of cloth is worth 24 dimes, what are 3 yards of the cloth worth?

2. If $\frac{2}{3}$ of a barrel of sugar cost \$12, what will 7 barrels cost?

3. I bought 5 yards of summer silk at the rate of \$10 for $3\frac{1}{3}$ yards what was the cost?

4. Mary bought 9 apples, at the rate of $\frac{3}{5}$ of an apple for $\frac{3}{4}$ of a cent; what did they cost?

5. Henry bought 25 peaches, at the rate of $\frac{3}{4}$ of a peach for $\frac{4}{5}$ of a cent; required the cost.

6. Three-fifths of what I paid for a cow was \$24; what did I pay for a horse whose value was 5 times that of the cow?

WRITTEN EXERCISES.

2. If $\frac{2}{3}$ of the cost of A's farm is \$2480, what did the farm cost? *Ans.* \$3720.

3. If $\frac{4}{7}$ of a barrel of flour cost \$2.50, what will 5 barrels cost at the same rate? *Ans.* $\$21.87\frac{1}{2}$.

4. If $\frac{6}{7}$ of a yard of cloth cost \$2.10, what will enough for a suit containing 9 yards cost? *Ans.* \$22.05.

5. If $\frac{5}{6}$ of a cwt. of sugar cost \$12.45, what will be the cost of 20 cwt., or 1 ton of sugar? *Ans.* \$298.80.

6. What will 21 tons of hay cost, if $\frac{7}{8}$ of a ton of hay cost \$12.85? *Ans.* \$308.40.

7. If $\$46\frac{7}{8}$ is $\frac{5}{16}$ of the price I pay for a horse, what would be the cost of 4 pairs of horses at the same rate? *Ans.* \$1200.

8. If a lawyer's clerk could write 60 pages in $10\frac{2}{3}$ hours, how much would he write in a week, working 10 hours a day? *Ans.* $337\frac{1}{2}$ pages.

9. How much will it cost to transport 25 hundredweight of freight from Philadelphia to Pittsburgh at \$4 for $2\frac{2}{11}$ hundredweight? *Ans.* \$$45\frac{5}{6}$.

10. How much will I pay for 15 acres of land, at the rate of $2\frac{1}{2}$ acres for \$425.25? *Ans.* \$2551.50.

11. A bought 40 bushels of grain, at the rate of $3\frac{1}{3}$ bushels for \$3.20; what did it cost? *Ans.* \$38.40.

12. B sold 77 gallons of milk, at the rate of \$1.25 for $5\frac{1}{2}$ gallons; required the cost. *Ans.* \$17.50.

13. C bought 39 cords of wood, at the rate of \$12.45 for $3\frac{1}{4}$ cords; how much did it cost? *Ans.* \$149.40.

14. D bought 81 tons of coal, at the rate of $5\frac{2}{5}$ tons for \$26.25; how much did it cost? *Ans.* \$393.75.

CASE III.

192. ***To pass from an integer to a fraction.***

1. If 1 barrel of apples cost \$8, what will $\frac{3}{4}$ of a barrel cost?

SOLUTION.—If 1 barrel of apples cost \$8, $\frac{1}{4}$ of a barrel cost $\frac{1}{4}$ of \$8, which is \$2; and if $\frac{1}{4}$ of a barrel cost \$2, $\frac{3}{4}$ of a barrel will cost 3 times \$2, which are \$6.

OPERATION.

1 cost \$8
$\frac{1}{4}$ " \$2
$\frac{3}{4}$ " \$6, *Ans.*

MENTAL EXERCISES.

1. What cost $\frac{5}{8}$ of a pound of beef, if 2 pounds cost 32 cents?

2. If a man can earn \$$5\frac{1}{4}$ in 3 days, how much will he earn in $\frac{1}{2}$ of a day?

3. If 2 barrels of flour cost \$$10\frac{5}{8}$, for what can I sell $\frac{3}{4}$ of a barrel at the rate of cost?

4. If Harry walks $6\frac{3}{8}$ miles in 2 hours, how far will he walk at this rate in $3\frac{1}{3}$ hours?

5. If 7 pounds of sugar cost 56 cents, for what can I sell $4\frac{1}{2}$ pounds, and neither gain nor lose?

6. A vessel sailed 60 miles in 12 hours; how far at this rate did it sail in $4\frac{3}{8}$ hours?

7. If a seamstress earns \$$1\frac{1}{3}$ in 2 days, how much will she earn in $10\frac{1}{2}$ days?

WRITTEN EXERCISES.

2. Henry paid $24.60 for a cow; how much would he have paid had he given $\frac{4}{5}$ as much? *Ans.* $19.68.

3. If 3 yards of cloth cost $7.50, what will $\frac{7}{8}$ of a yard of the same cloth cost? *Ans.* 2.18\frac{3}{4}$.

4. If 7 barrels of apples cost 19\frac{3}{5}$, what will $\frac{5}{6}$ of a barrel cost at the same rate? *Ans.* 2\frac{1}{3}$.

5. How much will 10$\frac{7}{8}$ tons of coal cost at the rate of 13\frac{5}{7}$ for 8 tons? *Ans.* 18\frac{9}{14}$.

6. A bought 9$\frac{8}{9}$ acres of land, at the rate of $1605 for 5 acres; required the cost. *Ans.* 3174\frac{1}{3}$.

7. How much must I pay for 18$\frac{15}{16}$ cords of wood, at the rate of 6\frac{2}{3}$ for 3 cords? *Ans.* 42\frac{1}{12}$.

8. If 9 tons of hay are worth 100\frac{4}{5}$, what must I give for 14$\frac{13}{14}$ tons? *Ans.* 167\frac{1}{5}$.

9. If 12 bushels of beans cost 11\frac{1}{4}$, how much will 15$\frac{14}{15}$ bushels cost? *Ans.* 14\frac{15}{16}$.

10. How much must I pay for 18$\frac{16}{18}$ tons of iron ore, at the rate of 17 tons for 33\frac{3}{4}$? *Ans.* 37\frac{1}{2}$.

11. A gave $1968 for 21 acres of land; for how much would he sell $\frac{18}{19}$ of an acre, at an advance of $4 an acre?

Ans. 92\frac{4}{7}$.

12. B gave $444 for 24 tons of hay; he sold his neighbor $\frac{19}{20}$ of a ton, at 2\frac{1}{2}$ less a ton; what did he receive?

Ans. $15.20.

CASE IV.

193. ***To pass from a fraction to a fraction.***

1. If $\frac{2}{3}$ of an acre of land cost $150, what will $\frac{4}{5}$ of an acre cost?

SOLUTION.—If $\frac{2}{3}$ of an acre cost $150, $\frac{1}{3}$ of an acre cost $\frac{1}{2}$ of $150, which is $75; and $\frac{3}{3}$ of an acre will cost 3 times $75, which are $225. If 1 acre cost $225, $\frac{1}{5}$ of an acre cost $\frac{1}{5}$ of $225, which is $45, and $\frac{4}{5}$ of an acre will cost 4 times $45, which are $180.

SOLUTION 2D.—If $\frac{2}{3}$ of an acre cost $150, $\frac{1}{3}$ of an acre cost $\frac{1}{2}$ of $150, and $\frac{3}{3}$ of an acre will cost $\frac{3}{2}\times$150; and $\frac{4}{5}$ of an acre will cost $\frac{4}{5}\times\frac{3}{2}\times$150, which by *cancellation* we find is $180.

OPERATION.

$\frac{4}{5}\times\frac{3}{2}\times$150=$180, *Ans.*

NOTE.—The 2d method, by cancellation, is preferred, especially with the more difficult *problems*.

6

MENTAL EXERCISES.

1. If $\frac{3}{4}$ of an apple cost $\frac{9}{10}$ of a cent, what will $\frac{5}{6}$ of an apple cost?

2. If $\frac{5}{6}$ of a barrel of flour is worth $\$6\frac{6}{7}$, what is $\frac{7}{8}$ of a barrel worth?

3. What is $\frac{5}{8}$ of an acre of land worth, if $\frac{7}{8}$ of an acre is worth $\$126$?

4. How much must I pay for $\frac{10}{11}$ of a cord of wood, if $\frac{5}{7}$ of a cord is worth $\$1.50$?

5. What must I pay for $\frac{5}{6}$ of a bushel of berries, if $\frac{3}{4}$ of a bushel is worth $\$6$?

6. What will $\frac{7}{8}$ of a yard of cloth cost, if $\frac{5}{6}$ of a yard cost $\$7\frac{1}{2}$?

7. I paid $\$1400$ for $4\frac{2}{3}$ acres of land; how much will $5\frac{3}{4}$ acres cost at the same rate?

8. A bought $7\frac{3}{5}$ yards of cloth for $\$10\frac{6}{7}$; what would he have paid for $8\frac{3}{4}$ yards at the same rate?

9. How many apples can you get for $53\frac{1}{3}$ cents, if $2\frac{1}{2}$ apples are worth $6\frac{2}{3}$ cents?

10. How much will $7\frac{1}{2}$ barrels of flour cost, at the rate of $\frac{3}{16}$ of a barrel for $\$1\frac{1}{2}$?

WRITTEN EXERCISES.

2. When $\$189\frac{1}{5}$ are paid for $8\frac{3}{5}$ acres of land, how much must be paid for $10\frac{5}{6}$ acres? *Ans.* $\$238\frac{1}{3}$.

3. A gave $6\frac{5}{7}$ yards of cloth for $13\frac{3}{7}$ cords of wood; how many cords could be got for $9\frac{3}{5}$ yards? *Ans.* $19\frac{1}{5}$ cords.

4. B gave $12\frac{1}{2}$ tons of iron for $32\frac{13}{16}$ barrels of flour; how many barrels would he get for $14\frac{3}{8}$ tons? *Ans.* $38\frac{1}{2}$.

5. C gave $5\frac{3}{8}$ tons of hay for $28\frac{2}{3}$ tons of coal; how much coal would he get for $7\frac{3}{4}$ tons of hay? *Ans.* $41\frac{1}{3}$ tons.

6. If 6 horses eat $7\frac{1}{5}$ bushels of oats in a day, how many horses would eat $33\frac{3}{5}$ bushels in a day? *Ans.* 28.

7. If 9 pigs cost $\$12\frac{3}{8}$, how many pigs can you buy for $115\frac{1}{2}$ dollars? *Ans.* 84.

8. If $2\frac{1}{2}$ barrels of flour cost $\$24\frac{3}{8}$, how many barrels can you buy for $\$87\frac{3}{4}$? *Ans.* 9.

9. How much must I pay for $6\frac{3}{5}$ pounds of sugar, at the rate of $\$0.44\frac{4}{9}$ for $2\frac{2}{7}$ pounds? *Ans.* $\$1.28\frac{1}{3}$.

10. How much must I pay for $10\frac{6}{7}$ bushels of rice at the rate of $\$15\frac{3}{8}$ for $5\frac{3}{7}$ bushels? *Ans* $\$30\frac{3}{4}$.

11. If $19\frac{3}{5}$ tons of iron cost $\$42\frac{6}{7}$, how many tons can you buy for $\$357\frac{1}{7}$? *Ans.* $163\frac{1}{3}$ tons.

12. How many bushels of wheat can I buy for \$114$\frac{2}{5}$, at the rate of 29$\frac{2}{3}$ bushels for \$28$\frac{3}{5}$? *Ans.* 118$\frac{2}{3}$ bushels.

13. If 35$\frac{3}{7}$ yards of cloth cost \$39$\frac{6}{7}$, how much will 5$\frac{7}{9}$ yards cost at the same rate? *Ans.* \$6.50.

14. How much will 57$\frac{3}{4}$ tons of hay cost, at the rate of 41$\frac{8}{9}$ tons for \$502$\frac{2}{3}$? *Ans.* \$693.

15. What must I pay for 14$\frac{12}{17}$ yards of cloth, if 47$\frac{8}{11}$ yards cost \$378$\frac{6}{7}$? *Ans.* \$117$\frac{6}{7}$.

16. What must I pay for 17$\frac{3}{4}$ tons of hay, if 5$\frac{5}{6}$ tons cost \$70$\frac{7}{9}$? *Ans.* \$215$\frac{11}{30}$.

17. What must I pay for 10$\frac{4}{5}$ barrels of rice, if 7$\frac{3}{5}$ barrels cost \$72$\frac{5}{6}$? *Ans.* \$103$\frac{1}{2}$.

18. If \$24$\frac{3}{4}$ will buy 25$\frac{2}{3}$ bushels of corn, how many dollars will buy 50$\frac{5}{9}$ bushels? *Ans.* \$48$\frac{3}{4}$.

19. If \$582$\frac{6}{7}$ will buy 13$\frac{5}{7}$ tons of iron, how many tons will \$1416$\frac{2}{3}$ buy? *Ans.* 33$\frac{1}{3}$.

20. If 26$\frac{8}{11}$ bushels of wheat are given for 6$\frac{8}{15}$ tons of hay, how many bushels must be given for 46$\frac{1}{5}$ tons of hay? *Ans.* 189 bushels.

21. If 7$\frac{2}{9}$ yards of cloth are given for 12$\frac{2}{15}$ bushels of corn, how many bushels of corn must be given for 19$\frac{1}{21}$ yards of cloth? *Ans.* 32 bushels.

CASE V.

194. ***Given a part and the remainder to find the whole.***

1. A man sold $\frac{3}{7}$ of his farm, and then had 216 acres remaining; how many acres in the entire farm?

OPERATION.

$\frac{7}{7}-\frac{3}{7}=\frac{4}{7}$

$\frac{4}{7}=216$

$\frac{1}{7}=54$

$\frac{7}{7}=378$, *Ans.*

SOLUTION.—If he sold $\frac{3}{7}$ of his farm, there remained $\frac{7}{7}-\frac{3}{7}$, which is $\frac{4}{7}$ of the farm, which equals 216 acres; if $\frac{4}{7}$ of the farm equals 216 acres, $\frac{1}{7}$ of the farm equals $\frac{1}{4}$ of 216 acres, which is 54 acres, and $\frac{7}{7}$ of the farm, etc.

MENTAL EXERCISES.

1. A merchant drew out of bank $\frac{1}{4}$ of his deposit, and then had \$360 in bank; what was his deposit?

2. A man willed $\frac{1}{5}$ of his property to his wife, and the remainder, \$4800, to his son; what was his property?

3. Mary gave $\frac{1}{3}$ of her money for a silk dress and $\frac{1}{5}$ for a cloak, and had \$10 remaining; how much had she at first?

4. Frank, after spending $\frac{3}{5}$ of his money, found that \$16 was $\frac{4}{5}$ of what he had remaining; how much money had he at first?

5. If $\frac{2}{5}$ of an army were killed, $\frac{3}{8}$ taken prisoners, and 800 men escaped, of how many men did the army consist?

6. What number is that which being increased by its $\frac{1}{3}$, and that sum diminished by $\frac{3}{5}$ of the number, the remainder is 50?

7. When E was married he was 27 years old, and $\frac{2}{3}$ of his age was 4 years more than $\frac{2}{3}$ of his wife's age; required the age of his wife.

8. A boy, after spending $\frac{1}{4}$ of his money for candies and $\frac{1}{3}$ for peaches, found that 20 cents was $\frac{4}{7}$ of what remained; how much money had he?

9. A thief stole $\frac{3}{5}$ of Harry's money, and before he was caught spent $\frac{2}{3}$ of it; the remainder, which was \$20 less than he stole, was given back; how much money had Harry?

WRITTEN EXERCISES.

2. A invested $\frac{3}{4}$ of his money in bank stock, and had \$6045 remaining; how much money had he? *Ans.* \$24,180.

3. A drover sold $\frac{2}{9}$ of his cows, and then had 210 remaining; how many had he before the sale? *Ans.* 270.

4. B sold $\frac{1}{3}$ of his gas stock to C and $\frac{1}{5}$ to D, and had remaining 1463 shares; how many shares did he sell to each, and how many had he at first? *Ans.* 3135.

5. A gave $\frac{3}{4}$ of his fortune for a farm, and paid $\frac{3}{5}$ of the remainder for building a house, and then had \$4671 remaining; how much had he at first? *Ans.* \$46,710.

6. A father gave $\frac{1}{4}$ of his fortune to his daughter, $\frac{1}{3}$ to his wife, and the remainder, which was \$6720, to his son; what did the wife and the daughter receive?

Ans. W., \$5376; D., \$4032.

7. One-third of the trees in an orchard bear apples, $\frac{1}{4}$ bear peaches, and the remainder, which is 100, bear plums; required the number of trees in the orchard. *Ans.* 240.

8. A boy lost $\frac{2}{3}$ of his kite string, and then added $53\frac{3}{4}$ feet, when he found that it was just $\frac{3}{4}$ of its original length; what was its original length? *Ans.* 129 feet.

MENTAL EXERCISES.

MISCELLANEOUS EXAMPLES.

1. If a barrel of apples cost $\$5\frac{1}{4}$, how many "quarters" does it cost?

2. If 4 boys can earn $\$6\frac{2}{3}$ in a week, how much does each boy earn?

3. If I buy at the mint 125 "quarters" for change, how many dollars must I pay?

4. What cost 8 barrels of apples, at the rate of $\$3\frac{3}{8}$ per barrel?

5. $\frac{5}{8}$ of $40 is 2 times what A gave for a bureau; what was the cost of the bureau?

6. $\frac{5}{7}$ of $\frac{14}{15}$ of a dollar is $\frac{1}{10}$ of the cost of a watch; how much did the watch cost?

7. A man owning $\frac{3}{4}$ of the stock of a bank sold $\frac{2}{3}$ of his share; how much did he sell?

8. Jane bought a slate for $\frac{1}{4}$ of a dollar, and an arithmetic for $\frac{5}{8}$ of a dollar; what was the cost of both?

9. If $\frac{1}{2}$ of a number, increased by $\frac{1}{3}$ of the number, equals 50, what is the number?

10. Maria owes a store bill of $\$\frac{3}{5}$; if she hands the clerk $\$\frac{3}{4}$, how much change would she receive?

11. If $\frac{1}{3}$ of a pole is in the air, $\frac{1}{2}$ in the water, and the rest in the mud, how much is in the mud?

12. How much will $4\frac{2}{3}$ yards of muslin cost, at the rate of 12 cents a yard?

13. How much will 10 pounds of sugar cost, at $\frac{3}{5}$ of a dime a pound?

14. If a yard of flannel costs $\$\frac{3}{5}$, how many yards can you buy for $12?

15. Mrs. Brown exchanged 10 pounds of butter, at 15 cents a pound, for calico worth $6\frac{1}{4}$ cents a yard; how many yards did she receive?

16. How much will 14 pounds of cloves cost, if $3\frac{1}{2}$ pounds can be bought for 50 cents?

17. If $2\frac{1}{2}$ yards of muslin cost 30 cents, how much will 10 yards cost, at the same rate?

18. How far will a man drive in $5\frac{1}{4}$ hours, at the rate of 21 miles in $3\frac{1}{2}$ hours?

19. If $\frac{2}{3}$ of a ton of hay is worth $\$4\frac{4}{5}$, what cost $7\frac{3}{8}$ tons of hay at the same rate?

20. John found 60 cents, which is $\frac{3}{4}$ of $\frac{1}{2}$ of what he then had; how much had he at first?

21. Henry gave his sister 20 cents, which is $\frac{2}{5}$ of what he had at first, and $\frac{1}{3}$ of what *his sister now* has; how much had each at first?

22. Amanda, having 50 pins, lost $\frac{4}{5}$ of them, and then found $\frac{3}{5}$ as many as remained; how many had she then?

23. A gave B 48 cents, and $\frac{5}{6}$ of this is 4 times as much as he had remaining; how much had he at first?

24. If 4 horses eat 2 tons of hay in 8 weeks, how long will it require 5 horses to eat the same?

25. Fanny's number of roses, increased by $\frac{1}{2}$ and $\frac{1}{3}$ of her number, equals 55; how many roses has she?

26. A gave $24 for a watch, and $\frac{1}{2}$-| $\frac{3}{4}$ of this is 4 times what he paid for the chain; required the cost of the chain.

27. A boy having 36 marbles, lost $\frac{3}{4}$ of them, and then found $\frac{5}{6}$ as many as he had at first; how many had he then?

28. $60 is $\frac{3}{5}$ of what B gave for a horse, and the cost of the horse, increased by its $\frac{3}{5}$, is 5 times what he paid for a sleigh; required the cost of the sleigh.

29. A has 40 fruit trees, $\frac{4}{10}$ of which bear apples, $\frac{1}{3}$ of the remainder bear pears, and the rest bear peaches; how many trees of each kind has he?

30. Janson's age, diminished by its $\frac{1}{4}$ and $\frac{1}{5}$, is 22 years, and his age is $\frac{4}{5}$ of his uncle's age; required the age of each.

31. Annie, having $\frac{3}{5}$ of a pound of candies, shared them equally with 5 of her schoolmates; what part of a pound did each receive?

32. Peter gave $\frac{1}{5}$ of his marbles to Samuel, and 2 times $\frac{1}{4}$ of them to Anson; how many had he at first, if he gave away 14 marbles?

33. A boy divided 14 apples equally among his companions, giving to each $3\frac{1}{2}$ apples; required the number of his companions.

34. Harry bought 8 bushels of potatoes, worth $\$\frac{3}{8}$ a bushel, and paid for them with eggs, worth $\$\frac{3}{4}$ a dozen; how many eggs did it take?

35. If $3\frac{1}{3}$ boxes of starch cost $\$5\frac{1}{4}$, how much will 10 boxes cost at the same rate?

36. Hannah's wedding dress cost $50, and $\frac{4}{5}$ of this is 4 times the cost of her bonnet, and also $\frac{2}{3}$ of the cost of her cloak; required the cost of the bonnet and cloak respectively.

37. Mary gave $\frac{3}{4}$ of her money to the poor, and then found $\frac{2}{3}$ as much as she gave away, and then had $30; how much had she at first?

38. William borrowed $\frac{4}{5}$ of Emily's money, and after spending $\frac{3}{4}$ of it, returned the remainder, which was $20; how much money had Emily?

39. A kite in the air fell $\frac{1}{2}$ of the distance to the ground, then arose $\frac{1}{3}$ of the distance it was from the ground, and then fell $\frac{1}{4}$ of the *distance it* arose; what part of the whole distance was it from the *ground?*

WRITTEN EXERCISES.

MISCELLANEOUS PROBLEMS.

1. Reduce $45\frac{7}{8}$ to an improper fraction. *Ans.* $\frac{367}{8}$.

2. Reduce $385\frac{9}{13}$ to an improper fraction. *Ans.* $\frac{5014}{13}$.

3. Reduce $\frac{2078}{12}$ to a mixed number. *Ans.* $173\frac{1}{6}$.

4. Reduce $\frac{4785}{125}$ to a mixed number. *Ans.* $38\frac{7}{25}$.

5. Reduce $\frac{65520}{102960}$ to its lowest terms. *Ans.* $\frac{7}{11}$.

6. Reduce $\frac{12543}{12654}$ to its lowest terms. *Ans.* $\frac{113}{114}$.

7. Reduce $\frac{4}{5}$ of $\frac{15}{16}$ of $\frac{18}{20}$ of $\frac{25}{36}$ to a simple fraction.
Ans. $\frac{15}{32}$.

8. Reduce $\frac{14}{15}$ of $\frac{18}{25}$ of $\frac{20}{21}$ of $\frac{24}{35}$ of $\frac{33}{88}$ to a simple fraction.
Ans. $\frac{432}{1875}$.

9. Reduce $\frac{2}{3}$, $\frac{3}{4}$, $\frac{5}{6}$, $\frac{7}{8}$, $\frac{9}{10}$ to a common denominator.
Ans. $\frac{80}{120}$, $\frac{90}{120}$, $\frac{100}{120}$, $\frac{105}{120}$, $\frac{108}{120}$.

10. Reduce $\frac{4}{5}$, $\frac{5}{6}$, $\frac{7}{8}$, $\frac{11}{12}$, $\frac{13}{14}$, $\frac{15}{16}$ to a common denominator.
Ans. $\frac{1344}{1680}$, $\frac{1400}{1680}$, $\frac{1470}{1680}$, $\frac{1540}{1680}$, $\frac{1560}{1680}$, $\frac{1575}{1680}$.

11. Reduce $\frac{13}{20}$ of $\frac{17}{19}$ of $\frac{35}{49}$ of $\frac{72}{99}$ of $\frac{57}{85}$ of $\frac{88}{72}$ to a simple fraction. *Ans.* $\frac{8}{85}$.

12. What is the sum of $\frac{4}{5}$, $\frac{11}{14}$, $\frac{9}{16}$, and $\frac{13}{28}$? *Ans.* $2\frac{49}{80}$.

13. Add $\frac{7}{18}$, $\frac{11}{15}$, $\frac{15}{22}$, $\frac{13}{20}$, and $\frac{17}{30}$. *Ans.* $3\frac{41}{1980}$.

14. Subtract $\frac{7}{8}$ of $3\frac{3}{7}$ from $\frac{3}{4}$ of $\frac{16}{21}$ of $9\frac{3}{4}$. *Ans.* $2\frac{4}{7}$.

15. Multiply $\frac{3}{4}$ of $3\frac{3}{7}$ by $\frac{3}{4}$ of $\frac{7}{8}$ of $\frac{16}{21}$ of $\frac{16}{27}$. *Ans.* $\frac{16}{21}$.

16. Divide $\frac{5}{9}$ of $\frac{27}{40}$ of $5\frac{4}{5}$ by $\frac{3}{4}$ of $\frac{2}{9}$ of $\frac{10}{12}$. *Ans.* $15\frac{33}{50}$.

17. Add $\dfrac{8\frac{4}{9}}{16\frac{8}{9}}$ to $\frac{1}{2}$ of $\frac{4}{5}$. *Ans.* $\frac{9}{10}$.

18. Add $\dfrac{3\frac{1}{2}}{7}$, $\dfrac{4\frac{1}{3}}{17\frac{1}{3}}$, and $\dfrac{10}{26\frac{1}{4}}$. *Ans.* $1\frac{11}{84}$

19. Subtract $\frac{45}{270}$ from $\dfrac{8\frac{1}{3}}{25}$. *Ans.* $\frac{1}{6}$.

20. From 1000 take $\frac{1}{1000}$. *Ans.* $999\frac{999}{1000}$.

21. Multiply $\frac{2}{7}$ of $\frac{21}{38}$ by $\frac{5}{8}$ of $15\dfrac{\frac{4}{5}}{16}$. *Ans.* $1\frac{295}{608}$.

22. Multiply $\dfrac{\frac{1}{4}-\frac{1}{9}}{\frac{1}{2}+\frac{1}{3}}+\frac{1}{2}$ by $12\frac{1}{7}$. *Ans.* $8\frac{2}{21}$.

23. Divide $3\frac{2}{3}$ by $\frac{1}{2}$ of $\dfrac{9\frac{1}{2}}{4}$. *Ans.* $3\frac{5}{57}$.

24. Divide $\frac{3}{4}$ of $\frac{8}{9}$ of $3\frac{1}{12}$ by $\frac{1}{7}$ of $\frac{14}{15}$ of $\frac{10}{27}$ *Ans.* $41\frac{5}{8}$.

25. What cost $45\frac{3}{4}$ yards of muslin at $12\frac{1}{2}$ cents a yard?
Ans. \$5.71$\frac{7}{8}$.

26. What cost $96\frac{7}{8}$ tons of coal, at \$12 per ton?
Ans. \$1162$\frac{1}{2}$.

27. What cost $196\frac{5}{8}$ pounds of meat, at $6\frac{3}{4}$ cents a pound?
Ans. \$13.27$\frac{7}{32}$.

28. What cost $327\frac{2}{3}$ yards of cloth, at $62\frac{1}{2}$ cents a yard?
Ans. \$204.79$\frac{1}{6}$.

29. What cost $\frac{5}{6}$ of a barrel of flour, if 1 barrel cost \$8?
Ans. \$6$\frac{2}{3}$.

30. What cost $\frac{7}{8}$ of a yard of cloth, if 2 yards cost \$18$\frac{2}{5}$?
Ans. \$8.05.

31. What cost $2\frac{2}{3}$ tons of hay, if $3\frac{1}{5}$ tons cost \$48?
Ans. \$40.

32. If 1 yard of cloth cost \$1$\frac{1}{8}$, what will $5\frac{7}{9}$ yards cost?
Ans. \$6.50.

33. If $2\frac{1}{2}$ tons of hay cost \$30, what will $57\frac{3}{4}$ tons cost?
Ans \$693.

34. If 1 yard of cloth cost \$8.01$\frac{3}{7}$, what will $14\frac{12}{17}$ yards cost? *Ans.* \$117$\frac{6}{7}$.

35. A has $\frac{3}{4}$ of \$8560, which is $2\frac{1}{2}$ times B's money; how much money has B? *Ans.* \$2568.

36. B has 725 acres of land; $2\frac{3}{5}$ times B's equals $3\frac{5}{7}$ times C's; how much has C? *Ans.* 507$\frac{1}{2}$ acres.

37. If 4 be added to both terms of the fraction $\frac{5}{6}$, will the value of the fraction be increased or diminished?

38. If 4 be subtracted from both terms of the fraction $\frac{5}{6}$, will the value of the fraction be increased or diminished?

39. Will the value of the fraction $\frac{5}{6}$ be increased or diminished if 4 be added to both terms? If 4 be subtracted from both terms?

40. Mr. Mann bought a tract of land for \$6500, and sold $\frac{3}{5}$ of it to Mr. Chase, and $\frac{5}{7}$ of the remainder to Mr. Colburn; what was the cost of the remainder? *Ans.* \$742$\frac{6}{7}$.

41. A man worked $25\frac{3}{4}$ days, and after paying his board and other expenses with $\frac{2}{5}$ of his earnings, he had \$12.36 *remaining*; what were his daily wages? *Ans.* 80 cents.

42. Mr. Bowman gave $12\frac{3}{4}$ bushels of potatoes, at 30 cents a bushel, for butter worth $18\frac{3}{4}$ cents a pound, how much butter did he get? *Ans.* $20\frac{2}{5}$ pounds.

43. James Allen bought land for $10,260, and sold it so as to gain $\frac{1}{19}$ of the cost, the gain being $3 an acre; how many acres did he buy? *Ans.* 180 acres.

44. Mr. Robinson bought 100 yards of double width sheeting; he sold $\frac{1}{4}$ of it to Mr. Brown and $\frac{2}{5}$ of the remainder to Mr. Jones; what is the value of the remainder at $25\frac{3}{4}$ cents a yard? *Ans.* 11.58\frac{3}{4}$.

45. If at a certain time of day a pole 63 feet long casts a shadow $36\frac{3}{4}$ feet long, what is the length of a pole which casts a shadow $33\frac{3}{5}$ feet long? *Ans.* $57\frac{3}{5}$ ft.

46. A cistern has a capacity of $289\frac{1}{11}$ gallons, and has a pipe discharging into it $25\frac{1}{3}$ gallons per hour, and there is a leak through which it loses $5\frac{1}{7}$ gallons per hour; how long will it take to fill the cistern? *Ans.* $14\frac{7}{22}$ hours.

47. Mr. Williams, laying in a stock of goods, invests $\frac{1}{3}$ of his money in flour, $\frac{1}{4}$ in sugar and molasses, $\frac{1}{12}$ in tea and coffee, and the remainder, $540, in sundry other groceries; what was the whole amount invested? *Ans.* $1620.

48. What is the value of $\dfrac{\frac{4}{7} \text{ of } \frac{3}{8}}{5\frac{1}{14}-4\frac{51}{56}}-\dfrac{\frac{5}{11} \text{ of } \frac{13}{15}}{7\frac{15}{22}-6\frac{1}{2}}$? *Ans.* 1.

49. What is the value of $\dfrac{\dfrac{1+\frac{1}{5}}{1-\frac{1}{5}}}{\dfrac{1\times\frac{1}{5}}{1\div\frac{1}{5}}}$? *Ans.* $37\frac{1}{2}$

50. Value of $\dfrac{6\frac{1}{4}+5\frac{1}{2}}{6\frac{1}{4}-5\frac{1}{2}}+\dfrac{6\frac{1}{4}\times5\frac{1}{2}}{6\frac{1}{4}\div5\frac{1}{2}}-\dfrac{7\frac{1}{2}}{3}$? *Ans.* $43\frac{5}{12}$

51. Value of $\left(\dfrac{\frac{1}{3}+\frac{3}{10}}{6\frac{1}{3}}+\dfrac{5\frac{5}{7}}{4\frac{4}{9}}\right)\div11\times\dfrac{15\frac{2}{5}}{8\frac{1}{12}}$? *Ans.* $\frac{8}{25}$.

52. Value of $\dfrac{1-\frac{1}{5}}{2}\times\dfrac{2-\frac{1}{5}}{3}\times\dfrac{3-\frac{1}{5}}{4}\div\dfrac{4-\frac{1}{5}}{5}$? *Ans.* $\frac{21}{95}$

INTRODUCTION TO DECIMALS.

MENTAL EXERCISES.

1. If a unit is divided into 10 equal parts, what is one of these parts called?

2. If one-tenth is divided into 10 equal parts, what is one part called? 2 parts? 3 parts?

3. If $\frac{1}{100}$ is divided into 10 equal parts, what is one part called? 2 parts? 4 parts?

4. What part of 4 is 4 tenths? of 5 is 5 tenths? of 5 tenths is 5 hundredths? of 6 tenths is 6 hundredths?

5. What is $\frac{1}{10}$ of 3 hundredths? What part of 3 hundredths is 3 thousandths? of 6 thousandths is 6 ten-thousandths?

6. In the number 4444, the 4 units is what part of the 4 tens? the 4 tens is what part of the 4 hundreds?

7. A term in units place denotes what part of the value which it does in tens place?

8. A term in tens place denotes what part of the value which it does in hundreds place?

9. By the same law, a term to the right of units place would denote what part of its value in units place? *Ans.* One-tenth.

10. How may we indicate that a term is at the right of units place? *Ans.* By placing a dot (.) at the right of units place.

11. How will you express $2\frac{5}{10}$ in this manner? *Ans.* 2.5.

12. What does the dot between the 2 and the 5 denote? *Ans.* That 2 is in units place and 5 in tenths place.

13. In the expression 11.1, the 1 at the right of units denotes what part of a unit?

14. In the expression 11.11, the second term at the right of the period denotes what part of a tenth? what part of a unit?

15. How may we then write tenths, hundredths, etc., without a denominator? *Ans.* By writing them at the right of units.

16. What shall we call the first place at the right of units? Second place? Third? Fourth? Fifth?

17. Write without a denominator 2 tenths; 3 tenths; 4 tenths; 6 tenths; 8 tenths; 3 hundredths; 5 hundredths; 1 thousandth; 3 thousandths; 6 thousandths.

18. Read the following expressions: 4.6; 25.45; 26.34; 18.05; 25.235; 36.205; 46.008.

19. These fractions arising from the successive division by 10 are called *decimal fractions.* The term decimal is derived from *decem, meaning ten.*

SECTION V.

DECIMAL FRACTIONS.

195. A **Decimal Fraction** is a number of *tenths, hundredths, thousandths,* etc.

196. A **Decimal Fraction** is usually expressed by placing a point before the numerator and omitting the denominator; thus .5 expresses $\frac{5}{10}$; .05 expresses $\frac{5}{100}$; .005 $\frac{5}{1000}$, etc.

197. The **Symbol** of a decimal is the *period,* called the *decimal point,* or *separatrix.* It indicates the decimal, and *separates* decimals and integers.

198. The places at the right of the decimal point are called *decimal places.* The first place to the right of the decimal point is *tenths,* the second place is *hundredths,* etc. Thus, .245 expresses 2 *tenths,* 4 *hundredths,* and 5 *thousandths.*

199. The method of expressing decimal fractions arises from the method of notation for integers, and is a continuation of it. This beautiful law, as applied to integers and fractions, is exhibited in the following

NOTATION AND NUMERATION TABLE.

Hund.-millions.	Ten-millions.	Millions.	Hund.-thousands.	Ten-thousands.	Thousands,	Hundreds.	Tens.	Units.	*Decimal Point.*	Tenths.	Hundredths.	Thousandths.	Ten-thousandths.	Hund.-thousandths.	Millionths.	Ten-millionths.	Hund.-millionths.	Billionths.
6	6	6	6	6	6	6	6	6	.	6	6	6	6	6	6	6	6	6
9th.	8th.	7th.	6th.	5th.	4th.	3d.	2d.	1st.		1st.	2d.	3d.	4th.	5th.	6th.	7th.	8th.	9th.
Integers.										Decimals								

200. A **Decimal** is a decimal fraction expressed by the method of decimal notation; as, .5, .25, etc.

201. A **Pure Decimal** is one which consists of decimal figures only; as, .345.

202. A **Mixed Decimal** is one which consists of an integer and a decimal; as 4.35.

203. A **Complex Decimal** is one which contains a common fraction at the right of the decimal; as, .45$\frac{1}{8}$.

NOTES.—1. The first treatise upon decimals was written by Stevinus, and published in 1585.

2. The decimal point, Dr. Peacock thinks, was introduced by Napier, the inventor of logarithms, in 1617, though De Morgan says that Richard Witt made as near an approach to it as Napier.

EXERCISES IN NUMERATION.

1. Read the decimal .45.

SOLUTION.—This expresses 4 tenths and 5 hundredths, or since 4 tenths equals 40 hundredths, and 40 hundredths plus 5 hundredths equal 45 hundredths, it may also be read 45 hundredths. Hence the following rules:

Rule I.—*Begin at tenths, and read the terms in order towards the right, giving each term its proper denomination.*

Rule II.—*Read the decimal as a whole number, and give it the denomination of the last term at the right.*

NOTE.—In the second method we may determine the denominator by numerating *from* the decimal point, and the numerator by numerating *towards* the decimal point.

Read the following decimals:

2. .35	7. .507	12. 5.0708
3. .76	8. .605	13. 8.75006
4. .84	9. .7605	14. 43.30027
5. .031	10. .0576	15. 756.00279
6. .162	11. .00312	16. 372.000086

EXERCISES IN NOTATION.

1. Express 25 thousandths in the form of a decimal.

SOLUTION.—25 thousandths equal 20 thousandths plus 5 thousandths, or 2 hundredths and 5 thousandths; hence we write the 5 in the third or thousandths place, 2 in the second or hundredths place, and fill the vacant tenths place with a cipher, and we have .025. Hence the following rules:

Rule I.—*Place the decimal point, and then write each term so that it may express its proper denomination, using ciphers when necessary.*

Rule II.—*Write the numerator, and then place the decimal point so that the right-hand term shall express the denomination of the decimal, using ciphers when necessary.*

NOTE.—To avoid ambiguity, where integers and decimals occur in the same written number, a comma should be placed between them; thus, three hundred and seven ten-thousandths should be written .0307, but three hundred, and seven ten-thousandths, 300.0007.

Express the following in decimal form:

2. Twenty-five hundredths.

3. 2 tenths and 8 hundredths.

4. 7 tenths and 9 hundredths.

5. Twenty-five thousandths.

6. 4 tenths and 7 thousandths.

7. Seven tenths and 8 thousandths.

8. Five hundred, and 25 thousandths.

9. Three tenths and 7 ten-thousandths.

10. Seven hundredths and 9 ten-thousandths.

11. Three hundred, and 78 ten-thousandths.

12. Five tenths, 6 hundredths, and 7 hundred thousandths.

13. Four hundredths, seven ten-thousandths, and 6 hundred-thousandths.

14. Nine hundred and sixty-nine hundred-thousandths.

15. Two tenths and three millionths.

16. Five hundredths, six thousandths, and eight millionths.

17. Thirty-five thousand, and eight millionths.

18. Ninety-three hundred and seven ten-millionths.

19. Eighteen thousand and one hundred-millionths.

20. Two million, and 6 thousand and 9 hundred-millionths.

Express the following as decimals:

21. $\frac{7}{10}$. *Ans.* .7.

22. $\frac{93}{100}$. *Ans.* .93.

23. $\frac{76}{1000}$. *Ans* .076.

24. $\frac{302}{10000}$. *Ans.* .0302.

25. $124\frac{7}{100}$. *Ans.* 124.07.

26. $967\frac{104}{1000}$. *Ans.* 967.104.

27. $96\frac{209}{10000}$. *Ans.* 96.0209.

28. $75\frac{3017}{1000000}$. *Ans.* 75.003017.

PRINCIPLES.

1. *Moving the decimal point one place to the right, multiplies the decimal by* 10; *two places, multiplies by* 100; *etc.*

For, if the point be moved one place to the right, each figure will express ten times as much as before, hence the whole decimal will be ten times as great; etc.

2. *Moving the decimal point one place to the left, divides the decimal by* 10; *two places, divides by* 100; *etc.*

For, if the point be moved one place to the left, each figure will express 1 tenth of its previous value, hence the whole decimal will be only 1 tenth as great; etc.

3. *Placing a cipher between the decimal point and the decimal, divides the decimal by* 10.

For, this moves each figure one place to the right in the scale, in which case they express 1 tenth as much as before, and hence the decimal is only 1 tenth as great.

4. *Annexing ciphers to the right of a decimal does not change its value.*

For, each figure retains the same place as before, and hence expresses the same value as before, and consequently the value of the decimal is unchanged.

MENTAL EXERCISES.

1. How many tenths in 1? in $\frac{1}{2}$? in $\frac{1}{5}$? in $\frac{3}{5}$?
2. How many hundredths in $\frac{1}{2}$? in $\frac{3}{5}$? in $\frac{4}{5}$? in $\frac{9}{10}$? in $\frac{13}{50}$?
3. How many hundredths in $\frac{1}{4}$? in $\frac{3}{4}$? in $\frac{1}{25}$? in $\frac{8}{25}$? in $\frac{7}{50}$?
4. How many halves in .5? in .50? 5ths in .2? in .4?
5. How many 4ths in .25? in .75? eighths in .125? in .375?
6. Express as a common fraction and reduce to lowest terms, .5; .4; .8; 25; .50; .75; .80; .125; .625.

REDUCTION OF DECIMALS.

204. There are two cases of the reduction of decimals:

1st. To reduce decimals to common fractions.

2d. To reduce common fractions to decimals.

CASE I.

205. ***To reduce a decimal to a common fraction.***

1. Reduce .45 to a common fraction.

SOLUTION.—.45 expressed in the form of a common fraction is $\frac{45}{100}$, which reduced to its lowest terms equals $\frac{9}{20}$. Hence the following

OPERATION.

$.45=\frac{45}{100}=\frac{9}{20}$, *Ans.*

Rule.—*Write the denominator under the decimal, omitting the decimal point, and reduce the common fraction to its lowest terms.*

WRITTEN EXERCISES.

Reduce the following decimals to common fractions:

2.	.35.	*Ans.* $\frac{7}{20}$.	7.	9.75.	*Ans.* $9\frac{3}{4}$.
3.	.48.	*Ans.* $\frac{12}{25}$.	8.	12.725.	*Ans.* $12\frac{29}{40}$.
4.	.125.	*Ans.* $\frac{1}{8}$.	9.	16.075.	*Ans.* $16\frac{3}{40}$.
5.	.625.	*Ans.* $\frac{5}{8}$.	10.	5.064.	*Ans.* $5\frac{8}{125}$.
6.	3.25.	*Ans.* $3\frac{1}{4}$.	11.	17.0125.	*Ans.* $17\frac{1}{80}$.

12. Reduce the complex decimal $.2\frac{2}{3}$ to a common fraction.

SOLUTION.—$.2\frac{2}{3}$ is $2\frac{2}{3}$ tenths, which, by writing the denominator, becomes $\frac{2\frac{2}{3}}{10}$, which equals $\frac{\frac{8}{3}}{10}$, or $\frac{8}{30}$, which reduced to its lowest terms equals $\frac{4}{15}$. Therefore, etc.

OPERATION.

$$.2\frac{2}{3} = \frac{2\frac{2}{3}}{10} = \frac{\frac{8}{3}}{10} = \frac{8}{30} = \frac{4}{15} \textit{ Ans.}$$

Reduce the following to common fractions or mixed numbers:

13. $.8\frac{1}{3}$.	*Ans.* $\frac{5}{6}$.	**18.** $4.8\frac{1}{3}$.	*Ans.* $4\frac{5}{6}$.
14. $.16\frac{2}{3}$.	*Ans.* $\frac{1}{6}$.	**19.** $12.18\frac{3}{4}$.	*Ans.* $12\frac{3}{16}$.
15. $.25\frac{3}{5}$.	*Ans.* $\frac{32}{125}$.	**20.** $22.03\frac{1}{3}$.	*Ans.* $22\frac{1}{30}$.
16. $.45\frac{5}{6}$.	*Ans.* $\frac{11}{24}$.	**21.** $50.08\frac{5}{6}$.	*Ans.* $50\frac{53}{600}$.
17. $\$0.16\frac{2}{3}$.	*Ans.* $\$\frac{1}{6}$.	**22.** $\$6.66\frac{2}{3}$.	*Ans.* $\$6\frac{2}{3}$.

CASE II.

206. ***To reduce a common fraction to a decimal.***

1. Reduce $\frac{5}{8}$ to a decimal.

SOLUTION.—$\frac{5}{8}$ equals $\frac{1}{8}$ of 5. 5 equals 50 tenths; $\frac{1}{8}$ of 50 tenths is 6 tenths and 2 tenths remaining; 2 tenths equal 20 hundredths; $\frac{1}{8}$ of 20 hundredths equals 2 hundredths and 4 hundredths remaining; 4 hundredths equal 40 thousandths; $\frac{1}{8}$ of 40 thousandths is 5 thousandths. Therefore, $\frac{5}{8}$ equals .625.

OPERATION.

$\frac{5}{8} = \frac{1}{8}$ of 5

$= 8\overline{)5.000}$

.625

Rule.—I. *Annex ciphers to the numerator, and divide by the denominator.*

II. *Point off as many decimal places in the quotient as there are ciphers used.*

NOTES.—1. In many cases the division will not terminate; the common fraction cannot then be exactly expressed by a decimal. Such decimals are called *interminate* or *infinite* decimals.

2. The symbol + annexed to a decimal indicates that it contains other decimal terms. The symbol — annexed to a decimal indicates that the last decimal term is increased by 1. This is often done when the next term is greater than 5.

WRITTEN EXERCISES.

Reduce the following common fractions to decimals:—

2. $\frac{1}{4}$.	*Ans* .25.	**8.** $\frac{11}{15}$.	*Ans.* .733+.
3. $\frac{3}{4}$.	*Ans.* .75.	**9.** $\frac{15}{16}$.	*Ans.* .9375.
4. $\frac{3}{8}$.	*Ans.* .375.	**10.** $\frac{16}{25}$.	*Ans.* .64.
5. $\frac{7}{8}$.	*Ans.* .875.	**11.** $\frac{16}{17}$.	*Ans* .9412—.
6. $\frac{3}{16}$.	*Ans.* .1875.	**12.** $\frac{18}{19}$.	*Ans.* .947368+.
7. $\frac{5}{16}$.	*Ans.* .3125.	**13.** $\frac{113}{256}$.	*Ans.* .44140625

ADDITION OF DECIMALS.

207. Addition of Decimals is the process of finding the sum of two or more decimals.

MENTAL EXERCISES.

Add

1. $\frac{3}{10}$ and $\frac{5}{10}$.
2. .3 and .5.
3. .04 and .03.
4. .6 and .04.
5. 25 and .5.
6. .07 and .08.
7. .05 and .136
8. 2.5 and 4.25.

WRITTEN EXERCISES.

1. What is the sum of 11.96, 25.075, 84.306, 90.728?

OPERATION.

```
 11.96
 25.075
 84.306
 90.728
-------
212.069
```

SOLUTION.—We write the numbers so that terms of the same order shall stand in the same column, and begin at the right to add. 8 thousandths, plus 6 thousandths, plus 5 thousandths, are 19 thousandths, which equals 1 hundredth and 9 thousandths; we write the 9 thousandths, and add the 1 hundredth to the next column: 2 hundredths, plus 7 hundredths, plus 6 hundredths, equal 15 hundredths, and the 1 hundredth added is 16 hundredths, which equals 1 tenth and 6 hundredths; we write the 6 hundredths. etc.

Rule.—I. *Write the numbers so that terms of the same order stand in the same column.*

II. *Add as in whole numbers, and place the decimal point between the units and tenths of the sum.*

2. Find the sum of 79.76, 85.08, 36.125, 140.309.
Ans. 341.274.

3. Find the sum of 87.09, 58.37, 95.42, 237.675.
Ans. 478.555

4. Add 18.79, 147.072, 856.709, 185.8761, 397.05784.
Ans. 1605.50494

5. Add 59.874, 435.095, 672.328, 976.309, 8467.500843.
Ans. 10611.106843.

6. What is the sum of \$25$\frac{3}{4}$, \$37$\frac{1}{5}$, \$28.37$\frac{1}{2}$, \$50.06$\frac{1}{4}$, \$15.37$\frac{1}{2}$, \$57$\frac{3}{8}$, \$15$\frac{7}{8}$, and \$23.87$\frac{1}{2}$? *Ans.* \$253.88$\frac{3}{4}$.

7. Add 9 and 7 tenths, 41 and 8 hundredths, 75 and 54 hundredths, 128 and 187 thousandths. *Ans.* 254.507.

8. Add 187 and 5 thousandths, 49 and 9 hundred-thousandths, 1876 and 245 millionths, 187 ten-thousandths, and 999 ten-millionths. *Ans.* 2112.0241349.

9. Add 798 and 9 ten-thousandths, 17 millionths, 18 thousandths and 98 ten-millionths, 67 hundred-thousandths and 95 ten-millionths. *Ans.* 798.0196063.

10. Find the sum of 3 decimal units of the 1st order, $6\frac{1}{2}$ decimal units of the 2d order, $4\frac{1}{4}$ of the 3d order, $3\frac{1}{8}$ of the 4th order, and $6\frac{1}{5}$ of the 5th order. *Ans.* .3696245.

SUBTRACTION OF DECIMALS.

208. Subtraction of Decimals is the process of finding the difference between two decimals.

MENTAL EXERCISES.

1. Subtract $\frac{4}{10}$ from $\frac{9}{10}$.
2. Subtract .4 from .9.
3. Subtract .06 from 08.
4. Subtract .05 from .55.
5. Subtract $\frac{6}{10}$ from .9.
6. Subtract .05 from .5.
7. Subtract .25 from .63.
8. Subtract .75 from 1.08.

WRITTEN EXERCISES.

1. From 972.163 take 856.235.

OPERATION.

$$\begin{array}{r} 972.163 \\ \underline{856.235} \\ 115.928 \end{array}$$

SOLUTION.—We write the numbers so that terms of the same order stand in the same column, and begin at the right to subtract. We cannot subtract 5 thousandths from 3 thousandths, hence we add ten thousandths to 3 thousandths, which equals 13 thousandths; 5 thousandths from 13 thousandths leaves 8 thousandths, which we write in the order of thousandths: since we have added 10 thousandths or 1 hundredth to the minuend, we must add 1 hundredth to the subtrahend; 1 hundredth and 3 hundredths are 4 hundredths; 4 hundredths from 6 hundredths leaves 2 hundredths, etc.

Rule.—I. *Write the subtrahend under the minuend, so that terms of the same order stand in the same column.*

II. *Subtract as in whole numbers, and place the decimal point between the units and tenths of the remainder.*

2. From 707.325 take 623.452. *Ans.* 83.873.

3. From 826.438 take 734.936. *Ans.* 91.502.

4. From 78.3057 take 29.084. *Ans.* 49.2217.

5. From 1230.207 take 384.1231. *Ans.* 846.0839.

6. From 2.07 take 1.432765. *Ans.* .637235.

7. From .3 take 3 hundred-millionths. *Ans.* .29999997.

8. From 1 and .001 take .01 and .000001. *Ans.* .990999.

9. From $2\frac{1}{2}$ take $2\frac{1}{2}$ thousandths and $2\frac{1}{2}$ billionths. *Ans.* 2.4974999975

MULTIPLICATION OF DECIMALS.

209. Multiplication of Decimals is the process of finding the product when one or both factors are decimals.

MENTAL EXERCISES.

What is the value of

1. 2 times $\frac{3}{10}$? 3×.4?
2. 2 times .6? 5×.6?
3. $\frac{5}{10}$ times $\frac{7}{10}$? .5×.7?
4. $\frac{4}{10}$ times $\frac{8}{10}$? .4×.8?
5. $\frac{8}{10}$×.6? .6×.7?
6. 5×$\frac{8}{100}$? 5× 08?
7. $\frac{3}{10}$×$\frac{12}{100}$? .3×.12?
8. .6×.15? .12×.12?
9. How many places in the product of *units* multiplied by *tenths*? *Tenths* by *tenths*? *Units* by *hundredths*? *Tenths* by *hundredths*? *Hundredths* by *hundredths*?
10. How many decimal places in the product if there are two in the multiplicand and one in the multiplier? If there are two in each factor? If three in multiplicand and two in multiplier?

WRITTEN EXERCISES.

1. Multiply 7.23 by .46.

OPERATION.

$$\begin{array}{r} 7.23 \\ .46 \\ \hline 4338 \\ 2892 \\ \hline 3.3258 \end{array}$$

SOLUTION.—7.23 multiplied by 46 equals 332.58, and multiplied by 46 *hundredths* the product is 1 hundredth as great, which, by removing the decimal point two places to the left, becomes 3.3258. Hence, 7.23 multiplied by .46 equals 3.3258.

SOLUTION 2D.—$7.23\times.46=\frac{723}{100}\times\frac{46}{100}=\frac{33258}{10000}=\frac{1}{10000}\times33258=3.3258$. From either of these solutions we derive the following

Rule.—*Multiply as in whole numbers, and from the right of the product point off as many decimal places as there are in both factors, prefixing ciphers when necessary.*

2. Multiply 27.13 by .67. *Ans.* 18.1771.
3. Multiply 43.08 by 2.36. *Ans.* 101.6688.
4. Multiply 79.52 by .019. *Ans.* 1.51088.
5. Multiply 8.534 by 20.074. *Ans.* 171.311516.
6. Multiply 123.107 by 1.52. *Ans.* 187.12264.
7. Multiply 512.073 by 35.08. *Ans.* 17963.52084.
8. Multiply 54.0079 by 7.072. *Ans.* 381.9438688.
9. Multiply 1.08096 by 3.5702. *Ans.* 3.859243392.
10. Multiply .03507 by .005873. *Ans.* .00020596611.
11. Multiply 2.0709 by .000246. *Ans.* .0005094414.
12. Value of $9\frac{1}{2}\times.08\frac{3}{4}\times10$? Ans. 8.3125.

13. Value of .5 of $\frac{7}{11} \times .05\frac{1}{2}$? *Ans.* .0175.

14. Value of .07 of $\frac{5}{6} \times 300 \times .09\frac{3}{7}$? *Ans.* 1.65.

15. Value of $\$25.6\frac{7}{8} \times .05\frac{1}{3} + \$15 \times 6.03\frac{1}{4}$?

Ans. \$91.8575.

16. $\overline{200 + .02 - \frac{1}{2}} \times .8\frac{1}{4} + (.8\frac{1}{4} - .12\frac{7}{10} \times \frac{3}{4}$ of $7\frac{1}{5})$?

Ans. 164.7432.

17. Multiply 1 hundredth by 1 thousandth, and add 1 tenth to the product. *Ans.* .10001.

18. What is the product of one-tenth by one-tenth? one hundred by one-hundredth? one million by one-millionth?

DIVISION OF DECIMALS.

210. Division of Decimals is the process of finding the quotient when one or both terms are decimals.

MENTAL EXERCISES.

What is the value of

1. $\frac{4}{10} \div 2$? $.4 \div 2$?
2. $\frac{24}{100} \div 6$? $.24 \div 6$?
3. $.50 \div 2$? $.5 \div 2$?
4. $.150 \div 2$? $.15 \div 2$?
5. $5 \div 4$? $.5 \div 4$? $.05 \div 4$?
6. $6 \div \frac{3}{10}$? $6 \div .3$? $6 \div .03$?
7. $.8 \div 4$? $8 \div .4$? $8 \div .04$?
8. $7.2 \div .6$? $.72 \div .6$? $.72 \div .06$?
9. $8.4 \div .7$? $.84 \div .7$? $.84 \div .07$?
10. $96 \div .8$? $9.6 \div .8$? $.96 \div .8$?

11. What is the product of *tenths* by *tenths*? What then is the quotient of *hundredths* by *tenths*? What is the product of *tenths* by *hundredths*? What is the quotient of *thousandths* divided by *tenths*?

12. If there are three decimal places in the multiplicand and two in the multiplier, how many in the product? If there are five in the dividend and three in the divisor, how many in the quotient?

WRITTEN EXERCISES.

1. Divide 12.3084 by 2.34.

OPERATION.

```
2.34)12.3084(5.26
     1170
      608
      468
      1404
      1404
```

SOLUTION.—Dividing as in whole numbers, we obtain the quotient 526; now, since the dividend is the product of the divisor and quotient, the number of decimal places in the dividend should equal the number in the divisor, plus the number in the quotient; hence the number of decimal places in the quotient equals the number of decimal places in the dividend, minus the number of places in the divisor; hence there should be *four* minus *two*, or *two* decimal places in the quotient, therefore the quotient is 5.26.

SOLUTION 2d.—$12.3084 \div 2.34 = \frac{123084}{10000} \div \frac{234}{100} = \frac{123084}{10000} \times \frac{100}{234} = \frac{123084}{100 \times 234} = \frac{1}{100} \times \frac{123084}{234} = \frac{1}{100}$ of $526 = 5.26$. From either of these solutions we derive the following

Rule.—I. *Annex ciphers to the dividend, if necessary to make the number of decimals equal to the number of decimal places in the divisor.*

II. *Divide as in whole numbers, annexing ciphers to the dividend when needed to continue the division.*

III. *Point off as many decimals in the quotient as the number of decimal places in the dividend exceeds the number in the divisor.*

NOTES.—1. When there are ciphers at the right of the divisor, cut them off, divide by the significant part, and then point off as many decimal places as before, plus the number of ciphers cut off.

2. Make complex decimals pure or divide them like common mixed numbers, or multiply both by the L. C. M. of the denominators, and then divide.

2. Divide 272.636 by 6.37. *Ans.* 42.8.
3. Divide 281.8585 by 3.85. *Ans.* 73.21.
4. Divide 57.85728 by 8.32. *Ans.* 6.954.
5. Divide 50.38218 by 67.8. *Ans.* .7431.
6. Divide 31.421154 by 85.43. *Ans.* .3678.
7. Divide 268.32615 by 4.705. *Ans.* 57.03.
8. Divide 2.3697792 by 34.56. *Ans.* .06857.
9. Divide 3411.3424 by 5.678. *Ans.* 600.8.
10. Divide 195.388698 by 32.185. *Ans.* 6.0708.
11. Divide 1678.543 by 185.79321. *Ans.* 9.03446+.

What is the value of

12.	.065÷.026?	*Ans.* 2.5.	17.	1÷.125.	*Ans.* 8.
13.	375÷.025?	*Ans.* 15.	18.	.51÷.015.	*Ans.* 34.
14.	.008÷.04?	*Ans.* .2.	19.	155÷.0625.	*Ans.* 2480.
15.	.012÷.06?	*Ans.* .2.	20.	.00625÷25.	*Ans.* .00025.
16.	.004÷2.5?	*Ans.* .0016.	21.	25÷.00625.	*Ans.* 4000.

22. $(2.04 \div 17 + \overline{47 \div 200} \times 5000) - \frac{7}{8}$? *Ans.* 1174.245.

23. $(789 - .789) \div (\overline{.75 - .075} \times .75 \text{ of } 8)$? *Ans.* 194.62.

24. $\overline{45 \times .181} + \overline{3.6 \times .3\frac{1}{3}} \div (3.5 - \frac{3.43}{7})$? *Ans.* $8.5436\frac{214}{301}$.

25. $(\$347.84 \div \$10.87) \times \overline{.0025 + .01\frac{3}{4}} \times 50$? *Ans.* 32.

26. $(\$1080 \times 3.27) \div \$10.90 - (\$790 \div \$3.95)$? *Ans.* 124.

WRITTEN PROBLEMS.

MISCELLANEOUS EXAMPLES.

1. What common fraction equals .00096? *Ans.* $\frac{3}{3125}$.

2. Add $\frac{.2}{2.7}$, $11\frac{\frac{1}{2}}{5.4}$, $2\frac{7}{45}$, and .1. *Ans.* $13\frac{19}{45}$.

3. Add $\frac{.4}{5}$, $\frac{2\frac{5}{6}}{5\frac{2}{3}}$, $\frac{21}{48}$, and $\frac{2\frac{4}{9}}{1\frac{7}{18}}$. *Ans.* $2\frac{311}{400}$.

4. Multiply .0075 cwt. by 4.008. *Ans.* .03006 cwt.

5. Divide $13\frac{22}{35}$ tons by $28\frac{\frac{1}{5}}{.7}$. *Ans.* $\frac{53}{110}$ tons.

6. Multiply 1.25 of $.8\frac{3}{4}$ of $9\frac{6}{35}$ by .8 of $\frac{\frac{7}{8}}{.2}$ of 8.75.
Ans. $307\frac{53}{256}$.

7. Divide $118\frac{1}{4}$ by $.04\frac{1}{2}$; also $2.4001\frac{3}{8}$ by $1.56\frac{1.25}{6\frac{1}{4}}$.
Ans. $2627\frac{7}{9}$; $1.53\frac{518}{781}$.

8. Divide $14\frac{\frac{2}{3}}{27}$ square yards by $11\frac{1\frac{3}{4}}{13\frac{1}{2}}$. *Ans.* $1\frac{469}{1803}$.

9. Divide £$240\frac{3}{4}$ by $\frac{3}{4}$ of $\frac{24\frac{1}{2}}{28}$ of $\frac{5}{7}$ of $1\frac{4}{5}$. *Ans.* £$285\frac{1}{3}$.

10. What number multiplied by $7\frac{3}{5}$ will give $6\frac{2}{3}$ for a product? *Ans.* $\frac{50}{57}$.

11. Divide seven millionths by twelve and a half ten-millionths. *Ans.* 5.6.

12. Add $\frac{1}{4}$, $\frac{8\frac{1}{2}}{36}$, .375, and .5, and multiply the sum by $6.2\frac{1}{2}$.
Ans. $8\frac{73}{144}$.

13. Subtract $\frac{3}{7}$ of 12 from $\frac{5}{8}$ of $\frac{8\frac{3}{4}}{\frac{2}{5}}$, and divide the remainder by $\frac{5}{7}$. *Ans.* $11\frac{301}{320}$.

14. Multiply the sum of .5 and $\frac{3\frac{1}{2}}{14}$ by the difference between $\frac{8\frac{1}{2}}{17}$ and $\frac{10}{26\frac{2}{3}}$. *Ans.* $\frac{3}{32}$.

15. Multiply $\frac{62\frac{1}{2}}{1000}$ by 25 millionths, and divide the product by 125 hundred-thousandths. *Ans.* .00125.

16. The product of two numbers is $\frac{4}{9}$, and one of them is $\frac{3}{8}$ of $\frac{5}{9}$ of 2; what is the other? *Ans.* $1\frac{1}{15}$.

17. Add 2.5 cwt., 1.25 cwt., 5 cwt., 4.375 cwt., 1.875 cwt., 2.5 cwt., 1.225 cwt., and 1.275 cwt. *Ans.* 20 cwt.

18. What number multiplied by $\frac{4}{9}$ of $\frac{13}{17}$ of $5\frac{1}{2}$ is equal to $\frac{2}{3}$ of $\frac{9}{11}$ of $27\frac{1}{2}$? *Ans.* $8\frac{7}{286}$.

19. Divide the sum of six thousandths and six millionths by their difference, to 6 decimal places. *Ans.* 1.002002+.

20. From the sum of $\frac{2}{11}$, $\frac{3}{5}$, and $\frac{1}{3}$, take the remainder obtained by subtracting $\frac{7}{35}$ from $\frac{4}{13}$ of $\frac{65}{8}$ of $\frac{1}{7}$. *Ans.* $1\frac{585}{616}$.

21. What number must be divided by $\frac{3}{8}$ of $2\frac{2}{3}$ of $\frac{5\frac{1}{2}}{1\frac{3}{8}}$, to make a quotient equal to the value of $6\frac{2}{3}$ of $\frac{1\frac{1}{2}}{10}$ of 6? *Ans.* 24.

22. John had $\$225\frac{3}{8}$ and earned $\frac{1}{5}$ of $3\frac{1}{3}$ times $\$88\frac{1}{3}$ more; having lost part of his money, he found that he had $\$186\frac{3}{8}$ remaining; what amount did he lose? *Ans.* $\$97\frac{8}{9}$.

23. A ton of iron ore from the mines of Cornwall yields .65 of a ton of pure iron; how much iron will 578.8 tons of ore yield? *Ans.* 376.22 tons.

24. An oil refiner has on hand 16,745 gallons of coal oil; how many casks, each containing 42.5 gallons, can be filled with it? *Ans.* 394 casks.

25. How many dress patterns of 12.25 yards each, can be cut from a piece of French percale, containing 50 yards? *Ans.* 4+ patterns.

26. An engine pumped 41.25 barrels of 31.5 gallons each from a tank containing 1500 gallons; how many gallons remained? *Ans.* 200.625 gal.

27. I bought 3 loads of wood, the first containing 1.02 cords, the second 1.05 cords, the third .945 cords; what did it cost at $3.60 a cord? *Ans.* $10.854.

28. A dealer bought 1086 bushels of wheat; how many bins, each holding 20.25 bushels, will it fill, and how many bushel-bags can be filled from what remains? *Ans.* 53 bins; 12.75 bags.

29. Two speculators bought 4320 acres of Western land,

which they divide so that one has .37½ and the other .62½ of it; how many acres had each? *Ans.* 1620; 2700.

30. How many lengths of 6-inch stove-pipe can be made from 87.48 pounds of Norway iron, if one length requires 3.24 pounds? *Ans.* 27.

31. Mr. Newlin paid $4000 for a truck farm, giving $76.25 an acre for 27.25 acres, and $85.75 for the remainder; how many acres did he buy? *Ans.* 49.66+ acres.

32. A man bought 345.75 tons of hay, at $16.25 a ton, ⅓ of which he sold at $17.75 a ton, and the rest at cost; how much was the gain? *Ans.* $172.87½.

33. Mr. Hartman bought a lot of wheat and sold .15 of it to one man, and .25 of it to another, and kept 572.85 bushels; how much did he buy? *Ans.* 954.75 bu.

34. A man devotes .12 of his income to charity, .25 for educating his children, .45 for household expenses, and saves the remainder, which is $284.76; required his income.

Ans. $1582.

35. How many cords of wood, at $5.12½ a cord, must I give for 91.25 bushels of wheat, at $1.40 a bushel, and 85 bushels of rye, at $1.25 a bushel? *Ans.* 45.66 cords.

36. The circumference of the fore wheel of a wagon is 12.75 feet, and of the hind wheel 14.25 feet; how much oftener does one turn than the other in going 5280 feet, or one mile? *Ans.* 43.59+ times.

37. A ship whose cargo was worth $25,000, being disabled in a storm, .45½ of the whole cargo was thrown overboard; what would a merchant lose who owned .25 of the cargo? *Ans.* $2843.75.

38. A grain dealer expended $6210 for grain, ⅓ of it being for wheat, at $1.25 a bushel, ¼ of it for corn, at $0.75 a bushel, and the remainder for rye, at $1.12½ a bushel; how many bushels of each kind did he purchase?

Ans. Wheat, 1656 bu.; Corn, 2070 bu.; Rye, 2300 bu.

UNITED STATES MONEY.

211. **United States Money,** or the currency of the United States, is expressed in the decimal system.

212. The several **denominations** and their relation to each other are presented in the following table:

TABLE.

10 mills equal 1 cent.	10 dimes equal 1 dollar.
10 cents " 1 dime.	10 dollars " 1 eagle.

NOTE.—$\frac{1}{4}$ of a dollar = 25 cents; $\frac{1}{2}$ of a dollar = 50 cents; $\frac{3}{4}$ of a dollar = 75 cents; $\frac{1}{2}$ of a cent = 5 mills.

213. The **dollar** is the unit and is indicated by the symbol $; the eagle and dollar are read as a number of dollars. Thus $245 is read 245 dollars, instead of 24 eagles, 5 dollars.

214. The **dime** is one tenth of a dollar, and is expressed as tenths, the decimal point being placed between dimes and dollars. Thus $2.3 expresses 2 dollars and 3 dimes.

215. The **cent** is one tenth of a dime or one hundredth of a dollar, and is written in hundredths place. Thus $6.75 indicates 6 dollars 7 dimes and 5 cents. Dimes and cents, however, are usually read as a number of cents. Thus $6.75 is read 6 dollars and 75 cents.

216. Since dimes and cents are regarded as a number of cents, when the number of cents is less than 10, a cipher must be written in tenths place. Thus 3 dollars and 4 cents are written $3.04.

217. The **mill** is one tenth of a cent or one thousandth of a dollar, and is written in thousandths place. Thus $8.375 is read 8 dollars 37 cents and 5 mills.

NOTES.—1. In checks, notes, drafts, etc., cents are usually written as hundredths of a dollar in the form of a common fraction, as $\$12\frac{75}{100}$.

2. When the final result of a business computation contains mills, if 5 or more they are reckoned 1 cent, and if less than 5 they are rejected. Thus $7.187 would be reckoned as $7.19 and $3.162 as $3.16.

3. We used dollars and cents in treating the fundamental rules; we now give a more formal treatment of decimal currency, involving problems that pupils were not then prepared to solve.

EXERCISES IN NUMERATION.

Read the following:

1. $15.65.	5. $28.05.	9. $100.001.
2. $14.753.	6. $10.50.	10. $202.202.
3. $28.284.	7. $105.105.	11. $370.037.
4. $132.125.	8. $125.005.	12. $1030.001.

EXERCISES IN NOTATION.

Write on the slate or board

1. Seven dollars and twenty-five cents.
2. Twelve dollars, thirty cents, and five mills.
3. Twenty-five dollars, fifty-four cents, and five mills.
4. Thirty-four dollars, seven cents, and seven mills.
5. Nine eagles, six dollars, six cents, and eight mills.
6. Two hundred dollars, seven and one-half cents.
7. Forty-nine dollars, six dimes, and $7\frac{1}{2}$ cents.
8. Five hundred and thirty-eight dollars, $62\frac{1}{2}$ cents.

REDUCTION OF UNITED STATES MONEY.

218. Reduction is the process of changing a number from one denomination to another without altering its value.

219. From the explanation given we have the following

PRINCIPLES.

1. *To reduce cents to mills, annex one cipher.*
2. *To reduce dollars to cents, annex two ciphers.*
3. *To reduce dollars to mills, annex three ciphers.*
4. *To reduce cents to dollars, place the point two places from the right.*
5. *To reduce mills to dollars, place the point three places from the right.*

NOTE.—In reducing a number of dollars and cents to cents, etc., the separatrix should be removed; thus, $5.25 = 525 cents, and $8.755 = 8755 mills.

WRITTEN EXERCISES.

1. Reduce 5 dollars to cents.

SOLUTION.—In 1 dollar there are 100 cents, and in 5 dollars there are 5 times 100 cents, or 500 cents, or we annex two ciphers, as above directed.

2. Reduce 25 dollars to cents. *Ans.* 2500.

3. Reduce 24 cents to mills. *Ans.* 240.

4. Reduce 8 dollars to mills. *Ans.* 8000.

5. Reduce 20 dollars to mills. *Ans.* 20000.

6. Reduce 6 cents and 5 mills to mills. *Ans.* 65.

7. Reduce 12 dollars and 25 cents to cents. *Ans.* 1225.

8. Reduce 23 dollars and 5 cents to cents. *Ans.* 2305.

9. Reduce 1450 cents to dollars. *Ans.* $14.50.

10. Reduce 2700 cents to dollars. *Ans.* $27.

11. Reduce 5425 cents to dollars. *Ans.* $54.25.

12. Reduce 4170 mills to dollars. *Ans.* $4.17.

13. Reduce 250 mills to cents. *Ans.* 25 cents.

14. Reduce 865 mills to cents. *Ans.* 86½ cents.

15. Reduce 13875 mills to dollars, cents, and mills. *Ans.* $13.875.

16. Reduce 185326 mills to dollars, cents, and mills. *Ans.* $185.326.

17. How many dollars, cents, and mills in 235 dollars, 2 cents, and 5 mills? *Ans.* $235.025.

18. How many dollars, cents, and mills in 150 dollars, 10 cents, and 5 mills? *Ans.* $150.105.

FUNDAMENTAL OPERATIONS.

220. Since **United States Money** is expressed in the decimal scale, all the operations may be performed as in decimals.

Rule.—*To add, subtract, multiply, or divide in United States money, proceed according to the corresponding operations in decimals.*

WRITTEN EXERCISES.

1. Find the sum of $62½ and 62½ cents. *Ans.* $63.12½.

2. From $25 take 25 cents and 5 mills. *Ans.* $24.74½.

3. Require the sum of $18¾, 18¾ dimes, 18½ cents, and 18 mills. *Ans.* $20.828.

4. From 1 dollar take 1 mill; from $505 take 505 cents. *Ans.* $0.999; $499.95.

5. Subtract $6\frac{1}{4}$ cents from $6\frac{1}{4}$ dollars, and add the remainder to $6\frac{1}{4}$ dimes. *Ans.* \$7.08$\frac{1}{4}$.

6. What must a stock-dealer pay for 27 young cows at \$18$\frac{3}{4}$ apiece? *Ans.* \$506.25.

7. A drover sold the government 33 horses at the rate of \$125$\frac{1}{4}$ apiece; what did he receive? *Ans.* \$4133.25.

8. If a cattle-feeder received \$250.25 for the sale of 13 cows, what did they bring him a head? *Ans.* \$19.25.

9. I paid \$36.25 for 5 barrels of flour; what would 17 barrels have cost me at the same rate? *Ans.* \$123.25.

10. A country merchant paid \$96 for a lot of muslins, at 16 cents a yard; how many yards did he buy? *Ans.* 600.

11. How many cords of wood can be bought for \$1553.25, at the rate of \$4.75 per cord? *Ans.* 327 cords.

12. A carpenter earned in 4 months \$240; how many days did he labor if his wages were \$2.50 a day? *Ans.* 96.

13. Divide \$16$\frac{3}{4}$ by 12 cents; divide \$48 by 6 dimes; divide 18$\frac{3}{4}$ dimes by 15 mills. *Ans.* 139$\frac{7}{12}$; 80; 125.

14. A lady bought a watch for \$55$\frac{1}{2}$, a chain for \$22$\frac{3}{4}$, a key for \$6$\frac{1}{4}$, and sold them all at a gain of \$6$\frac{3}{4}$; what did she receive for them? *Ans.* \$91.25.

15. A teacher bought a book for \$2$\frac{1}{2}$, an inkstand for 62$\frac{1}{2}$ cents, some paper for \$1$\frac{1}{4}$, a map for \$4$\frac{3}{4}$, a globe for \$5$\frac{1}{2}$, and handed the clerk a ten-dollar-bill and a five-dollar bill; how much change should he receive? *Ans.* \$0.37$\frac{1}{2}$.

16. Mr. Benton bought 12 hogsheads of molasses of 63 gallons each, at the rate of 42$\frac{1}{2}$ cents a gallon, and sold it at 50 cents a gallon; what was the gain? *Ans.* \$56.70.

17. A lady bought 3 yards of muslin at 6$\frac{1}{4}$ cents a yard, 7 yards of linen at 87 cents a yard, and handed out a \$10 bill; what was her change? *Ans.* \$3.72.

18. There was sold one day, in New York, Low Grade extra flour, amounting to \$15,000 at \$6.25 per barrel; how many barrels were sold? *Ans.* 2400 barrels.

19. A lady in furnishing her house bought 3 sets of chairs at \$7.25 a set, 2 tables at \$5.25 apiece, 3 rocking-chairs at

\$4.65 apiece, and 45 yards of carpet at \$1.75 a yard; what was the amount of the bill? *Ans.* \$124.95.

20. A dealer bought 8 barrels of turpentine, each containing 31.5 gallons, at \$1.12½ a gallon, and sold it for \$1.37½ a gallon; what did he gain? *Ans.* \$63.

21. A man bought a boat load of coal for \$250, and by retailing it at \$5.75 a ton, he gained \$37.50; how many tons in the load? *Ans* 50 tons.

22. The charge for sending a telegram from New York to Harrisburg is \$.40 for 10 words, and 5 cents for each additional word; what would a dispatch of 24 words cost me? *Ans* \$1.10.

23. Thomas Williams & Co. sold 420 bushels of new ungraded corn at 60 cents per bushel, and received in exchange 120 bushels of oats at 43 cents per bushel, and No. 1 Minnesota wheat, at \$1.67 per bushel; how many bushels did they receive? *Ans.* 120 bushels.

COMMERCIAL TRANSACTIONS.

221. In **Commercial Transactions** there are ordinarily three quantities considered, the *quantity*, the *price*, and the *cost*.

222. The **Quantity** is the amount bought or sold, estimated by the number of times it contains the *unit of measure*.

223. The **Price** is the value of one of the units of measure of any commodity. The *Cost* is the value of the whole quantity.

224. An **Aliquot Part** of a number is the whole or mixed number which will exactly divide that number.

ALIQUOT PARTS OF \$1.

5 cents = \$$\frac{1}{20}$.	16⅔ cents = \$⅙.
6¼ cents = \$$\frac{1}{16}$.	20 cents = \$⅕.
8⅓ cents = \$$\frac{1}{12}$.	25 cents = \$¼.
10 cents = \$$\frac{1}{10}$.	33⅓ cents = \$⅓
12½ cents = \$⅛.	50 cents = \$½.

225. The simple operations of finding price, cost, and quantity have already been sufficiently indicated, and we shall here consider only a few special cases.

CASE I.

226. ***To find the cost of a quantity, the price being an aliquot part of $1.***

1. What cost 48 yards of muslin at $12\frac{1}{2}$ cents a yard?

SOLUTION.—At $1 a yard, the cost would be $48; hence at $12\frac{1}{2}$ cents, which is $\frac{1}{8}$ of $1, the cost will be $\frac{1}{8}$ of $48, or $6. Hence the

OPERATION.

$$8\overline{)48}$$
$$6$$

Rule.—*Take such a fractional part of the given quantity as the price is of* $1.

2. What cost 12 pieces of calico, each containing 32 yards, at $6\frac{1}{4}$ cents a yard? *Ans.* $24.

3. What cost 4 bales of Sea Island cotton, each containing 300 lb., at $16\frac{2}{3}$ cents a pound? *Ans.* $200.

4. Mrs. Wilson bought a bag of Rio coffee, containing 40 lb., at $33\frac{1}{3}$ cents a lb., and 10 lb. of crushed sugar, at $12\frac{1}{2}$ cents a lb.; what was the cost of both? *Ans.* $$14.58\frac{1}{3}$.

5. Bought 20 yards of black cashmere, at $$1.37\frac{1}{2}$ a yard, 20 yards of paper muslin, at 10 cents a yard, and 3 dozen crocheted buttons, at 25 cents a dozen; what was my bill? *Ans.* $30.25.

CASE II.

227. ***To find the cost, the quantity and the price of* 100 *or* 1000 *being given.***

1. What is the cost of 6501 feet of poplar boards, at $17.25 a thousand?

SOLUTION.—If 1000 feet cost $17.25, 1 foot will cost $\frac{1}{1000}$ of $17.25, and 6501 feet will cost 6501 times $\frac{1}{1000}$ of $17.25, which is the same as $\frac{1}{1000}$ of 6501 times $17.25, which by multiplying and cutting off three places in the product, we find is $112.14.

OPERATION.

17.25
6501
112.14225

Ans. $112.14.

Rule.—*Multiply the price by the quantity, and point off in the product two places for price per hundred, or three places for price per thousand.*

NOTE.—The price per hundred or per thousand is expressed thus ℔ C, or ℔ M.

2. What is the value of 12,500 shingles at $6.25 ℔ M?
Ans. $78.12½.

3. What is the cost of 7877 feet of hemlock scantling, at $3.37½ ℔ M? Ans. $26.58.

4. What is the cost of 17,225 bricks, at $7.75 ℔ M, and 1560 lb. of sheet lead, at $10.12½ ℔ C? Ans. $291.44.

5. A grocer sold one day 9760 pounds of wheat flour, at $4.25 ℔ C; what was the amount of the sale?
Ans. $414.80.

6. Messrs. Wood & Co. burn in their store in a year 62,560 cubic feet of gas, at $4 50 ℔ M; what is their gas bill for a year? Ans. $281.52.

7. If a compositor is paid 45 cents per thousand ems, what will he receive for setting up a book of 560 pages of 1120 ems each? Ans. $282.24.

8. Mr. Barr receives for making cigars $7.50 ℔ M; what will be his profit on 43,780 cigars, if he pays $5.75 for the amount of tobacco required to make a thousand?
Ans. $76.61½.

CASE III.

228. ***To find the price of* 100 *or* 1000, *the quantity and cost being given.***

1. If I sell 4256 feet of boards for $62.776, what is the price ℔ M?

SOLUTION.—If 4256 feet cost $62.776, 1 foot will cost $\frac{1}{4256}$ of $62.776, and 1000 feet will cost 1000 times $\frac{1}{4256}$ of $62.776, which, dividing and removing the decimal point three places to the right, we find is $14¾. Hence the

OPERATION.
$62.776 ÷ 4256 = $14¾.

Rule.—*Divide the cost by the quantity, and remove the decimal point in the quotient two places to the right for price per hundred, or three places for price per thousand.*

2. I paid $7.59⅜ for 675 pickets for the fence around my garden; what was the price ℔ C? Ans. $1.12½.

3. Mr. Shirk sold 10,540 cedar rails for $553.35; what was the price ℔ C? Ans. $5¼.

4. If the cost of stereotyping a book of 320 pages of 850 ems each is $340, what is the cost per 1000 ems?

Ans. $1.25.

5. I retail envelopes at 12 cents a pack, gaining 3 cents on each pack of 24; what did they cost me ℔ M?

Ans. $3.75.

CASE IV.

229. ***To find the cost, the quantity and the price of a ton of 2000 pounds being given.***

1. At $75.45 a ton, what will be the cost of 7247 lb. of railroad iron?

SOLUTION.—Dividing $75.45, the price of a ton, by 2, we have $37.72½, the price of 1000 lb.; and proceeding as in Case II., we have $273.39 as the price of 7247 lb.

OPERATION.

2)75.45
37.72½
7247
273.39307½

Rule.—*Multiply half the price of a ton by the quantity, and remove the decimal point three places to the left.*

NOTE.—To find the price of a ton, cost and quantity being given, *divide the cost by the quantity, multiply the quotient by* 2, *and remove the decimal point three places to the right.*

2. A farmer sold 7375 pounds of hay at $13½ a ton; what did he receive for it? *Ans.* $49.78.

3. A porcelain manufacturer bought 5620 pounds of clay at $17.25 a ton; what did it cost him? *Ans.* $48.47.

4. A peddler sold to a paper-mill 1728 pounds of rags at $48 per ton; what did he receive? *Ans.* $41.47.

5. If 7240 pounds of bone dust cost $93.215, what is the price per ton? *Ans.* $25.75.

6. A coal dealer received $20.88 for 5760 pounds of Lehigh red ash coal; what was the price per ton? *Ans.* $7.25.

7. Shipped from Pittsburgh, on the Panhandle R. R., 34,761 pounds of pig iron @ $3.75 per ton, and 97,486 pounds of steel rails at $2.50 per ton; what was the charge?

Ans. $187.03.

8. Messrs. Wilson & Co. bought 36,850 pounds of lime at $3.60 a ton, and sold it at 22½ cents per hundred; what was the profit on the transaction? *Ans.* $16.58.

BILLS AND ACCOUNTS.

230. A **Bill** is a written statement of goods sold, services rendered, etc., giving the place, date, names of the parties, and the price, quantity, and cost of each item.

231. An **Invoice** is a full statement, sent to a purchaser or agent, at the time the goods are forwarded, containing the marks, contents, and prices of each package, the charges paid, and the mode of forwarding. The terms *Invoice* and *Bill* are often used interchangeably.

232. The **Footing** of a bill is the amount of its items. To *extend* an item is to write its cost in the proper column. A bill is *receipted* when the person to whom it is due, or his agent, writes at the bottom of the bill "Received Payment," and signs his name.

233. An **Account Current** is a written statement of the business transactions between two parties for a given time. The party who owes is the *Debtor;* the party owed is the *Creditor*.

234. To **Balance an Account** we find the difference of the footings of the two sides and add it to the smaller side, so that the two amounts are equal.

NOTES.—1. The abbreviation *Dr.* signifies *debit* or *debtor;* *Cr.*, *credit*, or *creditor;* @ signifies *at*, and denotes the price of the *unit of measure;* ℅ stands for *account;* ¢ for *cents;* No. or # for *number;* pcs. for *pieces*, and 62^2, 18^3, 6^1, signify respectively $62\frac{1}{2}$, $18\frac{3}{4}$, $6\frac{1}{4}$; 3 pcs. sheeting 42, 42^1, 45^3 signifies that the pieces contain respectively 42, $42\frac{1}{4}$ and $45\frac{3}{4}$ yards.

2. *Accounts current* are frequently made out every month, and are then called *monthly statements*, and generally contain only the amounts bought at each date, bills of the several items having been furnished at the time of buying.

3. Deductions are often made in bills, sometimes from the retail price of the items, and sometimes from the amount of the bill. Deductions from the retail price are generally made to customers buying in considerable quantities, and deductions from the amount are made for cash payments or payments within a short specified time, the prices in such bills being mostly wholesale. The symbol %, meaning *hundredth*, is frequently used; thus, less 6% means 6 *hundredths* deducted.

Rule I.—*In making out a* BILL, *extend the several items and take their sum.*

Rule II.—*In an* ACCOUNT, *find the difference between the debit and credit amounts.*

235. Required the footings and balances of the following bills and accounts:

BALTIMORE, *July* 15, 1876.

JOHN B. WYLIE & CO.,

Bought of MACNEIL, WALCOTT & CO.

24 Prs. Men's French Calf Boots,	@	$10		
65 " Women's Enameled "		1.87½		
75 " " Fox'd Gaiters,		1.50		
88 " Boys' Kip Boots,		2.50		
110 " " Brogans,		1.25		
90 " Misses' Rubber Shoes,		.75		
60 " Women's Arctics,		2.25		
	Less ⅛			
			827	50

Rec'd Payment,
MACNEIL, WALCOTT & CO.

WILMINGTON, DEL., *April* 27, 1876.

MR. JAMES LANE,

To A. B. HAYES, DR.

For Setting Boiler and Excavating,		100	00
" 13½ Days Work of Bricklayer,	@$4		
" 5½ " " " Laborer,	1.12½		
" 1 Load Sand,		2	50
" 15 Bushels Lime,	.38		
" 500 Dark Stretchers,	@$14 ℔ M		
" 500 Salmon Bricks,	" 8 " "		
		179	39

PROVIDENCE, R. I., *June* 1, 1876.

MR. JAMES HEDGE,

Bought of GLADDING & BRO.,

1876							
Apr.	3	1 Webster Unabridged,	3/10	12			
	7	4 Leisure Day Rhymes,	¼	2	00		
May	2	3 Eight Cousins,	¼	1	50		
	6	2 Ticknor's Life and Letters,	¼	3	50		
	"	4 Mysterious Island,	⅛	3	00		
	29	3 Farm Legends,	3/10	2	50		
	"	2 Three Brides,	⅛	1	75		
						38	61

Rec'd payment,
JOHN B. TIBBETS,
for GLADDING & BRO.

PHILADELPHIA, *Oct.* 1, 1876.

MRS. GEORGE GORDON,

Bought of WALKER & BROUGHER.

Terms: 30 days.

2	Pr.	W. G. Ewers and Basins	1.40				
2		" Cov'd Soaps	50				
2		" Brush Trays	50				
½	Doz.	" Bowls	1.75				
1	"	" Mugs		1	50		
2	"	Stem Wines	1.00				
2	"	Glass Peppers	1.25				
2	"	Individual Salts	50				
3	"	Oval Glass Dishes $\frac{\text{1-3 in}}{100}$ $\frac{\text{1-6 in}}{150}$ $\frac{\text{1-7 in}}{200}$					
		Crate and Porterage		1	25	18	43

NOTE.— $\frac{\text{1-3 in}}{100}$ $\frac{\text{1-6 in}}{150}$ $\frac{\text{1-7 in}}{200}$ means that there are 1 dozen 3 inch dishes @ $1, 1 dozen 6 inch @ $1.50, and 1 dozen 7 inch @ $2.

STATEMENT OF ACCOUNT.

NEW YORK, *April* 7, 1876.

MR. HENRY CADY,

In % *with* BADEAU & LOCKWOOD,

1875		DR.					
Oct.	9	To 15 Drums of Figs,	2 25				
"	30	" 9 Boxes Oranges,	4 75				
Dec.	3	" 12 " Lemons,	5 25				
1876							
Jan.	29	" 4 Barrels Dried Apples,	6 75				
Feb.	"	" 10 " " Peaches,	5 25				
Mar.	18	" 30 lb. Leghorn Citron,	25				
		CR.					
1876							
Jan.	11	By 2 bbl. Superfine State Flour,	6 00				
"	"	" 200 lb. Buckwheat "	.03¼				
Feb.	1	" 1 Firkin Butter, 56 lb.,	28				
"	29	" 100 lb. Corn Meal, ℔ C.,		1	35		
Mar.	1	" Cash,		108	00		
		Balance due Badeau & Lockwood,				82	47

CINCINNATI, O., *Sept. 25th*, 1874.

MR. JOHN WALLER,

Bought of TIERNEY & GARDEN.

Terms: 30 days.

½	Dozen Gents Blk. Fur Hats,	16	50		
½	" " " " "	19	50		
1	" " " " "			18	00
½	" Boys " Sax. "	9	00		
½	" " " " "	12	00		
½	" Children's Fancy "	6	00		
½	" Ladies Trimmed "	15	00		
½	" " Untrimmed "	7	50	—	—
	Less 6%,			—	—
				57	11

Received payment,

TIERNEY & GARDEN,

Per J. W. M.

PITTSBURGH, *Nov.* 1, 1876.

Messrs. WILEY & MORTON,

Bought of J. G. HERR & CO.

2	Doz. Rubber Long Combs,	4	50		
¼	Gro. Horn " $\frac{3.50}{\text{※}8}$, $\frac{4.50}{10}$, $\frac{6.00}{12}$				
$\frac{10}{12}$	Doz. Tooth Brushes,	1	75		
$\frac{7}{12}$	" Jet Sets,	7	25		
1	Gro. Rubber Coat and Vest Buttons ea. 1.00 & 75,			1	75
½	Gt. " Black Suspender "	1	62		
40	Yd. Jaconet,		20		
14	" Irish Linen,		67		
3	Pcs. Silk Velvet Ribbon $\frac{65}{1\frac{3}{4}}$, $\frac{72^{2}}{2\frac{1}{2}}$, $\frac{1.35}{4}$, $\frac{2.00}{6}$,			—	—
	Less $\frac{1}{10}$,			—	—
				47	07

Rec'd payment,

J. G. HERR & Co.

NOTE.—This bill shows another method of abbreviating entries of similar items. ¼ Gro. Horn Combs $\frac{3.50}{\text{※}8}$, $\frac{4.50}{10}$, $\frac{6.00}{12}$, means that there are ¼ gross No. 8 @ $3.50, ¼ gross No. 10 @ $4.50, and ¼ gross No. 12 @ $6. In the last item $1\frac{3}{4}$, $2\frac{1}{2}$, etc., are the width of the ribbon, the upper numbers being the price as before.

J. T. WAY & Co

In % *with* WOOD, MARSH & HEYWARD,

DR. CR

1875					1875				
Sept.	25	To 1 pc. Blue Anchor Cloth 38^1 yd. @ \$5.62^2,			Oct.	2	By 1 pc. Irish Linen 17^1 yd. @57^2¢,		
"	30	" 2 pcs. Black Cloth 42 yd. @\$7.65,			"	20	" 3 pcs. Plaid Jaconets 60 yd. @37^2¢		
Oct.	8	" 1 pc. Doeskin Cassimere 24 yd. @ \$1.87^2,			Nov.	30	" 1 pc. Swiss Mull 20 yd. @32^2¢,		
Nov.	9	" 2 pcs. Black Rib'd Cassimere 63 yd. @\$1.63,			Jan.	2	" Note for Balance @30 da.,	739	32
Dec.	14	" 2 pcs. Fisher's Tweed 64^1 yd. @75¢,							
"	30	" 2 pcs. Brown Corduroy 41^1 yd. @ \$1.12^2,							

WRITTEN EXERCISES.

236. Let pupils make out bills, accounts, and invoices in proper form, from the following statements:

1. The Lafayette Restaurant ordered Oct. 10, 1876, from Thos. Bradley, the following: 67 lb. Ribs of Beef @ 20¢; 2 Kidneys, @ 15¢; 3 Lambs, @ \$5; 3 Calves' Heads, @ 75¢; 8 Sweetbreads, @ 30¢; 2 Calves' Livers, @ 70¢; 37 lb. Veal, @ 17¢; make out the bill. *Ans.* \$41.04.

2. Eugene G. Blackford delivered the following, June 12, 1876, at the Continental Hotel: 15 Salmon, @ 30¢; 37 Lobsters, @ 10¢; 5 Fresh Mackerel, @ 12¢; 7 Spanish Mackerel, @ 20¢; 2 Shad, @ 40¢; 12 Soles @ 10¢; 9 Seabass, @ 10¢; 15 Sheepheads, @ 12¢; 3 doz. Soft Crabs, @ 50 ¢; 1½ doz. Frogs, \$1.75; extend the items and find the footing of the bill. *Ans.* \$19.03.

3. James Thomson bought of William Wilson, Pittsburgh, Jan. 20, 1873, 1 lb. Citric Acid, \$1.55; 5 lb. Chloride Lime. @ 8¢; jar 20¢; 10 lb. Epsom Salts, @ 5¢; 3 lb. Gum Shellac, @ 60¢; 1 doz. Wilson's Cod Liver Oil, @ \$7; ½ doz. Mishler's Herb Bitters, @ \$8; 2 lb. Carbolic Acid Crystals, @ \$1.75; what was the amount of the bill? *Ans.* \$18.95.

4. Tyndale & Co., Philadelphia, sold to Mrs. John Smith, Dec. 31, 1874, the following articles: 1 Soup Tureen, @ \$3.50; 2 Sauce Tureens, @ \$1.25; 1 doz. Tulip Goblets, @ \$1.40; 1 doz. Individual Salts, @ 50¢; 2 Glass Pitchers, @ 62½¢; 3 Oval Glass Dishes, @ 50¢; 1 Glass Nappy, @ 75¢; 4 Cov'd Dishes, @ \$1.25; 3 doz. Stoneware Plates,—1 doz. 6 in., @ \$1.25, 1 doz. 7 in., @ \$1.40, 1 doz. 8 in., @ \$1.60; required the bill, receipted. *Ans.* \$20.65.

5. Mrs. Amelia Watson, Newark, N. J., presented the following bill to James Haven, March 1, 1875: Board for 4 weeks, @ \$9; fuel and light 4 weeks, @ \$1.50; washing 5½ doz. @ \$1. Mr. Haven presented the following bill to Mrs. Watson at the same date: February 5th, 15 lb. Tea, @ 75¢; 10 lb. Coffee, @ 35¢; February 9th, 25 lb. Granulated Sugar, @ 12½¢; 5 lb. Brown Sugar, @ 10¢; 1 barrel No. 1 Mackerel, \$25; February 26th, 5 lb. Butter, @ 50¢, and 1½ dozen Eggs, @ 42¢. Make out both bills, receipting the smaller and crediting the amount upon the other. *Ans.* Bal. \$1—.

6. August 25, 1875, Franklin S. Fuller, of Memphis, Tenn., purchased of Kuhn & Furst, Philadelphia, 25 boxes Raisins, @ $\frac{10}{\$3.25}$, $\frac{15}{\$2.55}$; 350 lb. Currants, @ $\frac{100}{7\frac{1}{2}¢}$, $\frac{250}{7¢}$; 150 lb. Dates, @ 5¼¢; 100 lb. Turkish Prunes, @ 8½¢; 125 lb. French Prunes, @ $\frac{50}{15¢}$, $\frac{75}{12\frac{1}{2}¢}$; 12 bunches of Baracoa Bananas @ \$1.75; and gave a due bill for amount. Make out bill and receipt it. *Ans.* \$150.

7. Mr. Thomas Walker, of Aiken, S. C., bought of Hess, Rogers and Chambers, Philadelphia, the following articles, Oct. 14, 1876: 3 doz. Ladies' Berlin Gloves, @ \$2.25; 1 doz. Ladies' White Silk do., @ \$5.50; 1 doz. do. Berlin ½ Gauntlets, @ \$3.75; 1 doz. Colored Buck do., @ \$15; 1 doz. Ladies' Black Jouvin Kid Gloves, @ \$16; ½ doz. White do., @ \$15; 2 doz. Gents' Buck Driving Gauntlets, @ \$16.50; 2 doz. Gents' White Kid Gloves, @ \$11: 1 doz. Child's White Cotton Hose, $\frac{80}{3}$, $\frac{90}{3'}$, $\frac{95}{4}$, $\frac{100}{4'}$, $\frac{110}{5}$; make out bill for amount, deducting 20%. *Ans.* \$91.40.

SECTION VI.

DENOMINATE NUMBERS.

237. A **Denominate Number** is a concrete number in which the unit is a *measure;* as 3 *feet,* 4 *pounds,* etc.

238. A **Measure** is a unit by which quantity of *magnitude* or *continuous* quantity is estimated numerically; as, a *yard,* a *pound,* etc.

239. A **Compound Number** is a number which expresses several different units of the same kind of quantity; as, 4 yd. 2 ft. 11 in.

240. The **Terms** of a compound number are the *numbers of its different units.* Thus the *terms* in the example given are 4 yd., 2 ft., and 11 in.

241. **Similar Compound Numbers** are compound numbers which express the same kind of quantity.

242. **Denominate Numbers** may be embraced under eight distinct classes, as follows:

1. Value.
2. Weight.
3. Length.
4. Surface.
5. Volume.
6. Capacity.
7. Time.
8. Angles.

NOTE.—Concrete numbers are of two classes: 1st, those in which the unit is *natural*; 2d, those in which it is *artificial.* Natural units are such as exist in nature, and artificial units are those which are agreed upon to measure quantity of magnitude. The latter are called *denominate numbers.*

MEASURES OF VALUE.

243. The **Value** of anything is its worth, or that property which makes it useful or estimable.

244. **Money** is the measure of the value of things. It is of two kinds, coin and paper money.

245. **Coin,** or **Specie,** is metal prepared and authorized by government to be used as money.

246. Paper Money consists of printed promises to pay the bearer a certain amount, duly authorized to be used as money.

247. Currency is whatever circulates as money. It is of two kinds, *specie currency* and *paper currency*.

248. Legal Tender is a term applied to money which is required by law to be accepted in payment of debts.

249. An **Alloy** is a baser metal compounded with either gold or silver for the purpose of rendering it harder and more durable. In coinage, the alloy is considered as having no value.

UNITED STATES MONEY.

250. United States, or **Federal Money,** is the legal currency of the United States.

TABLE.

10 mills (m)	. .	= 1 cent	.	ct.
10 cents	. .	= 1 dime	.	d.
10 dimes	. .	= 1 dollar	.	$.
10 dollars	. .	= 1 eagle	.	E.

I. NAME.—United States money is so called because it is the money of the United States. It is called *Federal Money* because it was the money of the Federal Union. It was adopted by Act of Congress, Aug. 8, 1786.

II. TERMS.—The term *dollar* is supposed to be from *Dale* or *Daleburg*, a town where it was first coined; *dime* is from the French *disme*, meaning *a tenth; cent* is from the Latin *centum, a hundred; mill* is from the Latin *mille, a thousand; eagle* is from the name of the national bird. The cent was proposed by Robert Morris, and named by Thomas Jefferson.

III. UNIT.—The *unit* is the *gold dollar.* The currency is founded upon the decimal system, dimes, cents, and mills being written as decimals. This gives great simplicity to the operations.

IV. COINS.—The coins are of *gold, silver, nickel,* and *bronze.* The *gold coins* are the *double eagle, eagle, half-eagle, quarter-eagle, three dollars,* and *one dollar.* The *silver coins* are the *trade dollar, half-dollar, quarter-dollar, twenty-cent piece,* and *dime.* The *nickel coins* are the *three-cent* and *five-cent pieces.* The *bronze coin* is the *cent.* The silver half-dime and three-cent piece, the bronze two-cent piece, the nickel cent, and the old copper cent and half-cent, although still seen in circulation, are no longer coined. The mill has never been a coin; it is merely a convenient name for the tenth part of a cent.

V. COMPOSITION.—The gold and silver coins consist of 9 parts of pure metal and 1 part alloy. The alloy of the silver coin consists of pure copper; the alloy of the gold coin consists of silver and copper, the silver not to exceed $\frac{1}{10}$ of the alloy. The nickel coins contain $\frac{1}{4}$ nickel and $\frac{3}{4}$ copper. The bronze coins consist of 95 parts copper, and 5 parts tin and zinc.

VI. Gold coins are a legal tender for any amount; silver coins, of the present coinage, for any amount not exceeding $5 in any one payment; bronze and nickel coins for any amount not exceeding 25 cents in any one payment.

MENTAL EXERCISES.

1. How many cents in $\$\frac{1}{2}$? $\$\frac{1}{10}$? $\$\frac{3}{4}$? $\$\frac{3}{5}$? $\$\frac{2}{5}$? $\$\frac{3}{8}$? $\$\frac{5}{8}$?

2. What part of a dollar is 10 cts.? $12\frac{1}{2}$ cts.? 20 cts.? 25 cts.? $16\frac{2}{3}$ cts.? $33\frac{1}{3}$ cts.? $37\frac{1}{2}$ cts.? 50 cts.? $62\frac{1}{2}$ cts.? 75 cts.? $83\frac{1}{3}$ cts.?

3. What part of 5 eagles is 15 dimes? what part of 12 cents is $\frac{4}{5}$ of a dime?

4. How many eagles in 60 dollars? in 400 dimes? in 8500 cents? in 25,000 mills?

ENGLISH, OR STERLING MONEY.

251. English, or Sterling Money, is the legal currency of England.

TABLE.

4 farthings (far. or qr.)	=	1 penny	d.
12 pence	=	1 shilling	s.
20 shillings	=	1 pound or sovereign,	£
21 shillings	=	1 guinea	G.

I. NAME.—The term *Sterling* is supposed to be derived from *Easterling*, the name given to early German traders, who came from the east to England. Their money was called Easterling Money, which was contracted into *Sterling Money*.

II. TERMS.—The term *farthing* is a modification of "four things," the old English penny being marked with a cross so deeply impressed that it could be broken into two or four pieces, called respectively *half-penny* and *four things*. The *Pound*, as a measure of value, was derived from the pound as a measure of weight, 240 pence formerly weighing a pound. The *guinea* is so called because the gold of which it was first made came from Guinea, in Africa.

III. SYMBOLS.—The symbols £., s., d., qr., are the initials of the Latin words *libra*, *solidus*, *denarius*, and *quadrans;* signifying respectively, pound, shilling, penny, and quarter.

IV. UNIT.—The *unit* is the *pound*, represented by the sovereign and £1 bank note. Its value by late act of Congress is fixed at $4.8665.

V. COINS.—The coins are of three classes: *gold*, *silver*, and *copper*. The *gold coins* are the *sovereign* (=£1), and *half sovereign* (=10 s.), *guinea* (=21 s.) and *half guinea* (=10 s. 6 d.). The *silver coins* are the *crown* (=5 s.), the *half crown* (=2 s. 6 d.), the *florin* (=2 s.), the *shilling*, and the *six-penny*, *four-penny*, and *three-penny* pieces. The *copper coins* are the *penny*, *half-penny*, and *farthing*.

VI. COMPOSITION.—The standard for gold coins is 22 carats fine, that is, 11 parts pure gold, and 1 part alloy. The standard for silver is 37 parts pure silver and 3 parts alloy, hence the silver coins are $\frac{37}{40}$ pure, and $\frac{3}{40}$ copper. Pence and half-pence are made of pure copper.

MENTAL EXERCISES.

How many

1. Farthings in 6 pence?
2. Pence in 8 shillings?
3. Shillings in 108 pence?
4. Shillings in 25 pounds?
5. Pounds in 480 shillings?
6. Pence in 124 guineas?
7. What part of 2 pence is 6 farthings? What part of 3 shillings is 5 pence?
8. What part of 16 pence is $\frac{2}{3}$ of a shilling? What part of a guinea is $\frac{3}{4}$ of a pound?

WRITTEN EXERCISES.

1. How many pence in £16 11 s. 10 d.?

SOLUTION.—In one pound there are 20 shillings, and in £16 there are 16 times 20 shillings, which, increased by 11 s., are 331 shillings. In one shilling there are 12 pence, and in 331 shillings there are 331 times 12 pence, which, increased by 10 pence, equals 3982 pence.

OPERATION.

```
 £    s.   d.
16   11   10
20
---
331
 12
---
3982d. Ans.
```

2. How many pounds, shillings, and pence in 4367 pence?

SOLUTION.—There are 12 pence in one shilling, hence in 4367 pence there are as many shillings as 12 is contained times in 4367, which are 363 shillings and 11 d. remaining. There are 20 shillings in one pound, hence in 363 shillings there are as many pounds as 20 is contained times in 363, which are £18 and 3 shillings remaining. Hence in 4367 d. there are £18 3 s. 11 d.

OPERATION.

```
      d.
12)4367
2|0)36|3 - 11 d.
   £18  - 3 s.
Ans. £18 3 s. 11 d.
```

3. How many farthings in 9 s. 8 d. 3 far.? *Ans.* 467 far.
4. Farthings in £19 17 s. 11 d. 2 far.? *Ans.* 19102 far.
5. Shillings, pence, and farthings, in 7859 far.? *Ans.* 163 s. 8 d. 3 far.
6. Pounds, shillings, etc., in 58763 far.? *Ans.* £61 4 s. 2 d. 3 far.

CANADA MONEY.

252. The **Currency of Canada** is nominally the same as that of the United States, the table and denominations being the same.

253. The decimal currency was adopted in 1858, the Act taking effect in 1859, previous to which their currency was the same as the English.

I. COINS.—The coins consist of silver and copper. The *silver coins* are the *50-cent piece*, the *25-cent piece*, the *shilling* or *20-cent piece*, the *dime*, the *half-dime*. The *copper coin* is the *cent*.

IV. VALUE.—The coins are nominally equal to the corresponding coins of United States money, but the intrinsic value is a little less. The *eagle* of the United States is the legal tender for sums of $10 and upwards.

III. COMPOSITION.—The silver coins consist of 925 parts silver and 75 parts copper; or 37 parts silver to 3 parts copper, the same as the English silver coins.

FRENCH MONEY.

254. French Money is the legal currency of France. The *unit* is the *franc*, whose value is 19.3 cents.

255. The **Franc** is divided into tenths and hundredths, called respectively *decimes* and *centimes*. The *decime*, like our dime, is not used in business calculations, but is expressed by *centimes;* thus, instead of 5 decimes we say 50 centimes.

GERMAN MONEY.

256. The **German Empire** has adopted a new and uniform system of coinage.

257. The **Unit** is the *mark* (*Reichsmark*) worth 23.85 cents, and this is divided into 100 *pfennige*.

WRITTEN EXERCISES.

1. How many dollars in 25.50 francs? *Ans.* $4.9215.
2. How many francs in $256? *Ans.* 1326.42+fr.
3. How many dollars in 753 marks 15 pfennige? *Ans.* $179.626+.
4. How many marks in $456.25? *Ans.* 1913—marks.
5. Bought an opera-glass for 20 francs; what did it cost in United States money? *Ans.* $3.86.

MEASURES OF WEIGHT.

258. Weight is the measure of the force by which bodies are naturally drawn towards the earth.

259. There are three kinds of weight in common use. *Troy Weight*, *Apothecaries' Weight*, and *Avoirdupois Weight*.

TROY WEIGHT.

260. Troy Weight is used in weighing gold, silver, jewels, liquors in philosophical experiments, etc.

TABLE.

24 grains (gr.)	= 1 pennyweight, . .	pwt.
20 pennyweights	= 1 ounce, ...	oz.
12 ounces	= 1 pound, ...	lb.

I. Name.—The term *Troy* is said to be derived from *Troyes*, the name of a town in France, where this weight was first used in Europe. It was brought from Cairo in Egypt during the crusades of the 12th century.

II. Terms.—The term *pound* is from the Latin *pendo*, to *bend* or *weigh*. The term *ounce* is from the Latin *uncia*, a *twelfth part*, the ounce being one-twelfth part of a pound. The *pennyweight* was the weight of the old English silver penny. The term *grain* originated in the custom of using a number of grains of wheat for the weight of a penny. These grains were taken from the middle of the ear, and well dried, thirty-two at first, and afterwards twenty-four, being used to make a pennyweight.

III. Symbols.—The symbol *oz.* is thought to be from the Spanish word *onza*, signifying *ounce*; though Webster derives it from the use of the termination ʒ to express abbreviations, which was afterwards changed to *z*, *lb.* is from *libra*, the Latin for pound. *Pwt.* is a combination of *p.* for *penny*, and *wt.* for *weight*.

IV. Unit.—The *standard unit* of weight is the *Troy pound*. It is equal the weight of 22.794377 cubic inches of distilled water, at the temperature of 39.83° Fahrenheit, barometer at 30 inches.

MENTAL EXERCISES.

How many

1. Grains in 6 pwt.? 5 pwt.?
2. Pwt. in 72 gr.? in 144 gr.?
3. Pwt. in 7 oz.? 9 oz.?
4. Ounces in 100 pwt.? 60 pwt?
5. Ounces in 9 lb.? 11 lb.?
6. Pounds in 120 oz.? 144 oz.?
7. What part of a pound Troy are 4 oz.? 6 oz.? 8 oz.?
8. How many parts pure gold in a ring 18 carats fine?

WRITTEN EXERCISES.

How many

1. Grains in 7 pwt. 15 gr.? *Ans.* 183 gr.
2. Grains in 5 oz. 14 pwt. 13 gr.? *Ans.* 2749 gr.
3. Lb., oz., and pwt. in 758 pwt.? *Ans.* 3 lb. 1 oz. 18 pwt.
4. Oz., pwt., and gr. in 6275 gr.? *Ans.* 13 oz. 1 pwt. 11 gr.
5. Grains in 19 lb. 11 oz. 17 pwt. 19 gr.? *Ans.* 115147 gr.
6. Pounds, oz., etc., in 18759 gr.?
Ans. 3 lb. 3 oz. 1 pwt. 15 gr.

APOTHECARIES' WEIGHT.

261. Apothecaries' Weight is used in prescribing and mixing dry medicines. Medicines are bought and sold by Avoirdupois Weight.

TABLE.

20 grains (gr.)	. . . . =	1 scruple, ...	℈.
3 scruples	. . . . . =	1 dram, ...	ʒ.
8 drams	. . . . . =	1 ounce, ...	℥.
12 ounces	. . . . . =	1 pound, ...	℔.

I. Name.—The name arises from the weight being used by *apothecaries.*

II. Terms.—The term *scruple* is from the Latin *scrupulus, a little stone.* The term *dram* is from the Greek *drachma, a piece of money.*

III. Symbols.—The symbols have been supposed to be modifications of the figure 3, suggested by there being 3 scruples in a dram. Another supposition is that they are from inscriptions upon the ancient monuments of Egypt.

IV. Unit.—The *Unit* is the *pound,* and is identical with the Troy pound, as are also the ounce and grain, the ounce being differently divided.

V. Physicians use the Roman notation in writing prescriptions, using the small letters, preceded by the symbols, and writing *j* for *i* when it terminates a number. Thus, 12 gr. is written gr.xij.; 2 scruples, ℈ij. ℞ is an abbreviation for *recipe, take;* ā or āā (from the Greek ἀνά) means, *of each; ss.* for *semis* or *half,* as ℈ivss. means 4½ scruples; *P.* for *particula,* or *little part; P. aeq.* for *equal parts; q. p., quantum placet, as much as you please.*

MENTAL EXERCISES.

How many

1. Grains in 7℈? 11℈?
2. Scruples in 9ʒ? 16ʒ?
3. Drams in 24℈? 96℈?
4. Drams in 5℥? 7℥?
5. Ounces in 88ʒ? 96ʒ?
6. Pounds in 108℥? 168℥?
7. How many doses of 8 grains each, can be made from an ounce of quinine?

WRITTEN EXERCISES.

How many

1. Drams in 7℔. 5℥? *Ans.* 712ʒ.
2. Pounds and ounces in 239℥? *Ans.* 19℔. 11℥.
3. Scruples in 19℔. 8℥ 5ʒ 2℈? *Ans.* 5681℈.
4. Grains in 47℔. 10℥ 7ʒ 2℈ 19gr.? *Ans.* 275999gr.
5. Pounds, etc., in 92375gr.? *Ans.* 16℔. 3ʒ. 1℈ 15gr.
6. How many pills of 4gr. each, can be made from 3ʒ 2℈ of calomel? *Ans.* 55.

AVOIRDUPOIS WEIGHT.

262. Avoirdupois Weight is used for weighing everything except jewels, precious metals, etc.

TABLE.

16 ounces . .	= 1 pound, . .	lb.
100 pounds . .	= 1 hundredweight,	cwt.
20 hundredweight .	= 1 ton, . .	T.

I. NAME.—The term *Avoirdupois* is probably from the French *avoir du poids, to have weight.*

II. TERMS.—The term *ton* is from the Saxon *tunne*, a *cask*. The origin of the other terms has already been given. The symbol *cwt.* is from *centum, hundred*, and *weight*. The term *dram* has been used for $\frac{1}{16}$ of an ounce, but is obsolete, fractions of an ounce being used in business transactions.

III. UNIT.—The *unit* is the *pound*. It is derived from the Troy pound, and contains 7000 grains Troy. It is equal to the weight of 27.7015 cubic inches of water at 39.83° Fah., the barometer being at 30 inches.

IV. In Great Britain 28 lb. equal 1 qr., 112 lb. equal 1 cwt., and 2240 lb. equal 1 ton. These are called the *long hundred* and *long ton;* they were formerly used in this country, but are now used only at the custom-houses in invoices of English goods, in the wholesale iron and plate trade, and in wholesaling and freighting coal from the coal mines of Pennsylvania.

V. The following denominations are frequently used:

25 lb. of powder	make 1 barrel.	100 lb. of raisins	make 1 cask.
56 " butter	" 1 firkin.	196 " flour	" 1 barrel.
84 " "	" 1 tub.	200 " pork, beef, or fish	1 barrel.
100 " grain or flour	" 1 cental.	240 " lime	" 1 cask.
100 " dry fish	" 1 quintal.	280 " salt at N.Y.S.works	1 barrel.
100 " nails	" 1 keg.	600 " rice	1 barrel.

MENTAL EXERCISES.

How many

1. Ounces in 8 lb.? 12 lb.?
2. Pounds in 112 oz.? 9 cwt.?
3. Cwt. in 4800 lb.? 1000 T.?
4. Tons in 2200 cwt.? 2500 cwt.?
5. How many cwt. in $\frac{1}{2}$ of a ton? in $\frac{1}{4}$ of a ton?
6. How many pounds in 50 cwt.? in 75 cwt.? in 80 cwt.?

WRITTEN EXERCISES.

How many

1. Pounds in 5 T. 16 cwt. 75 lb.? *Ans.* 11675 lb.
2. Tons, etc., in 7897 pounds? *Ans.* 3 T. 18 cwt. 97 lb.
3. Ounces in 29 cwt. 73 lb. 15 oz.? *Ans.* 47583 oz.
4. Ounces in 6 T. 83 lb. 13 oz.? *Ans.* 193341 oz.
5. Tons in 117309 ounces? *Ans.* 3 T. 13 cwt. 31 lb. 13 oz.
6. Which is heavier, and how much, a pound of lead or a pound of gold?

7. Which is heavier, and how much, an ounce of feathers or an ounce of silver?

MEASURES OF LENGTH.

263. Measures of Length are used in measuring length, breadth, height, distance, etc.

264. A Line is that which has length without breadth or thickness. It is estimated by ascertaining how many times it contains some definite length, regarded as a unit of measure.

265. An **Angle** is the opening between two lines which diverge from a common point. Thus ACD and DCB are angles.

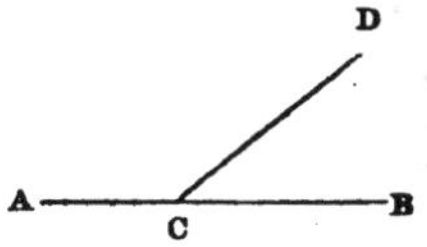

266. The **Vertex** of an angle is the point from which the two lines diverge; thus, C is the vertex of the angle BCD.

267. A **Right Angle** is formed by one line perpendicular to another; as ABC or CBD. One line is *perpendicular* to another when it makes the two adjacent angles equal.

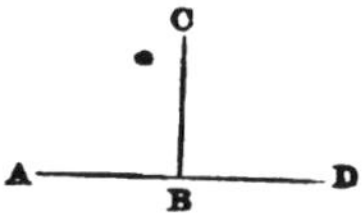

LONG MEASURE.

268. Long Measure is used for the general purposes of measuring length and distances.

TABLE.

12 inches (in.) . . .	=1 foot, ...	ft.
3 feet	=1 yard, ...	yd.
$5\frac{1}{2}$ yards, or $16\frac{1}{2}$ feet .	=1 rod, ...	rd.
320 rods	=1 statute mile, .	mi.
3 miles	=1 league, . . .	lea.
69.16 miles ($69\frac{1}{6}$ nearly)	=1 degree of latitude,	deg. or °,
	or of longitude at the equator.	

1 TERMS.—The term *inch* is from *uncia*, a *twelfth*; *foot* is from the human foot; *yard* was a *rod* or *shoot*; *rod* is from a *measuring stick* or *rod*; *furlong*, which has now become obsolete, is from *fur*, *furrow*, and

lang, long, the *length of a furrow; mile* is from *mille passuum*, 1000 paces; *span* is the space measured from the end of the thumb to the end of the little finger extended; *cubit*, from the elbow to the end of the middle finger; *fathom*, the length of the two arms extended. The ancient yard of England is said to have been determined by the length of the arm of King Henry I.

II. UNIT.—The *standard unit* of length is the *yard*, from which all other measures of length, and also those of capacity, weight, etc., are derived. It is identical with the Imperial yard of Great Britain, which, under William IV., was declared to be fixed by dividing a pendulum, which vibrates seconds in a vacuum, at the level of the sea, at 62° Fah., in the latitude of London, into 391393 equal parts, and taking 360000 of these parts for the yard. Subsequent scientific experiments have proved that such a standard is impracticable.—*See Brooks's Philosophy of Arithmetic.*

III. THE MILE.—The *geographic* or *nautical mile* is equal to 1 minute of one of the great circles of the earth; hence it equals $\frac{1}{60}$ of $\frac{1}{360}$ of the circumference of the earth, which equals about 1.15 statute miles. The English mile is the same as that of the United States. 3 statute miles make a *land league;* 3 nautical miles make a *nautical league.*

IV. OTHER MEASURES.—The following denominations are frequently used: in clock-making, 6 *points* = 1 *line*, and 12 *lines* = 1 *inch;* in measuring the foot, 3 *barleycorns* or *sizes* = 1 inch; in measuring the height of horses, 4 inches = 1 *hand*, the measure being taken directly over the fore-shoulder; 1 *span* = 9 inches; 1 *common cubit* = 18 inches, and 1 *sacred cubit* = 21.888 inches; 1 *pace* = 3.3 feet; a *knot* is equal to a nautical mile. Formerly 40 rods made 1 *furlong* and 8 furlongs one mile, but these are seldom used.

MENTAL EXERCISES.

How many

1. Feet in 72 in.? 14 yd.?
2. Yards in 36 ft.? 10 rd.?
3. Rods in 22 yd.? 2 fur.?
4. Furlongs in 200 rd.? 7 mi.?
5. Miles in 320 fur.? 9 lea.?
6. Yards in 36 inches? 72 in.?
7. What part of 1 yard are 2 ft.; and what part of 2 furlongs are $5\frac{1}{2}$ yards?
8. How high is a horse which measures $15\frac{1}{4}$ hands in height?

WRITTEN EXERCISES.

1. Reduce 27 ft. 8 in. to inches. *Ans.* 332 in.
2. 3 rods 4 yd. 2 ft. to feet. *Ans.* $63\frac{1}{2}$ ft.
3. 567 inches to rods. *Ans.* 2 rd. 4 yd. 2 ft. 3 in.
4. 243 rd. 7 in. to inches. *Ans.* 48121 in.
5. 19870 ft. to miles. *Ans.* 3 mi. 244 rd. 1 yd. 1 ft.
6. 277 rd. 4 yd. to yards. *Ans.* $1527\frac{1}{2}$ yd.
7. 70287 in. to miles. *Ans.* 1 mi. 34 rd. 5 yd. 1 ft. 3 in.
8. 3 mi. 265 rd. 7 in. to inches. *Ans.* 242557 in.
9. If the Atlantic is 3000 miles wide, how many feet is that, and how many steps, 3 feet long, would a person take to walk the distance? *Ans.* 5280000 steps.

SURVEYORS' LINEAR MEASURE.

269. Surveyors' Linear Measure is used by surveyors and engineers in measuring the dimensions of land, distances, etc.

TABLE.

7.92 inches (in.)	=	1 link,	li.
100 links	=	1 chain,	ch.
80 chains	=	1 mile,	mi.

I. NAME.—*Gunter's Chain* is named after the reputed inventor, Edmund Gunter, an English mathematician, born 1581.

II. UNIT.—The unit is a chain called *Gunter's Chain*, which consists of 100 links, and is 4 rods, 66 feet, or 792 inches long.

III. The denomination *rods* is seldom used by surveyors, distances being represented in chains and links. Since each link is $\frac{1}{100}$ of a chain, the number of links is generally expressed as a decimal; thus, 5 chains and 47 links are written 5.47 chains. Engineers generally use a chain 100 feet long.

MARINERS' AND CLOTH MEASURES.

270. Mariners' Measure is used by seamen in measuring distances, the depth of the sea, etc. *Cloth Measure* is used for measuring cloth, ribbons, etc.

MARINERS' MEASURE.

6 feet	=	1 fathom.
120 fathoms	=	1 cable length.
880 fathoms	=	1 mile.

CLOTH MEASURE.

1 yard	=	36 inches.
$\frac{1}{2}$ yard	=	18 inches.
$\frac{1}{4}$ yard	=	9 inches.
$\frac{1}{8}$ yard	=	$4\frac{1}{2}$ inches.

I. The *foot* and *yard* of these two measures are the linear foot and yard. The *nail* in Cloth Measure is obsolete. At the custom-houses the yard is divided into tenths, hundredths, etc.

II. In the old table of Cloth Measure there were given 3 qrs. = 1 Ell Flemish; 5 qrs. = 1 Ell English; 6 qrs. = 1 Ell French; 6 qrs. $1\frac{1}{2}$ in. = 1 Ell Scotch.

MENTAL EXERCISES.

How many

1. Inches in a chain?
2. Feet in a chain? a fathom?
3. Feet in 6 fathoms?
4. Feet in a cable length?

WRITTEN EXERCISES.

1. Reduce 4 miles to links. *Ans.* 32000 li.
2. Reduce 7 mi. 63 ch. to links. *Ans.* 62300 li.
3. Reduce 7856 fathoms to miles. *Ans.* 8 mi. 816 fath.
4. Reduce 98763 in. to chains. *Ans.* 124 ch. 70 li. $\frac{3}{5}$ in.

MEASURES OF SURFACE.

271. A **Surface** is that which has length and breadth without thickness.

272. A **Square** is a plane surface which has four equal sides and four right angles.

273. All **Surfaces** are *measured* by ascertaining the number of times they contain a small square regarded as the unit of measure.

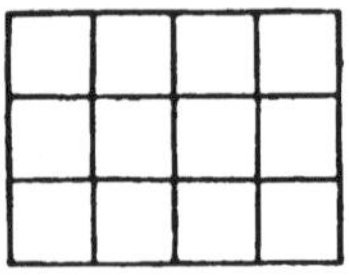

Thus, in the surface in the margin there are three rows of squares, each row containing 4 squares; hence there are 3 times 4 or 12 squares in all; and since these make up the entire surface, the measure of the surface, called its *area*, is 12 square units.

SURFACE OR SQUARE MEASURE.

274. **Surface or Square Measure** is used in measuring surfaces, as land, boards, amount of painting, papering, plastering, paving, etc.

TABLE.

144 square inches (sq. in.)	=	1 square foot,	sq. ft.
9 square feet	=	1 square yard,	sq. yd.
30¼ square yards, or 272¼ square feet,	=	1 perch or sq. rod,	P.
160 perches	=	1 acre,	A.
640 acres	=	1 square mile,	sq. mi.

I. Terms.—*Perch* is from the French *perche*, a *pole; acre* was primarily an open plowed or sowed field.

II. Unit.—The unit for land is the *acre:* for other surfaces it is usually the *square yard.*

III. The *Perch* is a surface equal to a *square rod.* The *rood* is found now only in old title-deeds and surveys; it is equal to 40 perches.

IV. A square piece of land, measuring 209 feet, or about 70 paces on each side, equals very nearly one acre.

MENTAL EXERCISES.

How many

1. Square inches in 3 sq. ft.? 5 sq. ft.?
2. Square feet in 720 sq. in.? 7 sq. yd.?
3. Square yards in 144 sq. ft.? 5 P.?
4. Perches in 121 sq. yd.? 12 R.?

5. Roods in 160 P.? 15 A.?

6. Acres in 84 R.? 3 sq. mi.

7. What is the difference between 3 feet square and 3 square feet? between 4 inches square and 4 square inches?

WRITTEN EXERCISES.

1. Reduce 9870 P. to acres. *Ans.* 61 A. 110 P.

2. 313 P. to square yards. *Ans.* 9468¼ sq. yd.

3. 3 A. 133 P. to square feet. *Ans.* 166889¼ sq. ft.

4. 109 P. 17 sq. yd. to sq. in. *Ans.* 4295268 sq. in.

5. 238975 sq. ft. to acres.
Ans. 5 A. 77 P. 211 sq. ft. 180 sq. in.

6. 9876543 sq. in. to acres.
Ans. 1 A. 91 P. 252 sq. ft. 51 sq. in.

SURVEYORS' SQUARE MEASURE.

275. Surveyors' Square Measure is used by surveyors in computing the area of land.

TABLE.

10,000 square links (sq. li.)	=	1 square chain,	sq. ch.
10 square chains . .	=	1 acre, . .	A.
640 acres . . .	=	1 square mile,	sq. mi.
36 sq. mi. (6 miles square)	=	1 township,	Tp.

Also 625 sq. li. = 1 perch; 16 perches = 1 sq. chain; 10 sq. ch. = 1 acre; or 40 perches = 1 rood; 4 roods = 1 acre. The *perch* and *rood* are not so much used as formerly, the contents of land being commonly estimated in square miles, acres, and hundredths.

MENTAL EXERCISES.

How many

1. Square links in 4 sq. ch.?

2. Square chains in 60000 sq. li.?

3. Acres in 60 sq. ch.? 2 sq. mi.?

4. Sq. chains in 65 acres? 42 acres?

5. Square miles in 1280 A.? 2 Tp.?

6. Townships in 72 sq. mi.? 144 sq. mi.?

WRITTEN EXERCISES.

1. Reduce 21 sq. mi. 65 A. 9 sq. ch. to sq. links.
Ans. 1350590000 sq. li.

2. Reduce 1,000,000 sq. li. to acres. *Ans.* 10 A.

3. 2 Tp. 21 sq. mi. 47 A. to sq. ch. *Ans.* 595670 sq. ch.

4. 8700 sq. chains to sq. miles. *Ans.* 1 sq. mi. 230 A.

5. 3 Tp. 250 A. 9 sq. ch. to sq. links. *Ans.* 6937090000.

MEASURES OF VOLUME.

276. A **Volume** is that which has length, breadth and thickness, or height. A volume is also called a *solid.*

277. A **Cube** is a volume bounded by six equal squares.

278. All **Volumes** are *measured* by ascertaining the number of times they contain a *small cube* regarded as a *unit of measure.*

Thus, in the cube in the margin, it will be seen that there are 3 times 3, or 9 cubes upon one surface, and since there are three such layers, there are 3 times 9, or 27 little cubes in all; and since these make up the entire volume, the measure of the cube, called its *contents,* is 27 cubic units.

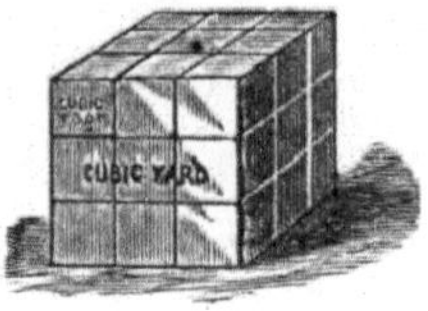

CUBIC, OR SOLID MEASURE

279. **Cubic, or Solid Measure** is used in measuring things which have length, breadth, and thickness.

TABLE.

1728 cubic inches (cu. in.)	=	1 cubic foot	cu. ft.
27 cubic feet . .	=	1 cubic yard,	cu. yd.
16 cubic feet . .	=	1 cord foot,	cd. ft.
8 cord feet, or 128 cubic feet . .	=	1 cord of wood,	cd.

I. A *cord of wood,* so named from being originally measured by a *cord.* or *string,* is a pile 8 ft. long, 4 ft. wide, and 4 ft. high. A *cord foot* is a part of this pile 1 ft. long; it equals 16 cubic feet. *See Art.* 358.

II. The *ton* of 40 ft. for *round,* or 50 ft. for *hewn* timber, is seldom used.

MENTAL EXERCISES.

How many

1. Inches in 2 cu. ft.? 4 cu. ft.?
2. Feet in 3 cu. yd.? 5 cd. ft.?
3. Cubic yards in 81 cu. ft.?
4. Feet in 6 cd.? in 4 cd.?
5. Cords in 256 cu. ft.? 72 cd. ft.?
6. Cubic yards in 135 cu. ft.?

7. What is the difference between a 2 inch cube and 2 cubic inches? between a 3 *inch cube* and 3 cubic inches?

WRITTEN EXERCISES.

Reduce

1. 3 cd. 7 cd. ft. to cubic feet. *Ans.* 496 cu. ft.
2. 7856 cd. ft. to cords. *Ans.* 982 cd.
3. 16 cd. 14 cu. ft. to cubic feet. *Ans.* 2062 cu. ft.
4. 19876 cu. ft. to cords. *Ans.* 155 cd. 36 cu. ft.
5. In 78 cords of wood how many cubic inches? *Ans.* 17252352.
6. In 5876543 cubic inches how many cords of wood? *Ans.* 26 cd. 72 cu. ft. 1343 cu. in.

MEASURES OF CAPACITY.

280. Measures of Capacity are volumes used to determine the quantity of fluids and many dry substances.

281. Measures of capacity are therefore of two kinds; *Measures of Liquids* and *Measures of Dry Substances.*

282. Liquid Measures are of two kinds,—*Liquid* or *Wine Measure* and *Apothecaries' Fluid Measure.*

LIQUID OR WINE MEASURE.

283. Liquid or **Wine Measure** is used for measuring all kinds of liquids.

TABLE.

4 gills (gi.)	=	1 pint,	pt.
2 pints	=	1 quart,	qt.
4 quarts	=	1 gallon,	gal.
$31\frac{1}{2}$ gallons	=	1 barrel,	bar.
63 gallons, or 2 bar.	=	1 hogshead,	hhd.

I. NAME.—It is called *Wine Measure* because wine was measured by it, while beer was measured by another measure.

II. TERMS.—*Gill* is from Low Latin *gilla*, a *drinking-glass*; *pint* is from the Anglo-Saxon *pyndan*, *to shut in*, *to pen*, or from the Greek *pinto*, *to drink*; *quart* is from the Latin *quartus*, a *fourth*. The derivation of *gallon* is not clear; in the French, a *galon* is a *grocer's box*.

III. UNIT.—The *standard unit* of wine measure is the *gallon*, which contains 231 cubic inches, and will hold a little more than $8\frac{1}{3}$ lb. Av. of distilled water. This is called the *Winchester gallon*, from the standard having been formerly kept at Winchester, England.

IV. *Barrels* and *hogsheads* are of variable capacity. The table values are used in estimating the capacity of wells, cisterns, vats, etc. In Massachusetts, the barrel is estimated at 32 gallons. A pint of water weighs nearly one pound, hence the old adage, "A pint's a pound, the world around."

V. Ale, beer, and milk, were formerly sold by a *gallon* of 282 cu. in., the subdivisions being *quarts* and *pints*. The measure was greater than wine measure, as beer was less costly than wine. This measure is now seldom used.

APOTHECARIES' FLUID MEASURE.

284. Apothecaries' Fluid Measure is used for measuring liquids in preparing medical prescriptions.

TABLE.

60 minims (♏)	=	1 fluidrachm	f℥.
8 fluidrachms	=	1 fluidounce	f℥.
16 fluidounces	=	1 pint	O.
8 pints	=	1 gallon	Cong.

I. TERMS.—*Minim* is from the Latin *minimus*, *the least*, the minim being the smallest fluid measure used. Several of the other terms are formed by prefixing *fluid* to the terms of Apothecaries' Weight.

II. SYMBOLS.—*Cong.* is the abbreviation of *congius*, the Latin for gallon. *O.* is the initial of *octarius*, the Latin for *one-eighth*, the pint being one-eighth of a gallon. Drops are indicated in a physician's prescription by *gtt.*, for the Latin *guttæ*.

III. In estimating the quantity of fluids, 45 drops equal about a fluidrachm; a common teaspoon holds about one fluidrachm; a common tablespoon, about ½ a fluidounce; a wineglass about 1½ fluidounces; a common teacup about 4 fluidounces. The minim is equivalent to a drop of water, but the drops of different fluids vary in size according to the tenacity of the liquid.

DRY MEASURE.

285. Dry Measure is used in measuring dry substances, such as grain, fruit, salt, coal, etc.

TABLE.

2 pints (pt.)	=	1 quart,	qt.
8 quarts	=	1 peck,	pk.
4 pecks	=	1 bushel,	bu.

I. TERMS.—*Peck* is supposed to be a corruption of *pack*, or to be derived from the French *picotin*, a *peck*.

II. UNIT.—The *unit* of dry measure is the Winchester bushel, formerly used in England, and named from the place where the standard was preserved. Its form is a cylinder, 18½ in. in diameter and 8 in. deep. Its volume is 2150.42 cu. in., and it contains 77.627413 lb. Av. of distilled water. The New York bushel is nearly identical with the imperial bushel of Great Britain, containing 2218.192 cu. in.

III. The *Chaldron*, consisting in some places of 36 bu. and in others of 32 bu., is used in some parts of the United States for measuring coal, but is being discontinued here, as it has been in England. One-half of a peck, or 4 quarts, is called a *dry gallon*.

IV. The *Cental* of 100 lb. is a standard recently recommended by the Boards of Trade in New York, Cincinnati, Chicago, and other large cities, for estimating grain, seeds, etc.

286. The **Weight of a Bushel** of the principal kinds of grain and seeds has been fixed by statute in many of the States, as shown by the following

TABLE.

	Cal.	Conn.	Del.	Ill.	Ind.	Iowa.	Ky.	La.	Me.	Mass.	Mich.	Minn.	Mo.	N. C.	N. H.	N. J.	N. Y.	Ohio.	Or.	Penn.	Vt.	W. T.	Wis.
Barley,	50			48	48	48	48	32		46	48	48	48	48		48	48	48	46	47	46	45	48
Buckwheat,	40	45		40	50	52	52			46	42	42	52	50		50	48		42	48	46	42	42
Clover Seed,				60	60	60	60				60	60	60			64	60	60	60			60	60
Indian Corn,	52	56	56	52	56	56	56	56		56	56	56	52	54		56	58	56	56	56	56	56	56
Oats,	32	28		32	32	35	$33\frac{1}{3}$	32	30	30	32	32	35	30	30	30	32	32	34	32	32	36	32
Rye,	54	56		54	56	56	56	32		56	56	56	56			56	56	56	56	56	56	56	56
Timothy Seed				45	45	45	45						45				44						46
Wheat,	60	56	60	60	60	60	60	60		60	60	60	60	60		60	60	60	60	60	60	60	60

MENTAL EXERCISES.

How many

1. Pints in 7 qt.? 45 gi.?
2. Quarts in 12 gal.? 15 gal.?
3. Quarts in 3 pk.? 7 pk.?
4. Gills in 7 pt.? 12 pt.?
5. Pecks in 4 bu.? 9 bu.?
6. Bushels in 20 pk.? 40 pk.?
7. Fluidrachms in 120 ♏?
8. Pints in 64 f℥? in 5 Cong.?
9. What part of 2 quarts is 2 gills? What part of 3 gallons is 6 pints?
10. At 10 cents a peck, how many bushels of apples can be bought for 12 dollars?

WRITTEN EXERCISES.

How many

1. Pints in 16 gal. 3 qt. 1 pt. of wine? *Ans.* 135 pt.
2. Gallons in 7635 gills of vinegar?
   Ans. 238 gal. 2 qt. 3 gi.
3. Minims in 5 Cong. 3 O. 12 f℥. 5 fʒ? *Ans.* 336300 ♏.
4. Cong. in 724475 ♏?
   Ans. 11 Cong. 6 O. 5 f℥. 2 fʒ. 35 ♏.
5. Pints in 26 bu. 3 pk. 7 qt. 1 pt. of berries?
   Ans. 1727 pt.
6. Bushels in 37891 pints of clover seed? *Ans.* $592\frac{3}{64}$ bu.
7. What cost 5 gal. 3 qt. 1 pt. of vinegar at 3 cents a pint?
   Ans. $1.41.

8. What cost 217 bu. 3 pk. 5 qt. of canary seed at $18\frac{3}{4}$ cents a quart? *Ans.* 1307.43\frac{3}{4}$.

CIRCULAR MEASURE.

287. Circular Measure is used to measure angles and directions, latitude and longitude, difference of time, etc.

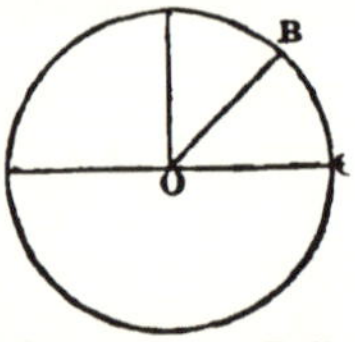

288. A **Circle** is a plane figure bounded by a curved line, every point of which is equally distant from a point within, called the *centre*.

289. The **Circumference** of a circle is the bounding line; any part of the circumference, as BC, is an *arc*. An arc of one-fourth of the circumference is called a *quadrant*.

290. For the purpose of measuring angles, the circumference is divided into 360 equal parts, called *degrees*; each degree into 60 equal parts, called *minutes;* each minute into 60 equal parts, called *seconds*.

291. Any angle having its vertex at the centre, is measured by the arc included between its sides; thus, the angle COB is measured by the arc BC. A right angle is measured by 90 degrees, or a quadrant; half a right angle, by 45 degrees; etc.

TABLE.

60 seconds (″) . .	=	1 minute . .	′
60 minutes . . .	=	1 degree . .	°
30 degrees . . .	=	1 sign . .	S.
12 signs, or 360° . .	=	1 circumference	C.

I. Terms.—The term *minute* is from the Latin *minutum*, which signifies a *small part*. The term *second* is an abbreviated expression for *second* minutes, or minutes of the *second order*. *Signs* are used in astronomy as a measure of the zodiac.

II. Unit.—The *unit* is the *degree*, which is $\frac{1}{360}$ of the circumference of a circle. A *quadrant* is one-fourth of a circumference, or 90°. A *minute* of the earth's circumference is called a *geographic mile*.

III. Divisions.—The divisions of the circumference are not of absolute length; they are merely *equal parts*, used to indicate the size of angles. Thus, a *quadrant*, whether the circle is large or small, measures a *right angle*.

WRITTEN EXERCISES.

Reduce

1. 25′ 30″ to seconds. *Ans.* 1530″.
2. 6000″ to degrees. *Ans.* 1° 40′.
3. 4S. 17° 18′ 42″ to seconds. *Ans.* 494322″.
4. 195600″ to signs. *Ans.* 1S. 24° 20′.
5. 3 quadrants 15° to seconds. *Ans.* 1026000″.
6. If Venus, at a certain time, is 44° 27′ 30″ east of Mercury, how many seconds are they apart? *Ans.* 160050″.

TIME.

292. Time is a portion of duration. The *measures* of time are fixed by the revolution of the earth on its axis and around the sun.

TABLE

60 seconds (sec.) .	=	1 minute,	min.
60 minutes . .	=	1 hour, . .	h.
24 hours . .	=	1 day, . .	da.
365 days . .	=	1 common year, .	yr.
366 days . .	=	1 leap year, .	yr.
100 years . .	=	1 century, .	cen.

ALSO,

7 days . . .	=	1 week,	wk.
4 weeks . . .	=	1 lunar month,	mo.
13 lunar months, 1 d., 6 h., or 12 calendar months,	=	1 year, .	yr.

I. TERMS.—*Second* and *minute* are parts of an hour, corresponding to the parts of a degree in Circular Measure. *Hour* is derived from the Latin *hora*, originally a definite space of time fixed by natural laws; a *day*, derived from the Saxon *daeg*, is the time of the revolution of the earth upon its axis; a *week* is a period of uncertain origin, but which has been used from time immemorial in Eastern countries; a *month*, from Saxon *monadh*, from *mona*, the moon, is the time of one revolution of the moon around the earth; a *year*, from the Saxon *gear*, is the time of the earth's revolution around the sun; a *century* comes from the Latin *centuria*, a collection of a hundred things.

II. UNIT.—The *unit of time* is the *day*; it is determined by the revolution of the earth on its axis. The *Sidereal Day* is the exact time of the revolution of the earth on its axis. The *Solar Day* is the time of the apparent revolution of the sun around the earth. The *Astronomical Day* is the solar day, beginning and ending at noon. The *Civil Day* is the average length of all the solar days of the year; it begins at 12 o'clock midnight, and consists of two periods of 12 hours each.

THE CALENDAR.

293. **The Calendar** is a division of time into periods adapted to the purposes of civil life.

294. The **Year** is divided into 12 calendar months, three of which constitute a period called a *Season*.

295. The seasons, months, and number of days in each, are given in the following table:

No. of Mo.	Month.	Season.	No. of Days.
1	January,	Winter,	31
2	February,		28 or 29
3	March,	Spring,	31
4	April,		30
5	May,		31
6	June,	Summer,	30
7	July,		31
8	August,		31
9	September,	Autumn,	30
10	October,		31
11	November,		30
12	December,	Winter,	31

296. The time from any day of one month to any day of another month in the same year is readily found by the following table:

TABLE

SHOWING THE NUMBER OF DAYS FROM ANY DAY OF ONE MONTH TO THE SAME DAY OF ANY OTHER MONTH IN THE SAME YEAR.

From any day of	To the same day of											
	Jan.	Feb.	Mar	Apr	May	J'ne	July	Aug	Sep.	Oct.	Nov	Dec.
January	365	31	59	90	120	151	181	212	243	273	304	334
February	334	365	28	59	89	120	150	181	212	242	273	303
March	306	337	365	31	61	92	122	153	184	214	245	275
April	275	306	334	365	30	61	91	122	153	183	214	244
May	245	276	304	335	365	31	61	92	123	153	184	214
June	214	245	273	304	334	365	30	61	92	122	153	183
July	184	215	243	274	304	335	365	31	62	92	123	153
August	153	184	212	243	273	304	334	365	31	61	92	122
September	122	153	181	212	242	273	303	334	365	30	61	91
October	92	123	151	182	212	243	273	304	335	365	31	61
November	61	92	120	151	181	212	242	273	304	334	365	30
December	31	62	90	121	151	182	212	243	274	304	335	365

Method of using the Table.—Suppose we wish to find the number of days from March 10th to November 16th. We find March in the vertical column, and November at the top, and at the intersection we find 245, to which adding 6 days we have 251, the number of days required. The table being constructed for February 28 days, the proper allowance must be made for leap year.

I. **Terms.**—January is derived from *Janus*, the god of the year, to whom this month was sacred. February is from *februa*, the Roman festival of expiation, celebrated on the 15th of this month. January and February were added to the Roman calendar by Numa, Romulus having previously divided the year into 10 months. March is from *Mars*, the god of war and reputed father of Romulus. It was the first month of the Roman calendar. April is probably from the Latin *aperire*, to open, from the opening of the buds, or the bosom of the earth in producing vegetation. May is from *Maia*, the mother of Mercury, to whom the Romans offered sacrifices on the first day of this month. June is from *Juno*, the sister and wife of Jupiter, to whom it was sacred. July was named by Mark Antony after *Julius* Cæsar, who was born in this month. It was previously called Quintilis. August was named after *Augustus* Cæsar, who entered upon his first consulate in this month. It was formerly called Sextilis, or sixth month. September, October, November, and December, are respectively named from the Latin numerals, *Septem*, *Octo*, *Novem*, and *Decem*, as when the year began in March, they were the seventh, eighth, ninth, and tenth months, as their names indicate.

II. The number of days in each month is easily remembered by the following stanza :—

Thirty days hath September,
April, June, and November;
All the rest have thirty-one,
Excepting February alone;
To which we twenty-eight assign,
Till leap year gives it twenty-nine.

MENTAL EXERCISES.

1. How many seconds in 30 minutes? in 25 minutes?
2. How many minutes in 360 seconds? in 3 hours?
3. How many hours in 660 minutes? in 4 days?
4. How many days in 72 hours? in 4 weeks?
5. How many days in 3 common years? in 4 leap years?
6. How many common years in 730 days?
7. Name the months which have 30 days each ; those which have 31 days each.

WRITTEN EXERCISES.

1. How many seconds in one day? *Ans.* 86400 sec.
2. How many seconds in one week? *Ans.* 604800 sec.
3. How many minutes in 5 da. 7 h. 45 min.? *Ans.* 7665 min.
4. Reduce 56780 seconds to hours. *Ans.* 15 h. 46 min. 20 sec.

5. In one common year how many hours; how many minutes; how many seconds?

Ans. 8760 h.; 525,600 min.; 31,536,000 sec.

6. How many days from May 8th to October 19th?

Ans. 164 days.

7. How many days from March 28th to December 16th?

Ans. 263 days.

8. How many days from February 21st, 1860, to September 10th, 1860? *Ans.* 202 days.

ADJUSTMENT OF THE CALENDAR.

297. A **True** or **Solar Year** is the exact time in which the earth revolves around the sun. It consists of 365 da. 5h. 48 min. 49.7 sec. Now, since it is inconvenient to reckon the fractional part of a day each year, it is necessary to arrange a correct calendar in which each year may have a whole number of days. This is done by causing some years to consist of 365 days, and others of 366 days. The former are called common years, the latter, Bissextile, or Leap years.

298. The calendar is reckoned according to the following rule:

Rule.—*Every year that is divisible by* 4, *except the centennial years, and every centennial year divisible by* 400, *is a leap year; all the others are common years.*

NOTE.—The centennial years are those whose expressions in figures end in two ciphers.

EXPLANATION.—I. If we reckon 365 days as one year, the time lost in the calendar in one year is 5 h. 48 min. 49.7 sec., and the time lost in four years is 23 h. 15 min. 18.8 sec., that is, *one day*, lacking only 44 min. 41.2 sec.; hence the first error can be corrected by adding *one* day every *four* years, making the year to consist of 366 days.

II. If every fourth year be reckoned as leap year, since we add 44 min., etc., too much, the time *gained* in the calendar in four years is 44 min. 41.2 sec., and in 100 years it will be 18 h. 37 min. 10 sec., that is, *one* day, lacking 5 h. 22 min. 50 sec.; hence the second error may be corrected by deducting one day from each centennial leap year, thus calling each centennial year a common year of 365 days.

III. Again, if every centennial year be reckoned as a common year, since we do not add enough, the time lost in 100 years will be 5 h. 22 min. 50 sec., and in 400 years it will be 21 h. 31 min. 20 sec.; hence the time lost in 400 years will be 1 day lacking 2 h. 28 min. 40 sec., and this error may be *rectified by* making every 4th centennial year a leap year

In the same way we may make the calendar correct for any number of years.

NOTE.—The reckoning of time by the ancients was very inaccurate. The calendar was reformed by Julius Cæsar, 46 B. C., who made the year to consist of 365¼ days, adding one day every fourth year. In 1582, Pope Gregory corrected the error which resulted from the above correction, by striking out 10 days from the calendar, calling the 5th of October the 15th, and ordaining that henceforth only those centennial years should be leap years which are divisible by 400. This change was soon adopted by most Catholic countries, but Great Britain did not make the change till 1752, when the error amounted to 11 days. Some countries, as Russia, still adhere to the Julian Calendar, their dates being about 12 days behind ours. The dates are distinguished as Old Style and New Style.

MENTAL EXERCISES.

1. How many centuries is it since the birth of Christ?
2. When did the 18th century end and the 19th century begin?
3. How many leap years and how many common years in every century?
4. Which of the following are leap years: 1700? 1760? 1776? 1800? 1876? 1880? 1890? 1900? 2000?
5. My watch ticks 4 times in a second; how many times will it tick in a day?

MISCELLANEOUS TABLES.

299. The following tables are frequently used, the first in counting certain kinds of articles, and the second in the paper trade.

COUNTING.	PAPER.
12 units = 1 dozen.	24 sheets = 1 quire.
12 dozen = 1 gross.	20 quires = 1 ream.
12 gross = 1 great gross.	480 sheets = 1 ream.
20 units = 1 score.	

I. Two things of a kind are frequently called a *pair* and *six* a *set*.

II. Paper is sold at retail by sheets, quires, and reams, and at wholesale by reams.

BOOKS.

300. In printing books large sheets of paper are used, which are folded into leaves according to the size of the book. The terms *folio*, *quarto*, *octavo*, etc., as applied to printed books, are based on sheets about 18 × 24 in., about half the sizes now generally used, and indicate the number of leaves into which such a sheet is folded.

A sheet folded in 2 leaves is called a folio, makes 4 pages.					
A sheet folded in 4	"	"	"	a quarto or 4to, makes 8 pages.
A sheet folded in 8	"	"	"	an octavo or 8vo, makes 16 pages.
A sheet folded in 12	"	"	"	a 12mo, makes 24 pages.
A sheet folded in 16	"	"	"	a 16mo, makes 32 "
A sheet folded in 18	"	"	"	an 18mo, makes 36 "
A sheet folded in 24	"	"	"	a 24mo, makes 48 "
etc.				etc.

NOTE.—Printing paper is made of many sizes, according to the requirements of the printer. In book printing 24 × 38 inches, called *Double Medium*, is perhaps used most largely.

301. Clerks and copyists are often paid by the *folio* for making copies of legal papers, records, and documents.

72 words make 1 folio, or sheet of common law.
90 " " 1 " " " " chancery.

MENTAL EXERCISES.

1. How many dozens in ½ a gross? In 2½ gross? In ½ a great gross?

2. How many pairs in 40? Scores in 60? Scores in 70? Sets in 48?

3. How many sheets in 3 quires? In 2½ quires? In ½ a ream? In ¼ ream?

4. How many eggs in 2½ dozen? In half a dozen? In a dozen and a quarter?

5. How many years in 3 score? In 3 score and 10? In 4 score and a half?

6. How many sheets of paper will be required to make a 12mo. book of 360 pages? Of 480 pages?

7. How many sheets will be required to make an octavo book of 320 pages? Of 400 pages?

8. How many octavo books will the paper for a quarto book make, of the same number of pages?

WRITTEN EXERCISES.

1. How many fine black crayons are there in 42 boxes, each containing 1 gross? *Ans.* 6048.

2. Sold 63 boxes of Maynard's writing ink, each box containing 3 dozen bottles; how many gross?
Ans. 15 gross, 9 dozen.

3. What would be the cost of 3240 sheets of foolscap at 36 cents a quire? *Ans.* $48.60.

4. A lady copied in one month 795.5 chancery folios at 12¢ per folio; what did she receive? *Ans.* $95.46.

5. A printer used 3 reams 5 quires 19 sheets of paper for printing half-sheet posters; how many did he print, allowing 1 quire to a ream for waste? *Ans.* 3000.

6. How much paper would it require to print 5000 copies of this book, a 12mo., allowing 1 quire to each ream for waste?

REDUCTION OF COMPOUND NUMBERS.

302. Reduction is the process of changing a number from one denomination to another without altering its value.

303. There are **Two Cases**: *Reduction Descending* and *Reduction Ascending*.

These two cases have been considered in the examples under the tables, but we will present a few more problems under their proper heads.

REDUCTION DESCENDING.

304. Reduction Descending is the process of reducing a number to a lower denomination.

1. Reduce £6 5 s. 3 d. to pence.

SOLUTION.—In 1 pound there are 20 shillings, and in £6 there are 6 times 20 shillings, or 120 shillings; 120 shillings plus 5 shillings are 125 shillings; in 1 shilling there are 12 pence, and in 125 shillings there are 125 times 12 pence, or 1500 pence; 1500 pence plus 3 pence are 1503 d. Therefore, etc.

OPERATION.

£	s.	d.
6	5	3
20		
125 s.		
12		
1503 d., *Ans.*		

Rule.—I. *Multiply the number of the highest denomination given, by the number of units of the next lower denomination which equals one of this higher, and to the product add the number given, if any, of this lower denomination.*

II. *Multiply this result as before, and proceed in the same manner until we arrive at the required denomination.*

WRITTEN EXERCISES.

2. Reduce 15 lb. 4 oz. 3 pwt. 15 gr. to gr. *Ans.* 88407 gr.
3. 17 ℔. 7℥ 5ʒ 1℈ 18 gr. to gr. *Ans.* 101618 gr.
4. 3 cable lengths 100 fathoms to feet. *Ans.* 2760 ft.
5. In 18 T. 13 cwt. 75 lb. 14 oz. how many ounces? *Ans.* 598014 oz.

6. How many inches in 1 degree? *Ans.* 4381977.6 in.

7. How many links in 3 miles 20 chains? *Ans.* 26000 li.

8. Reduce 18 A. 25 P. 16 sq. yd. to sq. inches.
Ans. 113908356 sq. in.

9. How many square chains in one township?
Ans. 230400 sq. ch.

10. Reduce 56 cu. yd. 10 cu. ft. 1500 cu. in. to cubic inches. *Ans.* 2631516 cu. in.

11. Reduce Cong. vj. O. vij. f℥xv. fʒiv. ♏xxx. to minims.
Ans. 429870♏.

12. A hogshead of molasses contained 98 gallons; required the number of gills. *Ans.* 3136 gi.

13. Between the times of two observations, the planet Mars was found to have passed over 3 S. 25° 40″; how many seconds in this distance? *Ans.* 414040″.

14. Adam died at the age of 930 years; how many seconds old was he? *Ans.* 29,347,944,621 sec.

15. If the pulse beat 75 times a minute, how often does it beat in a week? *Ans.* 756000 times.

16. How many inches from Lancaster to Philadelphia, if it is just 68 miles? *Ans.* 4,308,480 in.

REDUCTION ASCENDING.

305. Reduction Ascending is the process of reducing a number to a higher denomination.

1. In 46743 pence how many pounds?

SOLUTION.—There are 12 pence in 1 shilling, hence in 46743 pence there are as many shillings as 12 is contained times in 46743, which are 3895s. and 3 d. remaining; there are 20 shillings in 1 pound, hence in 3895s. there are as many pounds as 20 is contained times in 3895, which are £194 and 15s. remaining. therefore, in 46743 pence there are £194 15s. 3d.

OPERATION.

12)46743
20)3895 + 3 d.
£194 + 15s.
Ans. £194 15s. 3 d.

Rule.—I. *Divide the given number by the number of units in that denomination which equals one of the next higher.*

II. *Divide the quotient in the same way, and thus proceed until we arrive at the required denomination.*

III. *The last quotient and the remainders, if any, will be the result required.*

WRITTEN EXERCISES.

2. Reduce 2579 pence to pounds. *Ans.* £10 14 s. 11 d.

3. Reduce 76942 grains to pounds Troy.
Ans. 13 lb. 4 oz. 5 pwt. 22 gr.

4. How many pounds in 75826 farthings?
Ans. £78 19 s. 8 d. 2 far.

5. Reduce 57642 gr. to ℔. *Ans.* 10 ℔. 2℈ 2 gr.

6. Reduce 65784 oz. to tons.
Ans. 2 T. 1 cwt. 11 lb. 8 oz.

7. How many miles in 179234 inches?
Ans. 2 mi. 265 rd. 1 yd. 8 in.

8. In 74325 li. how many miles?
Ans. 9 mi. 23 ch. 25 li.

9. Reduce 743750 cu. in. to cords.
Ans. 3 cd. 46 cu. ft. 710 cu. in.

10. In 67,437,842,351 sq. in. how many square miles?
Ans. 16 sq. mi. 511 A. 17 P. 17 sq. yd. 8 sq. ft. 59 sq. in.

11. Reduce 347236 ♏ to Cong.
Ans. Cong. v. O. v. f℥iij. f℥iij. ♏xvj.

12. In 6434 gi. how many gallons? *Ans.* 201 gal. 2 gi.

13. How many bushels in 2401 pints?
Ans. 37 bu. 2 pk. 1 pt.

14. How many common years in 47,424,496 seconds?
Ans. 1 yr. 183 da. 21 h. 28 min. 16 sec.

15. A ship changed her longitude in a storm 721 geographical miles; what was the difference in longitude?
Ans. 12° 1'.

16. A butter-dealer sold 6476 pounds of butter in a month; how many firkins did he sell? *Ans.* 115 firkins 36 lb.

17. A Pennsylvania miller ground a quantity of buckwheat and took 72 lb. of flour for toll, which was $\frac{1}{10}$ of the whole; how many bushels were there brought to mill if 4 lb. of grain produced 3 lb. of flour? *Ans.* 20 bushels.

MENTAL EXERCISES.

MISCELLANEOUS EXAMPLES.

1. How much will 9 eggs cost, at 20 cents a dozen?

2. How much will 8 pearl buttons cost, at 75 cents a dozen?

3. What will 4 table-spoons cost, at the rate of $11.52 a gross?

4. What will shoe-tacks cost a dozen, at the rate of $2.88 a great gross?

5. What will 8 sheets of paper cost, at 18 cents a quire?

6. What will 12 quires of paper cost, at $4 a ream?

7. What cost ½ a gross of lead-pencils, at 50 cents a dozen?

8. What will 2 reams of sand-paper cost, at 22 cents a quire?

9. If a ream of emery-paper costs $12, what does it cost a quire?

10. What will 4 gross of pens cost, if the wholesale rate is 2 pens for a cent?

11. A vessel was sunk in 8 fathoms of water; how many feet deep was it?

12. How many feet high is a horse which measures 15½ hands high?

13. How much will 3 pecks of potatoes cost, at 80 cents a bushel?

14. When apples sell at 20 cents a peck, how much are they worth a bushel?

15. What will 3 quarts of molasses cost, at 96 cents a gallon?

16. How many vials, holding a gill each, can be filled with half a gallon of alcohol?

17. How many half-pint bottles will 2 gallons of Arnold's writing-fluid fill?

18. If I pay $1.60 for a bushel of blackberries, how much is that a quart?

19. If it takes 6 buttons for 1 shirt, how many shirts will 2 gross of buttons trim?

20. The distance across a city park is 132 paces; how many rods is it?

21. What will 15 pounds of flour cost, at the rate of $4 a hundred weight?

22. A boy bought a bushel of chestnuts for $2.40, and sold them at 12 cents a quart; what did he gain?

23. I bought 30 lb. of flour for $1.50; how much is that a hundred-weight?

24. Albert's horse is 5 feet in height; how many hands high is his horse?

25. A vessel sails 4 leagues an hour; how many hours will it take to sail 120 miles?

WRITTEN EXERCISES.

MISCELLANEOUS EXAMPLES.

1. What will 5 tons of coal cost, at 37½¢ a hundred? *Ans.* $37.50.

2. How many guineas in 47 pounds and 5 shillings? *Ans.* 45 guineas.

3. How many pounds Avoirdupois in 105000 grains? *Ans.* 15 lb.

4. At 6¢ a pound, what will 5 cwt. 75 lb. of sugar cost? *Ans.* $34.50.

5. How much will 6 barrels of flour cost, at the rate of 3½ cents a pound? *Ans.* $41.16.

6. How many suits of clothes, containing 6¾ yd. each, can be cut out of 93 yd. of cloth? *Ans.* 13; 5¼ yd. over.

7. How many doses of medicine, each weighing 7 grains, can be made out 1 ℥ 2 ʒ 2 ℈ 11 gr.? *Ans.* 93 doses.

8. How many cannon-balls, each weighing 41 lb. 10⅔ oz., can be made out of a ton of iron? *Ans.* 48 balls.

9. How many times will a wheel, 15 ft. 4 in. in circumference, revolve in going 50 miles? *Ans.* $17217\frac{9}{23}$.

10. Which is greater, and how much, six dozen dozen or a half a dozen dozen? *Ans.* 1st, by 792.

11. How many kegs, each containing 5 gal. 2 qt. 1 pt., can be filled from a tun (4 hhd.) of wine? *Ans.* 44⅘.

12. How many lots of 5 A. 82 P. are there in a field containing 66 A. 24 P.? *Ans.* 12 lots.

13. At 13 cents a pound, how much rice can be bought for $81,250? *Ans.* 312½ tons.

14. How much time will a person lose in 50 years by taking an hour's nap each afternoon? *Ans.* 2 yr. $30\frac{5}{12}$ da.

15. If a comet pass through an arc of 7° 5′ a day, how long will it be in describing an arc of 270°? *Ans.* $38\frac{2}{17}$ da.

16. How many minutes longer was January, 1860, than February of the same year? *Ans.* 2880 min.

17. If a physician uses on an average 5℥ 7ʒ 1℈ 4 gr. of drugs daily, how many did he use during February, 1876? *Ans.* 14 ℔. 3℥ 6ʒ 1℈ 16 gr.

18. How long will it take to count a million at the rate of 80 per minute, working 12 hours a day? *Ans.* $17\frac{13}{36}$ da.

19. How many half-pint bottles will it take to put up 6 gallons of Arnold's writing fluid? *Ans.* 96.

20. How many more seconds were there in 1876 than in a solar year? *Ans.* 65470.3 sec.

21. How many boards, 12 ft. long, will inclose a lot 50 rd. long and 27 rd. wide, the fence being 3 boards high?
Ans. $635\frac{1}{4}$.

22. How many ounces of calomel will it take to make 980 pills of 5 grains each? *Ans.* 10℥ 1ʒ 2℈.

23. If a weekly newspaper has 5600 subscribers, how many reams of paper will it require in a year, making no allowance for waste? *Ans.* 606 reams, 13 quires, 8 sheets.

24. How many minutes more in the spring of every common year than in the autumn? *Ans.* 1440 min.

25. How many pages are there in an octavo book, the printing of which requires 20 fully printed sheets? *Ans.* 320.

26. Mr. A's income averages 4 cents a minute; what will it be during the three summer months? *Ans.* $5299.20.

27. If 32,400 steel pens are made by a factory in a day, how many gross will be made in the month of March?
Ans. 6975 gross.

28. If a grocer's weights are $\frac{1}{4}$ of an oz. in a pound below the legal standard, how much does he gain fraudulently from the sale of 2 bags of Rio coffee, 116 lb. each, true weight, at $18\frac{3}{4}$¢ a pound? *Ans.* $\$0.69\frac{1}{21}$.

29. I bought 5 T. 14 cwt. of hay in a stack, but before it was all delivered, 1 T. 3 cwt. 56 lb. were spoiled by the weather. The price originally agreed upon was $88.35; what should I pay for what I received? *Ans.* $70.091.

30. A New Jersey truck-raiser sold 15 bu. 3 pk. 1 qt. of strawberries, at 11¢ a quart, and agreed to take flour in payment at $3\frac{1}{2}$¢ a pound, as far as it would make an exact number of barrels, and the rest to be paid in cash; how many barrels and how much cash did he receive?
Ans. 8 barrels and 67¢

ADDITION OF COMPOUND NUMBERS.

306. Addition of Compound Numbers is the process of finding the sum of two or more similar compound numbers.

1. Find the sum of £7 6 s. 8 d.; £5 9 s. 3 d.; £14 13 s. 10 d.; and £11 19 s. 11 d.

SOLUTION.—We write the numbers so that similar units shall stand in the same column, and begin at the right to add. 11 d. plus 10 d., plus 3 d., plus 8 d., are 32 d., which by reduction we find equals 2 s. and 8 d.; we write the 8 d. in the pence column, and reserve the 2 s. to add to the column of shillings: 2 s. plus 19 s., plus 13 s., plus 9 s., plus 6 s., are 49 s., which by reduction we find equals £2 and 9 s.; we write the 9 s. in the column of shillings, and reserve the £2 to add to the column of pounds; £2 plus £11, plus £14, plus £5, plus £7, equal £39, which we write under the pounds. Hence the following

OPERATION.

£	s.	d.
7	6	8
5	9	3
14	13	10
11	19	11
39	9	8

Rule.—I. *Write the compound numbers so that similar units stand in the same column.*

II. *Begin with the lowest denomination and add each column separately, placing the sum, when less than a unit of the next higher denomination, under the column added.*

III. *When the sum equals one or more units of the next higher denomination, reduce it to this denomination, write the remainder under the column added, and add the quotient obtained by reduction to the next column.*

IV. *Proceed in the same manner with all the columns to the last, under which write the entire sum.*

Proof.—The same as in addition of simple numbers.

NOTE.—In writing, if any places are wanting supply them with a cipher.

WRITTEN EXERCISES.

(2)

£	s.	d.
16	12	5
127	13	7
192	18	10
168	14	11
505	19	9

(3)

£	s.	d.
125	16	7
116	13	9
242	17	5
363	15	8

(4)

£	s.	d.
672	18	11
149	10	9
941	17	10
876	19	8

(5)

lb.	oz.	pwt.	gr.
16	10	18	12
14	9	17	23
15	7	15	20
17	11	14	19
23	8	16	22

(6)

cd.	cu. ft.	cu. in.
18	123	762
25	109	835
37	127	976
21	113	204
36	120	630

(7)

T.	cwt.	lb.	oz.
18	18	68	15
25	12	44	13
52	14	73	9
71	17	92	7
26	15	74	14

(8)

℔.	℥	ʒ	℈	gr.
28	11	7	2	16
25	10	6	1	15
19	9	5	2	3
27	8	3	1	17
24	7	2	1	18

(9)

hhd.	gal.	qt.	pt.
27	36	2	1
38	60	3	0
49	55	1	1
74	27	0	1
107	19	3	0

(10)

℔.	℥	ʒ	℈	gr.
14	10	6	2	14
26	9	3	1	18
37	8	2	2	16
15	0	0	0	2
46	7	5	1	17

11. Find the sum of 16 mi. 100 rd. 3 yd. 2 ft. 7 in., 12 mi. 309 rd. 2 yd. 1 ft. 6 in., 15 mi. 274 rd. 5 yd. 2 ft. 9 in., 18 mi. 227 rd. 4 yd. 1 ft. 8 in. *Ans.* 63 mi. 273 rd. 1 ft.

12. Find the sum of 132 sq. yd. 8 sq. ft. 120 sq. in., 246 sq. yd. 7 sq. ft. 137 sq. in., 546 sq. yd. 3 sq. ft. 129 sq. in., 765 sq. yd. 6 sq. ft. 105 sq. in., 382 sq. yd. 5 sq. ft. 126 sq. in. *Ans.* 2074 sq. yd. 6 sq. ft. 41 sq. in.

13. Find the sum of 16 A. 104 P. 18 sq. yd. 7 sq. ft., 25 A. 116 P. 28 sq. yd. 8 sq. ft., 18 A. 139 P. 17 sq. yd. 6 sq. ft., 27 A. 106 P. 30 sq. yd. 8 sq. ft., 24 A. 155 P. 26 sq. yd. 5 sq. ft. *Ans.* 113 A. 144 P. 1 sq. yd. 7 sq. ft.

14. A vintner sold to A 5 hhd. 59 gal. 3 qt. 1 pt. of wine, to B 20 hhd. 45 gal. 2 qt., to C 39 hhd. 58 gal. 1 pt., and had as much as he sold A remaining; how much had he at first? *Ans.* 72 hhd. 34 gal. $1\frac{1}{2}$ qt.

15. What is the sum of 126 yr. 10 mo. 5 wk. 17 h., 236 yr. 9 mo. 2 wk. 7 da. 18 hr. 41 min., 425 yr. 8 mo. 4 wk. 3 da. 20 h. 16 min., 198 yr. 7 mo. 6 wk. 19 h. 52 min., 385 yr. 5 wk. 40 min.? *Ans.* 1373 yr. 3 mo. 3 wk. 6 da. 4 h. 29 min.

16. Find the sum of 144 cu. yd. 18 cu. ft. 1329 cu. in., 275 cu. yd. 25 cu. ft. 1076 cu. in., 382 cu. yd. 17 cu. ft. 1521 cu. in., 420 cu. yd. 20 cu. ft. 1507 cu. in., 367 cu. yd. 21 cu. ft. 1473 cu. in. *Ans.* 1591 cu. yd. 23 cu. ft. 1722 cu. in.

SUBTRACTION OF COMPOUND NUMBERS.

307. Subtraction of Compound Numbers is the process of finding the difference between two similar compound numbers.

1. From 10 oz. 12 pwt. 20 gr. take 7 oz. 15 pwt. 16 gr.

OPERATION.

oz.	pwt.	gr.
10	12	20
7	15	16
2	17	4

SOLUTION.—We write the subtrahend under the minuend, placing similar units in the same column, and begin at the lowest denomination to subtract; 16 gr. subtracted from 20 gr. leaves 4 gr. which we write under the grains: 15 pwt. from 12 pwt. we cannot take; we will therefore take 1 oz. from the 10 oz., leaving 9 oz.; 1 oz. equals 20 pwt., which, added to 12 pwt. equals 32 pwt.; 15 pwt. subtracted from 32 pwt. equals 17 pwt., which we write under the pwt.; 7 oz. from 9 oz. (or, since it will give the same result, we may add 1 oz. to 7 oz., and say 8 oz. from 10 oz.) leaves 2 oz. Hence the following

Rule.—I. *Write the subtrahend under the minuend, so that similar units stand in the same column.*

II. *Begin with the lowest denomination and subtract each term of the subtrahend from the corresponding term of the minuend.*

III. *If any term of the subtrahend exceeds the corresponding term of the minuend, add to the latter as many units of that denomination as make one of the next higher, and then subtract; add 1 also to the next term of the subtrahend before subtracting.*

IV. *Proceed in the same manner with each term to the last.*

Proof.—The same as in the subtraction of simple numbers.

NOTE.—The pupil will notice that the general principle of addition and subtraction is the same as in simple numbers, the difference being in the irregularity of the scale, the units themselves being expressed in the decimal scale.

WRITTEN EXERCISES.

(2)

£	s.	d.	far.
143	11	10	2
115	14	6	3
27	17	3	3

(3)

£	s.	d.	far.
930	17	7	3
246	19	8	1

(4)

lb.	oz.	pwt.	gr.
16	10	16	18
13	11	17	15
2	10	19	3

(5)

lb.	oz.	pwt.	gr.
125	8	14	20
96	9	10	23

(6)

cwt.	lb.	oz.
112	92	12
37	44	13

(7)

T.	cwt.	lb.	oz.
236	13	68	12
127	12	22	10

(8)

hhd.	gal.	qt.	pt.
128	27	0	1
106	30	2	1

(9)

yr.	mo.	wk.	da.	h.
216	10	2	5	16
123	10	3	2	20

(10)

sq. yd.	sq. ft.	sq.in.
226	20	120
134	25	130

(11)

m.	rd.	yd.	ft.	in.
85	260	0	2	8
29	296	5	1	9
55	283	$\frac{1}{2}$	0	11
		$\frac{1}{2}$=1		6
55	283	0	2	5

(12)

A.	P.	sq.yd.	sq.ft.	sq.in.
179	150	16	7	135
134	155	18	8	106
44	154	27$\frac{1}{4}$	8	29
		$\frac{1}{4}$=2		36
44	154	28	1	65

13. A farmer had 200 bu. of wheat and sold 28 bu. 2 pk. 5 qt. 1 pt. to one man, and as much more to another; how much remained? *Ans.* 142 bu. 2 pk. 5 qt.

14. A California miner having 112 lb. of gold, sent his mother 17 lb. 10 oz. 15 pwt. 20 gr., and 3 lb. 16 pwt. less to his father; how much did he retain?

Ans. 79 lb. 3 oz. 4 pwt. 8 gr.

15. Subtract 16 mi. 223 rd. 3 yd. 1 ft. 8 in. from 36 mi. 271 rd. 3 yd. 1 ft. 11 in. *Ans.* 20 mi. 48 rd. 3 in.

16. Subtract \$16 57¢ 5$\frac{1}{2}$ mills from \$25 20¢ 7$\frac{1}{3}$ mills, and add 2 eagles and 25$\frac{3}{4}$ dimes to the result.

Ans. \$31.206$\frac{5}{6}$.

17. Subtract 125 A. 37 P. 29 sq. yd. 4 sq. ft. 140 sq. in. from 240 A. 85 P. 16 sq. yd. 96 sq. in.

Ans. 115 A. 47 P. 16 sq. yd. 6 sq. ft. 136 sq. in.

18. A man had a hogshead of molasses, from which there leaked away 11 gal. 3qt. 1 pt., and then after putting in 12 gal. he found it lacked 16 gal. 1 pt. of containing 63 gal.; how much was in at first? *Ans.* 46 gal. 3 qt.

MULTIPLICATION OF COMPOUND NUMBERS.

308. Multiplication of Compound Numbers is the process of finding the product when the multiplicand is a compound number.

1. Multiply £12 11 s. 7 d. by 8.

OPERATION.

£	s.	d.
12	11	7
		8
100	12	8

SOLUTION.—We write the multiplier under the lowest denomination of the multiplicand, and begin at the right to multiply. 8 times 7 d. are 56 d., which, by reduction, we find equals 4 s. and 8 d.; we write the 8 d. under the pence, and reserve the 4 s. to add to the next product: 8 times 11 s. are 88 s., which, added to the 4 s., equals 92 s., which we find by reduction equals £4 and 12 s.; we write the 12 s. under the shillings, and reserve the £4 to add to the next product; 8 times £12 are £96, plus the £4, equals £100, which we write under the pounds. Hence the following

Rule.—I. *Write the multiplier under the lowest denomination of the multiplicand.*

II. *Begin with the lowest denomination, and multiply each term in succession as in simple numbers, reducing as in addition of compound numbers.*

Proof.—The same as in multiplication of simple numbers

NOTE.—If the multiplier is a large composite number, it will be more convenient to multiply by its factors.

WRITTEN EXERCISES.

(2)

cwt.	lb.	oz.
18	96	9
		5

(3)

lb.	oz.	pwt.	gr.
16	8	15	17
			3

(4)

mo.	da.	h.	min.	sec.
50	10	20	30	40
				7

(5)

£	s.	d.	far.
13	12	9	2
			8

(6)

hhd.	gal.	qt.	pt.
21	35	3	1
			9

(7)

℔.	℥	ʒ	℈	gr.
12	8	7	2	20
				11

8. Multiply 12 L. 2 mi. 232 rd. by 12.

Ans. 154 L. 2 mi. 224 rd.

9. Multiply 23 ch. (36 bu.) 18 bu. 2 pk. 7 qt. 1 pt. by 13.

Ans. 305 ch. 27 bu. 2 pk. 1 qt. 1 pt.

10. A farmer sold 5 loads of hay, each containing 15 cwt. 90 lb.; how much did he sell? *Ans.* 79 cwt. 50 lb.

11. Multiply 13 yr. 10 mo. 3 wk. 5 da. by 15, using the factors of the multiplier. *Ans.* 208 yr. 7 mo. 3 wk. 5 da.

12. If a man walk 17 mi. 300 rd. in each of 21 days, how far will he walk in all? *Ans.* 376 mi. 220 rd.

13. If a farmer raise 60 bu. 3 pk. 6 qt. 1 pt. of grain on one acre, how much can he raise at the same rate on 48 acres?
Ans. 2925 bu. 3 pk.

14. If a pipe discharges 11 hhd. 40 gal. 2 qt. 1 pt. of water in an hour, how much will it discharge in 56 hours?
Ans. 652 hhd. 7 gal.

15. A had 1000 A. of land; he sold B 96 A. 150 P., and C 4 times as much; how much remained? *Ans.* 515 A. 50 P.

16. A farmer raised 4000 bu. of grain; he sold 50 bu. 3 pk. 7 qt. to A, 7 times as much to B, and to C 6 times as much as to A and B together; how much remained?
Ans. 1145 bu. 3 pk.

DIVISION OF COMPOUND NUMBERS.

309. Division of Compound Numbers is the process of finding the quotient when the dividend is a compound number.

310. There are two cases:—

1st. To divide a compound number into equal parts.

2d. To divide one compound number by a similar one.

CASE I.

311. ***To divide a compound number into a number of equal parts.***

1. Divide £103 7 s. 6 d. into 5 equal parts, that is, take $\frac{1}{5}$ of it.

SOLUTION.—We write the divisor at the left of the dividend, and begin at the highest denomination to divide. $\frac{1}{5}$ of £103 is £20 and £3 remaining; £3 equal 60 s., which added to 7 s. equals 67 s.; $\frac{1}{5}$ of 67 s. is 13 s. and 2 s. remaining; 2 s. equal 24 d., which added to 6 d. equals 30 d.; $\frac{1}{5}$ of 30 d. is 6 d. Hence the following

OPERATION.

	£	s.	d.
5)	103	7	6
	20	13	6

Rule.—I. *Begin with the highest denomination of the dividend and divide each term in succession, as in simple numbers*

II. *When there is a remainder, reduce it to the next lower denomination, add it to the term of that denomination, and divide the result as before.*

III *Proceed in the same manner until all the terms are divided.*

Proof.—The same as in division of simple numbers.

NOTE.—When the divisor is a *large number* and *composite*, the factors being not greater than 12, it is perhaps more convenient to divide by the factors.

WRITTEN EXERCISES.

(2)

	£	s.	d.
4)	61	18	4
	15	9	7

(3)

	lb.	oz.	pwt.	gr.
6)	76	10	14	12

(4)

	T.	cwt.	lb.
7)	112	16	66
	16	2	38

(5)

	cwt.	lb.	oz.
8)	125	94	12

(6)

	hhd.	gal.	qt.	pt.	gi.
9)	108	42	2	1	2

(7)

	mi.	rd.	yd.	ft.
11)	120	313	3	2

8. Five sons share 112 A. 144 P. 24 sq. yd. of land; how much does each receive? *Ans.* 22 A. 92 P. 29 sq. yd.

9. If 9 farmers raise 1137 bu. 3 pk. 4 qt. 1 pt. of grain, what is the average amount raised by each? *Ans.* 126 bu. 1 pk. 5 qt. $1\frac{2}{3}$ pt.

10. A miner sends 37 lb. 10 oz. 17 pwt. 16 gr. of gold to his 8 sisters; how much does each receive? *Ans.* 4 lb. 8 oz. 17 pwt. 5 gr.

11. A man walked 376 mi. 276 rd. in 22 days; what was the average distance each day? *Ans.* 17 mi. $41\frac{7}{11}$ rd.

12. If 26 casks contain 21 hhd. 11 gal. 2 qt. 1 pt., what is the capacity of each cask? *Ans.* 51 gal. 1 qt. $\frac{17}{26}$ pt.

CASE II.

312. ***To divide one compound number by a similar one.***

1. Divide £26 6 s. 2 d. by £4 15 s. 8 d.

SOLUTION.—£26 6 s. 2 d. equals 6314 pence; £4 15 s. 8 d. equals 1148 pence; and dividing 6314 d. by 1148 d. we obtain a quotient of 5½. From this solution we have the following

OPERATION.

£26 6 s. 2 d. = 6314 d.
£4 15 s. 8 d. = 1148 d.
1148)6314(5½, *Ans.*
6314

Rule.—*Reduce both dividend and divisor to the lowest denomination mentioned in either, and then divide as in simple numbers.*

Proof.—The same as in division of simple numbers.

NOTE.—The division may also be made without reducing to the lowest denomination, and this will be shorter when the quotient is integral.

WRITTEN EXERCISES.

2. Divide £48 7 s. 4 d. by £6 11 d. *Ans.* 8.

3. 69 bu. 3 pk. 6 qt. by 6 bu. 3 pk. 6 qt. *Ans.* $10\frac{8}{87}$.

4. If a man feeds his horse 1 pk. 6 qt. of oats a day, how long will 3 bu. 2 qt. last him? *Ans.* 7 days.

5. How many demijohns, each containing 2 gal. 3 qt. 1 pt., can be filled from a tank holding 71 gal. 3 qt. 1 pt. of wine? *Ans.* 25.

6. A drove of cattle ate 6 T. 15 cwt. 87 lb. of hay in a week; how long will 33 T. 19 cwt. 35 lb. last them?

Ans. 5 weeks.

DIFFERENCE BETWEEN DATES.

CASE I.

313. ***To find the difference of time between two dates.***

1. Washington was born Feb. 22d, 1732, and died Dec. 14th, 1799; what was his age?

SOLUTION.—Dates are expressed in the *number* of the year, the month, and the day; hence the date of his birth is 1732 yr. 2 mo. 22 da., and the date of his death is 1799 yr. 12 mo. 14 da.; and the difference of these two dates will equal his age, which we find to be 67 yr. 9 mo. 22 da.

OPERATION.

yr.	mo.	da.
1799	12	14
1732	2	22
67	9	22

Rule.—*Write the number of the year, month, and day of the earlier date under the year, month, and day of the later date, and take the difference of the numbers.*

NOTE.—In this method we reckon 30 days to the month; when greater accuracy is required, we reckon the actual number of days in each month. The exact time between two dates is found by the table, Art. 296.

WRITTEN EXERCISES.

2. What is the difference in time from Dec. 12th, 1850, to Jan. 5th, 1860? *Ans.* 9 yr. 23 da.

3. Milton was born Dec. 9th, 1608, and died Nov. 8th 1675; what was his age? *Ans.* 66 yr. 10 mo. 29 da.

4. Franklin was born Jan. 6th, 1706, and died April 17th, 1790; required his age. *Ans.* 84 yr. 3 mo. 11 da.

5. What is the exact time a note has to run, dated Dec 30th, 1862, and payable Jan. 16th, 1864? *Ans.* 1 yr. 17 da.

6. The Revolution was commenced the 19th of April, 1775, and terminated January 20th, 1783; how long did it continue? *Ans.* 7 yr. 9 mo. 1 da.

7. The Civil War began April 11, 1861, and closed April 9, 1865; how long did it last? *Ans.* 3 yr. 11 mo. 28 da.

8. How many years, months, and days, from your birthday to the present date?

9. A man was born the 29th of Feb., 1824, and died the 3d of March, 1860; how many birthdays did he see, and what was his exact age? *Ans.* 10; Age, 36 yr. 4 da.

10. Two men were born on the first day of January, one in the year 1756, the other in 1822; both died 37 years after they were born, upon the 1st of January; required the difference of their ages? *Ans.* 1 day.

11. What is the exact time from 45 min. and 25 sec past 12 o'clock noon, May 20, 1874, to 24 min. 36 sec. past 4 o'clock, P. M., July 4, 1876?

Ans. 2 yr. 1 mo. 15 da. 3 h. 39 min. 11 sec.

CASE II.

314. ***To find the day of the week upon which any given day of the month will fall, the day of the week of some other date being given.***

NOTE.—A common year ends on the same day it begins, hence each year begins one day later than the preceding year. A year following leap year begins two days later.

1. If the 3d of April, 1876, occur on Monday, on what day of the week will the next 29th of August be?

OPERATION.

148÷7=21,+1.

SOLUTION.—By the table we find the difference of time to be 148 days: dividing by 7, the number of days in a week, we have 21 weeks and 1 day; the 29th of August must therefore be one day after Monday or Tuesday.

Rule.—*Find the number of days between the two dates, reduce this number to weeks; the number of days remaining will be the number of days from the given day of the week to the required day.*

NOTE.—In solving the problems after the 2d, the date in a previous problem may be assumed as the starting-point.

WRITTEN EXERCISES.

2. If the year 1776 began on Monday, on what day of the week was the 4th of July following? *Ans.* Thursday.

3. What day of the week was the surrender of Burgoyne, Oct. 17, 1777? *Ans.* Friday.

4. On what day of the week was Daniel Webster born, Jan. 18, 1782? *Ans.* Friday.

5. On what day of the week was the battle of Waterloo fought, June 18, 1815? *Ans.* Sunday.

6. On what day of the week was President Lincoln assassinated, April 14, 1865? *Ans.* Friday.

7. On what day of the week was General Grant inaugurated the first time, March 4, 1869? *Ans.* Thursday.

8. The Centennial opened on Wednesday, May 10, 1876; if the next Centennial opens on the same date, on what day will it open? *Ans.* Monday.

9. Suppose a letter should be published purporting to have been written at New Orleans on the day of the battle, and dated Wednesday, Jan. 8, 1815; would it be genuine?

10. Let the pupils now determine, from the above principles, the day of the week upon which they were born.

LATITUDE AND LONGITUDE.

315. The **Latitude** of a place is its distance from the equator, north or south. It is reckoned in degrees, minutes, and seconds, and cannot exceed 90°, or a quadrant.

316. The **Longitude** of a place is its distance, east or west, from a given meridian. It is reckoned in degrees, minutes, and seconds, and cannot exceed 180°, or a semicircumference.

NOTE.—In adding two longitudes, if their sum exceed 180 degrees, it must be subtracted from 360 degrees for the correct difference of longitude.

MENTAL EXERCISES.

1. From what is latitude reckoned? From what is longitude reckoned?

2. What is the greatest latitude a place may have? What is the greatest longitude a place may have?

3. What places have no latitude? What places have no longitude? What place has neither latitude nor longitude?

4. What is the use of latitude and longitude? Has every place a meridian of longitude?

5. What is the latitude of the equator? the latitude of the poles? the longitude of the poles?

317. From the above principles, to find the difference of latitude or longitude, we have the following rule:

Rule.—*When the latitudes or longitudes are both of the same name, subtract the less from the greater; when they are of different names, take their sum.*

WRITTEN EXERCISES.

1. The latitude of Richmond, Va., is 37° 20′ north, and of Savannah 32° 4′ 56″ north; what is the difference of latitude? *Ans.* 5° 15′ 4″.

2. The latitude of Charleston, S. C., is 32° 46′ 33″ north, and of Quito, 13′ 27″ south; what is the difference of latitude? *Ans.* 33°.

3. The longitude of Portland is 70° 13′ 34″ W., and of Mobile, 88° 1′ 29″ W.; what is the difference of longitude? *Ans.* 17° 47′ 55″.

4. The longitude of New Orleans is 90° W. and of Geneva 6° 9′ 5″ E.; what is the difference of longitude? *Ans.* 96° 9′ 5″.

5. The longitude of St. Paul, Minn., is 95° 4′ 55″ W., and of Berlin 13° 23′ 45″ E.; what is the difference of longitude? *Ans.* 108° 28′ 40″.

6. The longitude of Paris is 2° 20′ E., and of New York 74° 3′ W.; what is the difference of longitude? *Ans.* 76° 23′.

LONGITUDE AND TIME

318. The earth revolves on its axis from west to east once in 24 hours, and this causes the sun to appear to revolve around the earth from east to west in the same time. Places on the east of a certain point have later time, those on the west earlier time, since the sun appears to those on the east first.

319. The circumference of a circle contains 360°, hence the sun appears to travel through 360° in 24 hours, and in 1 hour it travels $\frac{1}{24}$ of 360°=15°; in 1 minute it travels $\frac{1}{60}$ of 15°=15′, and in 1 second it travels $\frac{1}{60}$ of 15′=15″. Hence the following table:

TABLE OF LONGITUDE AND TIME.

15° of longitude	=	1 hour of time.
15′ of "	=	1 minute of time.
15″ of "	=	1 second of time.

MENTAL EXERCISES.

1. In what time does the earth revolve on its axis? What part of a revolution does it make in 12 hours? in 6 hours?

2. How many degrees of the earth's surface pass under the sun's rays in 24 h.? in 12 h.? in 4 h.?

3. How many degrees of longitude make a difference of 1 hour in time? 2 hours? 3 hours? 4 hours?

4. In what direction does the earth turn on its axis? In what direction does the sun appear to move?

5. Does the sun appear first to places east or west of a given point?

6. When it is noon with us, is it earlier or later east of us? west of us?

7. When it is noon at Boston, what is the time 15° east of Boston? 15° west? 30° east? 30° west?

8. What difference in longitude makes a difference of 1 hour of time? of 1 minute? of 1 second?

9. What is the difference of longitude between two cities, if the difference of time is 1 hour? 1 h. 30 min.? 2 h.? 2 h. 45 min.?

10. If I start at New York and travel until my watch is 1 h. 30 min. too fast, in what direction and how far do I go?

11. If I start at Chicago and travel until my watch is 2 h. 15 min. too slow, how far and in what direction do I travel?

CASE I.

320. ***To find the difference of time of two places when their difference of longitude is given.***

1. The difference of longitude between two places is 40°; what is their difference of time?

SOLUTION.—Since 15° of longitude correspond to 1 h. of time, and 15′ of longitude to 1 min. of time, $\frac{1}{15}$ of the number of *degrees* and *minutes* will equal the number of *hours* and *minutes* difference in time. Dividing by 15 we have 2 h. 40 min. Hence the following

OPERATION.

$$\begin{array}{r} 15)\underline{40° \quad 0'} \\ 2 \quad 40 \end{array}$$

Rule.—*Divide the difference of longitude expressed in* ° ′ ″ *by* 15*; the result will be the difference of time in* H. MIN. SEC.

WRITTEN EXERCISES.

2. The difference of longitude of two places is 35°; what is their difference of time? *Ans.* 2 h. 20 min.

3. The longitude of Portsmouth is 70° 45′, and of Washington 77° 0′ 15″; required their difference of time.

Ans. 25 min. 1 sec.

4. The long. of New York is 74° 3′ west, and of New Orleans 90° west; required the difference in time.

Ans. 1 h. 3 min. 48 sec.

5. The longitude of Philadelphia is 75° 9′ 5″ west, and of Cincinnati 84° 29′ 31″ west; what is the time at Cincinnati when it is 10 A. M. at Philadelphia?

Ans. 22 min. $38\frac{4}{15}$ sec. past 9 A. M.

6. Vienna is 16° 23′ east longitude; what is the time there at 9 A. M. in Philadelphia?

Ans. 3 h. 6 min. $8\frac{1}{3}$ sec. P. M.

7. The long. of Washington is 76° 56′ west of London; what change must we make in our watches in coming from London to Washington? *Ans.* Set back 5 h. 7 min. 44 sec.

8. Paris is about 2° 20′ east longitude; what change would a person have to make in his watch in going from New York to Paris? *Ans.* Set ahead 5 h. 5 min. 32 sec.

9. The longitude of Rome is 12° 27′ east, and San Francisco 122° 26′ 15″ west; what time is it in the latter place when it is 4 P. M. in the former? *Ans.* 27 sec. past 7 A. M.

CASE II.

321. ***To find the difference of longitude of two places when their difference of time is given.***

1. The difference of time between two places is 26 minutes; what is their difference of longitude?

SOLUTION.—Since 1 h. of time corresponds to 15° of longitude, and 1 min. of time to 15′ of longitude, 15 times the number of *hours* and *minutes* difference in time will equal the number of *degrees* and *minutes* difference in longitude. Multiplying by 15 we have 6° 30′. Hence the following

OPERATION.

h.	min.
0	26
	15
6°	30′

Rule.—*Multiply the difference of time expressed in* H. MIN. SEC. *by* 15; *the result will be the difference of longitude in* ° ′ ″.

WRITTEN EXERCISES.

2. The difference of time between Philadelphia and Cincinnati is about 37 min. 20 sec.; what is the difference of longitude? *Ans.* 9° 20′.

3. The time at St. Louis is about 53 minutes earlier than the time at Washington; what is the difference in longitude? *Ans.* 13° 15′.

4. When it is noon at London it is about 7 o'clock, A. M., in Philadelphia; required the difference of longitude. *Ans.* About 75°.

5. In traveling from New York to Cincinnati I find my watch is 41 min. 32 sec. too fast; required the difference of longitude. *Ans.* 10° 23′.

6. In coming from San Francisco to Philadelphia I find my watch is 3 h. 9 min. $8\frac{2}{3}$ sec. too slow; what is the longitude of San Francisco, that of Philadelphia being 75° 9′ 5″? *Ans.* 122° 26′ 15″.

7. The longitude of Cambridge, England, is 5′ 21″ east, and the difference of time between it and Cambridge, Mass., is 4 h. 44 min. $50\frac{4}{5}$ sec.; required the longitude of the latter place. *Ans.* 71° 7′ 21″ west.

8. The longitude of New York is 74° 3′ west, and of Jerusalem is 35° 32′ east; when it is $4\frac{1}{2}$ o'clock, A. M., at New York, what is the time at Jerusalem? *Ans.* 48 min. 20 sec. past 11 A. M.

DENOMINATE FRACTIONS.

322. A **Denominate Fraction** is one in which the unit of the fraction is denominate; as, $\frac{2}{3}$ of a pound.

323. **Denominate Fractions** may be expressed either as *common* fractions or as *decimals*.

REDUCTION OF DENOMINATE FRACTIONS.

324. **Reduction of Denominate Fractions** is the process of changing them from one denomination to another without altering their value.

325. There are **two general cases,** reduction ascending and descending, which, for convenience of operation, are subdivided into several other cases.

REDUCTION DESCENDING.

CASE I.

326. ***To reduce a common denominate fraction to a fraction of a lower denomination.***

1. Reduce $\frac{1}{96}$ of a shilling to farthings.

OPERATION.

$$\frac{1}{96}\times\frac{12}{1}\times\frac{4}{1}=\frac{1}{2}\text{ far.}$$

SOLUTION.—Since there are 12 pence in one shilling, 12 times the number of shillings equals the number of pence; and since there are 4 farthings in 1 penny, 4 times the number of pence equals the number of farthings; hence $\frac{1}{96}$ of a shilling equals $\frac{1}{96}\times\frac{12}{1}\times\frac{4}{1}$ farthings, which by cancelling and multiplying becomes $\frac{1}{2}$ of a farthing. Therefore, etc.

Rule.—*Express the multiplication by the required multipliers, and reduce by cancellation.*

WRITTEN EXERCISES.

Reduce

2. $\frac{1}{160}$ of a bu. to the fraction of a pint. *Ans.* $\frac{2}{5}$.
3. $\frac{1}{1440}$ of an oz. to the fraction of a grain. *Ans.* $\frac{1}{3}$.
4. $\frac{3}{1920}$ of a day to the fraction of a minute. *Ans.* $2\frac{1}{4}$.
5. $\frac{7}{288}$ of a gal. to the fraction of a gill. *Ans.* $\frac{7}{9}$.
6. $\frac{5}{1386}$ of a rod to the fraction of an inch. *Ans.* $\frac{5}{7}$.
7. $\frac{3}{22400}$ of a ton to the fraction of an ounce. *Ans.* $4\frac{2}{7}$.
8. $\frac{5}{64152}$ of a sq. rd. to the fraction of a sq. in. *Ans.* $3\frac{1}{18}$.
9. $\frac{1}{18720}$ of a mile to the fraction of an inch. *Ans.* $3\frac{5}{13}$.

CASE II.

327. ***To reduce a common denominate fraction to integers of lower denomination.***

1. What is the value of $\frac{5}{9}$ of a pound Troy?

SOLUTION.—There are 12 oz. in one pound, hence 12 times the number of pounds equals the number of ounces; 12 times $\frac{5}{9}$ equals $\frac{60}{9}$ or $6\frac{2}{3}$ oz.: there are 20 pwt. in one oz., therefore 20 times the number of oz. equals the number of pwt.; 20 times $\frac{2}{3}$ equals $\frac{40}{3}$, or $13\frac{1}{3}$ pwt., etc.

OPERATION.

$\frac{5}{9}\times12=\frac{60}{9}=6\frac{2}{3}$ oz.
$\frac{2}{3}\times20=\frac{40}{3}=13\frac{1}{3}$ pwt.
$\frac{1}{3}\times24=8$ gr.

SOLUTION 2d.—$\frac{5}{9}$ of a pound equals $\frac{1}{9}$ of 5 lb.; and $\frac{1}{9}$ of 5 lb. we find by dividing is 6 oz. 13 pwt. 8 gr.

OPERATION.

lb.	oz.	pwt.	gr.
9)5	0	0	0
	6	13	8

Rule I.—*Reduce the fraction until we reach an integer and a fraction of a lower denomination, set aside the integer and reduce the fraction as before, and thus continue as far as necessary.*

Rule II.—*Regard the numerator as so many units of the given denomination, and divide by the denominator.*

WRITTEN EXERCISES.

What is the value

2. Of $\frac{5}{6}$ of an ounce? *Ans.* 16 pwt. 16 gr.

3. Of $\frac{4}{5}$ of a bushel? *Ans.* 3 pk. 1 qt. $1\frac{1}{5}$ pt.

4. Of $\frac{2}{3}$ of a mile? *Ans.* 213 rd. 1 yd. $2\frac{1}{2}$ ft.

5. Of $\frac{7}{8}$ of a rod? *Ans.* 4 yd. 2 ft. $5\frac{1}{4}$ in.

6. Of $\frac{9}{10}$ of a mile? *Ans.* 288 rd.

7. Of $\frac{7}{8}$ of a sign? *Ans.* 26° 15′.

8. Of $\frac{5}{7}$ of an acre? *Ans.* 114 P. 8 yd. 5 ft. $113\frac{1}{7}$ in.

CASE III.

328. ***To reduce a denominate decimal to integers of lower denominations.***

1. Reduce .875 gal. to integers of lower denominations.

SOLUTION.—There are 4 quarts in one gallon, therefore 4 times the number of gallons equals the number of quarts; 4 times .875 equals 3 qt. and .5 qt.; there are 2 pints in one quart, therefore 2 times the number of quarts equals the number of pints: 2 times .5 equals 1 pt. Therefore .875 gal. equals 3 qt. 1 pt.

OPERATION

$$\begin{array}{r} .875 \\ 4 \\ \hline 3.500 \\ 2 \\ \hline 1.000 \end{array}$$

Rule.—*Reduce the decimal until we reach an integer and a decimal of a lower denomination, set aside the integer, and reduce the decimal as before, and thus continue as far as necessary.*

WRITTEN EXERCISES.

What is the value

2. Of .825 of a pound Troy? *Ans.* 9 oz. 18 pwt.
3. Of .675 of a rod? *Ans.* 3 yd. $2\frac{11}{80}$ ft.
4. Of .364 of an acre? *Ans.* 58 P. $7\frac{13}{50}$ sq. yd.
5. Of .3275 of a hogshead? *Ans.* 20 gal. 2 qt. $1\frac{3}{50}$ pt.
6. Of .9735 of a bushel? *Ans.* 3 pk. $7\frac{19}{125}$ qt.
7. Of .3218 of a ton of iron? *Ans.* 6 cwt. 43 lb. 9.6 oz.
8. Of 2.1365 of a tun (4 hhd.) of wine? *Ans.* 2 tuns, 34 gal. 1 qt. 1 pt+.

REDUCTION ASCENDING.

CASE I.

329. ***To reduce a common denominate fraction to a common fraction of a higher denomination.***

1. Reduce $\frac{1}{2}$ of a farthing to the fraction of a shilling.

OPERATION.

$\frac{1}{2}\times\frac{1}{4}\times\frac{1}{12}=\frac{1}{96}$ s.

SOLUTION.—There are 4 farthings in a penny, therefore $\frac{1}{4}$ of the number of farthings equals the number of pence: there are 12 pence in 1 shilling, therefore $\frac{1}{12}$ of the number of pence equals the number of shillings; hence $\frac{1}{2}$ far. equals $\frac{1}{12}\times\frac{1}{4}\times\frac{1}{2}=\frac{1}{96}$ of a shilling.

Rule.—*Express the division by the required divisors, and reduce by cancellation.*

WRITTEN EXERCISES.

Reduce

2. $\frac{3}{4}$ of a grain to the fraction of an ounce. *Ans.* $\frac{1}{640}$.
3. $\frac{9}{10}$ of a foot to the fraction of a mile. *Ans.* $\frac{3}{17600}$.
4. $\frac{8}{9}$ of a gill to the fraction of a gallon. *Ans.* $\frac{1}{36}$.
5. $\frac{5}{7}$ of an inch to the fraction of a rod. *Ans.* $\frac{5}{1386}$.
6. $\frac{7}{8}$ of a lb. to the fraction of a ton. *Ans.* $\frac{7}{16000}$.
7. $4\frac{2}{7}$ oz. to the fraction of a ton. *Ans.* $\frac{3}{22400}$.
8. What part of a cord of wood is a pile containing 48 cubic feet? *Ans.* $\frac{3}{8}$.

CASE II.

330. ***To reduce a compound number to a common fraction of a higher denomination.***

1. Reduce 3 s. 6 d. 2 far. to the fraction of a pound.

SOLUTION.—By reduction we find 3 s. 6 d. 2 far. equal to 170 far., and also £1=960 far.; one farthing is $\frac{1}{960}$ of a pound, and 170 far. equals 170 times $\frac{1}{960}=\frac{170}{960}$, which reduced to its lowest terms, equals $\frac{17}{96}$. Therefore, etc.

OPERATION.

3 s. 6 d. 2 far. = 170 far.
£1 = 960 far.
$\frac{170}{960}=\frac{17}{96}$, *Ans.*

Rule.—*Reduce the number to its lowest denomination, and write under it the number of units of this denomination which make a unit of the required denomination; and then reduce the resulting fraction to its lowest terms.*

WRITTEN EXERCISES.

2. Reduce 3 oz. 8 pwt. 12 gr. to pounds. *Ans.* $\frac{137}{480}$.

3. What part of 2 bushels is 1 bu. 4 qt.? *Ans.* $\frac{9}{16}$.

4. Of a bar. of beer (36 gal.) is 22 gal. 2 qt.? *Ans.* $\frac{5}{8}$.

5. Of a barrel of wine is 4 gal. 3 qt. 1 pt.? *Ans.* $\frac{13}{84}$.

6. Of 5 ℥. is 2ʒ 1℈ 10 gr.? *Ans.* $\frac{1}{16}$.

7. Of 3 years is 3 wk. 6 da. 20 h.? *Ans.* $\frac{167}{6570}$.

8. What part of 4 inches square is 4 square inches? What part of a $2\frac{1}{2}$ inch cube is $2\frac{1}{2}$ cu. in.? *Ans.* $\frac{1}{4}$; $\frac{4}{25}$.

CASE III.

331. ***To reduce a compound number to a decimal of a higher denomination.***

1. Reduce 1 bu. 2 pk. 4 qt. to the decimal of a bushel.

SOLUTION.—There are 8 qt. in 1 pk., hence $\frac{1}{8}$ of the number of quarts equals the number of pecks; $\frac{1}{8}$ of 4 equals .5, which, with 2 pk. equals 2.5 pk.; there are 4 pk. in a bushel, hence $\frac{1}{4}$ of the number of pecks equals the number of bushels; $\frac{1}{4}$ of 2.5 equals .625, which, with 1 bu., equals 1.625 bu.; hence 1 bu. 2 pk. 4 qt. equals 1.625 bu.

OPERATION.

8 | 4
4 | 2.5
  | 1.625 *Ans.*

Rule.—I. *Divide the lowest term by the number of units which equals one of the next higher, and annex the decimal quotient to the integer of the next higher denomination.*

II. *Proceed in a similar manner until the whole is reduced to the required denomination.*

Note.—It may also be done by reducing to a common fraction, and the common fraction to a decimal.

WRITTEN EXERCISES.

Reduce

2. 5 oz. 10 pwt. 12 gr. to the decimal of a lb.
Ans. .4604166 lb.+.

3. 3 pk. 4 qt. 1 pt. to the decimal of a bu.
Ans. .890625 bu.

4. 9℥ 1ʒ 2℈ 8 gr. to the decimal of a ℔. *Ans.* .76875 ℔.

5. 286 rd. 1 yd. 3$\frac{3}{5}$ in. to the decimal of a mile.
Ans. .894375 mi.

6. What decimal part of 5 gal. is 1 qt. 3$\frac{3}{5}$ gi.?
Ans. .0725.

7. Reduce 295 rd. 3 yd. 2 ft. 8 in. to the decimal of a mile.
Ans. .92408459 mi.+.

8. Reduce 91 lb. 12$\frac{7}{8}$ oz. to the decimal of a ton.
Ans. .04590234375 ton.

MENTAL EXERCISES.

MISCELLANEOUS EXAMPLES.

1. How many seconds are there in the circumference of a carriage wheel?

2. A lady put up 50 quart cans of strawberries; how many pecks did she put up?

3. When the sun has seemed to pass over 4½ signs, how many seconds has he appeared to move?

4. How many pages of an octavo volume can be printed on 3 quires of paper?

5. A hostler at a hotel stable fed out 48 half-peck measures of oats; how many bushels did he feed?

6. Sarah picked a peck of cherries, and sold them at 4 cents a pint; how much did she receive for them?

7. A sailor took the sounding, and found the water 75 feet deep; how many fathoms was it?

8. My grandfather's age is "three-score years and ten;" how many years old is he?

9. A merchant buys ½ dozen hand-saws, at $20 a dozen; how shall he sell them to gain 50 cents apiece?

10. I paid $3.20 for a bushel of huckleberries; how shall I sell them to gain 2 cents a quart?

11. A merchant bought $\frac{1}{2}$ dozen shovels, at $9.60 a dozen; how shall he retail them to gain 20 cents apiece?

12. What must I pay a gross for hair-pins, that I may sell them for 5 cents a dozen, and gain 2 cents a dozen?

13. How many cubic inches in a 2-inch cube? in a 4-inch cube? in a 6-inch cube?

14. In 6 francs how many centimes? How many in $8\frac{1}{2}$ francs? in $10\frac{1}{4}$ francs?

15. How many leap years in every century? At what time did the 19th century begin?

16. What part of a circumference is 90°? 60°? 180°? Number of degrees in a quadrant?

17. A grocer paid $6.40 for a bushel of cranberries; how shall he sell them a quart to gain 2 cents a pint?

18. A huckster woman sold one day a bushel of pea-nuts by the half-pint glass, at 5 cents a glass; how much did she receive?

19. A lady bought 2 pairs of vases, 3 sets of chairs, and a dozen knives and forks; how many individual articles did she buy?

20. If Henry Clay spoke 90 words in a minute, how many words would he speak in an address of one hour?

21. Bought paper collars at 40 cents a box, each box containing a dozen; how much will I gain on each collar by selling them for 5 cents apiece?

WRITTEN EXERCISES.

MISCELLANEOUS PROBLEMS.

1. If 8 barrels of flour cost £24 12 s. 8 d., what will 12 barrels cost at the same rate? *Ans.* £36 19 s.

2. If 7 bales of goods weigh 20 cwt. 75 lb., what will 56 bales of the same size weigh? *Ans.* 166 cwt.

3. What cost $8\frac{2}{3}$ cords of wood, at the rate of £11 5 d. for $5\frac{3}{4}$ cords? *Ans.* £16 12 s. $2\frac{2}{3}$ d.

4. What cost $8\frac{3}{4}$ yd. of cloth, at the rate of $4.50 per yard? *Ans.* 39.37\frac{1}{2}$.

5. Multiply $8\frac{1}{2}$ d. by $6.33\frac{1}{3}$, and from the product take 2 s. 6 d. *Ans.* 1 s. $11\frac{5}{6}$ d.

6. Add $.00\frac{7}{8}$ sq. yd., $.04\frac{3}{4}$ sq. ft., and .0008 sq. in. *Ans.* 18.1808 sq. in

7. Reduce 1 shilling, 11 pence and 1.0384 farthings to the decimal of a guinea. *Ans.* .0923 guinea.

8. How many cubic feet in 75.125 cords? How many cubic inches in the same? *Ans.* 9616 cu. ft.

9. How much gold may be obtained from a ton of quartz rock, if it yields .0016 of its weight in gold? *Ans.* 3.2 lb.

10. How much is the cost of 24 cwt. 87 lb. of sugar, at $6.50 per hundredweight? *Ans.* $161.65½.

11. A man bought 4 hhd. 28 gal. 3 qt. of wine at $4.50 a gallon; what did it cost? *Ans.* $1263.375.

12. A grocer shipped 6120 eggs to Philadelphia in 6 barrels; how many did he pack in a barrel? *Ans.* 85 doz.

13. An apothecary bought 16 lb. 10½ oz. of drugs, at $12.25 a pound; required the cost. *Ans.* $204.039.

14. Since 9 o'clock the sun has seemed to pass over 4° 23′ 24″; what time is it? *Ans.* 9 h. 17 min. 33⅗ sec.

15. A druggist purchased 20 lb. 8¾ oz. of opium at 48¾ cents an ounce; what did it cost? *Ans.* $160.265+.

16. What is the weight of $1,000,000 in gold dollars at 25.8 gr. each? What is the weight in silver half-dollars at 192.9 gr. each? *Ans.* 4479 lb. 2 oz.; 66979 lb. 2 oz.

17. If I start at St. Louis, latitude 38° 37′ 28″ N., and travel due north 1800 miles; what latitude do I reach?

Ans. 64° 39′ 3″+.

18. What cost 31 lb. 14 oz. of drugs, if 6 lb. 6 oz. cost $31.40? *Ans.* $157.

19. What cost 5 cwt. 65 lb. of sugar, if .96 of a cwt. cost $7.50? *Ans* 44.14\frac{1}{16}$.

20. What cost 10 cwt. 81 lb. of hay, if 5 cwt. 55 lb. cost £2 10 s. 6 d.? *Ans.* £4 18 s. 4 d.+.

21. If £124 16 s. 6 d. are worth $599.16, how many dollars are £136 10 s. 6 d. worth? *Ans.* $655.32.

22. In what time will a man walk 120 mi. 260 rd. 4 ft. if he goes 12 mi. in 3 h. 20 min.?

Ans. 33 h. 33 min. 33 sec.+.

23. If A travels 24 mi. 198 rd. 4 yd. in 6 h. 30 min., how far will he go in 9 h. 45 min.?

Ans. 36 mi. 298 rd. 1 ft. 6 in

24. How many centals of wheat are equivalent to 1200 bushels? How many centals in 150 bar.? *Ans.* 720; 294.

25. How many bushels of buckwheat in Kentucky are equivalent to 520 bu. in Illinois? To 650 bu. in Pennsylvania? *Ans.* 400; 600.

26. A merchant bought in Connecticut 32 bushels of oats at 2¢ a pound, and sold them in New York at 80¢ a bushel; what was his profit? *Ans.* $4.48.

27. If B digs 363 rd. 7 yd. of ditch in 35 wk. 5 da., how long will it take to dig 910 rd. $3\frac{3}{4}$ yd., working 12 h. a day, 6 da. a week, and 4 wk. a month? *Ans.* 22 mo. 1 wk. $3\frac{1}{2}$ da.

28. If a river current carries a raft of lumber at the rate of 4 mi. $265\frac{3}{4}$ rd. per hour, how long will it be in taking it 346 mi. $28.90\frac{10}{11}$ rd.? *Ans.* 2 da. 23 h. 38 min. $50\frac{1}{2}$ sec. +.

29. Mr. Owen sold 15 bu. 3 pk. 4 qt. of apples at $2.75 a bushel, and took his pay in flour at $3\frac{1}{2}$¢ a pound, receiving only an exact number of barrels, and in sugar at 12¢ a pound for what remained; how many barrels of flour and pounds of sugar did he receive? *Ans.* 6 barrels; $20\frac{77}{96}$ lb.

30. Two men start from different places on the equator, and travel towards each other till they meet; on comparing their watches with the time of the place of meeting, it is found that the first is 45 minutes slow and the second 1 h. 15 min. fast; how far apart were the starting points and in what direction did each travel? *Ans.* 2074.8 mi.

31. The distance from a certain toll-gate east to a tavern is $3\frac{4}{11}$ miles; from the toll-gate west to a school-house is $45\frac{10}{33}$ rods; half-way between the tavern and the school-house is a creek 100 yards wide; how far from the toll-gate to the middle of the creek? to the further bank of the creek?

Ans. 1 mi. 195 rd. 2 yd. 2 ft. 9 in.; 1 mi. 204 rd. 3 yd. 1 ft. 3 in.

32. A balloon started from Paris with dispatches for Tours, and alighted near Bourges, 119 mi. $266.66\frac{2}{3}$ rods from Paris. Its actual route was $1\frac{1}{5}$ times this distance, which it made at the rate of 51 mi. 80 rd. an hour. Starting at 4 A. M., when did it alight? *Ans.* 6 h. 48 min. $21\frac{8}{41}$ sec. A. M.

SECTION VII

PRACTICAL MEASUREMENTS.

332. The **Applications of Measures** to the farm, the household, the mechanic arts, etc., are so extensive that we now present a distinct treatment of the subject.

333. These **Practical Measurements** include Measures of Surface, Measures of Volume, Measures of Capacity, and Comparison of Weights and of Money.

MEASURES OF SURFACE.

334. A **Surface** is that which has length and breadth without thickness.

THE RECTANGLE.

335. A **Rectangle** is a plane surface having four sides and four right angles. A slate, a door, the sides of a room, etc., are examples of rectangles.

336. A **Rectangle** has two *dimensions*, length and breadth. A *Square* is a rectangle in which the sides are all equal.

337. The **Area** of a rectangle is the surface included within its sides. It is expressed by the number of times it contains a small square as a *unit of measure*.

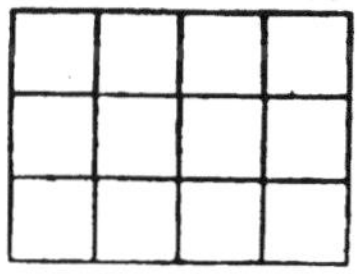

Rule I.—*To find the area of a square or rectangle, multiply its length by its breadth.*

For, in the rectangle above, the whole number of little squares is equal to the number in each row multiplied by the number of rows, which is equal to the number of linear units in the length multiplied by the number in the breadth.

Rule II.—*To find either side of a square or rectangle, divide the area by the other side.*

NOTES.—1. The sides multiplied must be of the *same denomination*, and the product will be *square units* of that denomination, which may be reduced, if necessary, to higher denominations.

2. In dividing, the *linear unit* of the side must be of the same name as the *square unit* of the area, and the quotient will be linear units of the same denomination.

WRITTEN EXERCISES.

1. How many square feet in a floor 32 ft. long by 21 ft. wide? how many square yards?

SOLUTION.—To find the area, we multiply the length by the breadth, and we have $32\times21=672$ sq. ft.; reducing this to square yards, we have $74\frac{2}{3}$ sq. yd.

2. How many square yards in the surface of a blackboard 27 ft. long by 4 ft. wide? *Ans.* 12 sq. yd.

3. How many square yards in a garden 215 ft. long by 109 ft. wide? *Ans.* $2603\frac{8}{9}$ sq. yd.

4. What is the width of a room 25 feet long, whose floor contains 500 sq. ft.? *Ans.* 20 ft.

5. A rectangle contains $466\frac{1}{8}$ sq. ft., and one side is 16 ft 6 in. long; how long is the other side? *Ans.* 28 ft. 3 in.

6. How many square feet in the sides of a room 18 ft long, 14 ft. 6 in. wide, and 9 ft. 6 in. high? *Ans.* $617\frac{1}{2}$.

7. A certain box is 3 ft. 6 in. long, 2 ft. 3 in. wide, and 1 ft. 4 in. high; how many square feet in its surface?
Ans. 31 sq. ft. 12 sq. in.

8. What is the surface of a cubical box, each of whose dimensions is 1 ft. 6 in.? *Ans.* $13\frac{1}{2}$ sq. ft.

THE TRIANGLE.

338. A **Triangle** is a plane surface having three sides and three angles; as, ABC.

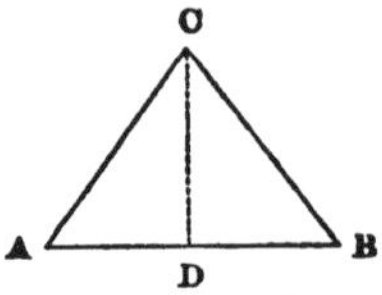

339. The **Base** is the side upon which it seems to stand, as AB. The **Altitude** is a line perpendicular to the base, drawn from the angle opposite; as, CD.

340. A triangle which has its three sides equal is called *equilateral;* when two sides are equal it is called *isosceles;* when its sides are unequal it is called *scalene.*

Rule I.—*To find the area of a triangle, multiply the base by one-half of the altitude.*

Rule II.—*To find the base or altitude of a triangle, divide the area by one-half the other dimension.*

WRITTEN EXERCISES

1. What is the area of a triangle whose base is 25 inches and altitude 18 inches?

SOLUTION.—To find the area, we multiply the base by one-half the altitude; $25 \times 9 = 225$; hence the area is 225 sq. in.

2. How many square feet in a triangle whose base is 18 ft. 6 in. and altitude 9 ft. 9 in.? *Ans.* 90 sq. ft. 27 sq. in.

3. What is the area of the gable end of a house 29 ft. wide the ridge being 12 ft. higher than the top of the wall? *Ans.* 174 sq. ft.

4. The area of a triangular bed of flowers is 25 sq. ft., and its base 10 ft.; what is the altitude? *Ans.* 5 ft.

5. The area of a triangular lot is 250 square yards, and its base is 250 ft.; what is its altitude? *Ans.* 18 ft.

6. The area of the gable of a house is 378 sq. ft., the base being 14 yards; what is the height of the ridge? *Ans.* 18 ft.

THE CIRCLE.

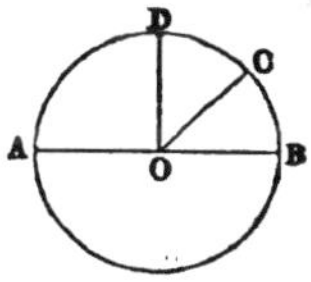

341. A **Circle** is a plane figure bounded by a curved line, every point of which is equally distant from a point within, called the *centre.*

342. The **Circumference** of a circle is the bounding line; any part of the circumference, as BC, is an *Arc.* An arc of one-fourth of the circumference is called a *Quadrant.*

343. The **Diameter** is a line passing through the centre and terminating in the circumference; as, AB. The *Radius* is a line drawn from the centre to the circumference; as, OD.

Rule I.—*To find the circumference of a circle, multiply the diameter by* 3.1416.

Rule II.—*To find the diameter of a circle, multiply the circumference by* .3183.

Rule III.—*To find the area of a circle, multiply the circumference by one-fourth of the diameter, or multiply the square of the radius by* 3.1416.

WRITTEN EXERCISES.

1. The diameter of a circle is $12\frac{1}{2}$ feet; what is its circumference?

Solution.—To find the circumference, we multiply the diameter by 3.1416; $3.1416 \times 12\frac{1}{2}$ equals 39.27; hence the circumference equals 39.27 ft.

2. What is the circumference of the planet Venus, its diameter being about 7800 miles? *Ans.* 24504.48 miles.

3. The distance round a circular pond is 500 feet; what is the distance across the pond? *Ans.* 159.15 ft.

4. How many times will a carriage wheel 4 ft. 6 in. in circumference revolve in driving 10 miles? *Ans.* $11733\frac{1}{3}$.

5. I have a circular flower-bed 50 feet in circumference; what is the area of the bed? *Ans.* 198 sq. ft. 135 sq. in.

6. A cow is fastened to a stake by a rope 16 feet long; what space can she graze over? *Ans.* 89.36+ sq. yd.

7. If the equatorial diameter of the earth is 7925.75 miles, what are its circumference and the length of a degree of longitude at the equator?

Ans. 24899.536+ miles; 69.16+ miles.

8. A circular flower-bed being divided into four equal parts by lines drawn from the centre, one section was planted with tulips; what was the area of the tulip-bed, its outer edge being 7 feet? *Ans.* 15.5967 sq. ft.

MEASUREMENT OF LAND.

344. The **Unit of Measure** of land is the *Acre*, which is sometimes divided into *square rods* and sometimes into *square chains*. Hundredths of an acre are also frequently used.

Government lands are divided by parallels and meridians into *townships*, which contain 36 square miles or *sections*, and each section is subdivided into *quarter-sections*. Hence, 640 acres make a *section*, and 160 acres a *quarter-section*. The quarter-sections are still further subdivided into *half-quarter-sections*, *quarter-quarter-sections*, and *lots*. Lots are often of irregular form on account of natural boundaries, but contain, as near as may be, a quarter-quarter-section.

NOTE.—The pupil will remember that *rods* multiplied by *rods* give *square rods*, *chains* by *chains* give *square chains;* also, that 1 acre = 10 square chains or 160 square rods.

WRITTEN EXERCISES.

1. How many square rods in a grass plat 65 ft. long and 15 ft. wide?

SOLUTION.—The area equals 65×15, or 975 sq. ft.; reducing to square rods, we have $3\frac{211}{363}$ sq. rd.

2. How many acres in a rectangular meadow 725 rods long and 400 rods wide? *Ans.* 1812 A. 80 P.

3. What is the value of a farm 208.7 rods long and 120 rods wide, at 81\frac{3}{4}$ an acre? *Ans.* 12795.91\frac{7}{8}$.

4. Mr. A bought 64 A. 116 P. of land for $3.50 per square rod, and sold it for $3.75 per square rod; what did he gain? *Ans.* $2589.

5. A rectangular pond is 200 rd. 17 yd. long, and 150 rd. 15 yd. wide; required its area. *Ans.* 193 A. 137 P. 15$\frac{3}{4}$ sq. yd.

6. I have a field 16.5 ch. long and 9.75 ch. wide; how much land does it contain? *Ans.* 16 A. 14 P.

7. Mr. Wilson's farm contains 163 A. 3 ch., and its length is 71 ch.; how many rods of fence would be required to surround it? *Ans.* 752 rd.

8. If a township is equally divided among 480 families, how many acres does each receive, and what part of a section? *Ans.* 48 acres; $\frac{3}{40}$ of a section.

9. How many rails are required to fence a quarter-quarter-section, the fence being 5 rails high, and each rail 8 ft. long; and what will be the cost at $35 per thousand rails? *Ans.* 3300 rails; $115.50.

10. A field 80 rods long contains 15 acres, while another field of the same width contains 9 acres; what is the length of the latter field? *Ans.* 48 rods.

11. How much less will it cost to fence a field 72 rods square than a rectangular field 3 times as long and $\frac{1}{3}$ as wide, if fencing cost $2.50 a rod? *Ans.* $480.

12. A mechanic having a lot of ground 50 rods square, planted 3 acres with corn, 200 square rods with vegetables, 15 rods square with flowers, and the remainder he kept to pasture his cow; how much of the lot was pasture?

Ans. 9 A. 155 P.

COST OF ARTIFICERS' WORK.

345. By **Artificers' Work** we mean plastering, painting, papering, paving, stone-cutting, etc.

346. Plastering, painting, papering, paving, and ceiling are estimated by the *square foot* or *square yard.* Roofing, flooring, partitioning, slating, etc., generally by the *square,* which consists of 100 *square feet,* but sometimes by the square foot or yard.

347. Shingles, which commonly measure 18 in. by 4 in., are estimated by the *thousand* or *bundle.* 1000 are generally allowed to a *square* of 100 sq. ft.

WRITTEN EXERCISES.

1. What will be the expense of paving a sidewalk 303 ft. long and $7\frac{1}{2}$ ft. wide, at $2.25 per square yard?

SOLUTION.—The area equals $303\times7\frac{1}{2}$, or $2272\frac{1}{2}$ sq. ft., which equals $252\frac{1}{2}$ sq. yd.; hence the cost is $\$2.25\times252\frac{1}{2}$, or $\$568.12\frac{1}{2}$.

2. What will it cost to plaster a school-room 40 ft. long, 20 ft. wide, and 10 ft. high, at $0.36 a square yard?

Ans. $80.

3. What is the cost of wainscoting a room 28 ft. long by 15 ft. 4 in. wide, to a height of 4 ft. 3 in. at $0.45 per square yard? *Ans.* $\$18.41\frac{2}{3}$.

4. What is the cost of slating a roof 52 ft. 10 in. long, each side being 20 ft. wide, at $15.25 per square?

Ans. $\$322.28\frac{1}{3}$.

5. A frame house is 50 ft. long, 28 ft. wide, and 35 ft. high; what will be the expense of outside painting at $12.25 per square? *Ans.* $668.85.

6. What will it cost to shingle a roof 64 ft. long and 32 feet from eaves to ridge, the first course along the eaves being double, at \$14.87½ a thousand? *Ans.* \$614.992.

7. What will be the expense of papering a room 40 ft. long, 32 ft. 4 in. wide, and 15½ ft. high, allowing 815 square feet for doors, windows, and washboards, at 25¢ per square foot; ceiling not included? *Ans.* \$356.83⅓.

8. A cistern 7 ft. 5 in. long, 4 ft. 6 in. wide, and 6 ft. 3 in. deep, is to be lined with zinc costing 12¢ a pound, allowing 5 lb. to the square foot; what will be the expense?

Ans. \$109.40.

CARPETING, PAPERING, ETC.

348. In **Carpeting, Papering, etc.**, it is frequently necessary to find the quantity of material of a given width required to cover or line a given surface. We do this by the following

Rule.—*Divide the surface we wish to cover by the area contained in a yard of the material.*

WRITTEN EXERCISES.

1. How many yards of carpeting, 1 yard wide, are required to cover a floor 18 ft. 8 in. by 15 ft. 9 in.?

SOLUTION.—18 ft. 8 in. equals 18⅔ ft.; 15 ft. 9 in. equals 15¾ ft.; 18⅔ × 15¾ equals 294 sq. ft., which equals 32⅔ sq. yd.; the area of 1 yard of the carpet is 1 sq. yd., and dividing 32⅔ by 1 we have 32⅔, the number of yards of carpet required.

2. How many sods, each 16 inches square, will be required to sod a grass plat 25 ft. long by 10 ft. 8 in. wide?

Ans. 150 sods.

3. How many planks, 6 ft. long by 1 ft. 6 in. wide, will it require to floor a room 27 ft. long by 17 ft. wide?

Ans. 51 planks.

4. A lady bought 15 yd. of velvet ½ of a yard wide; how much silk ⅞ of a yard wide must she buy to line it?

Ans. 8$\frac{4}{7}$ yd.

5. A housekeeper wishes to cover a floor 28 ft. long by 20 ft. 3 in. wide, with matting 4 ft. wide, at \$1.28 a yard; what will be the cost? *Ans.* \$60.48.

6. Miss Hartman wishes to carpet a room 18 ft long by 15 ft. 6 in. wide, with Brussels carpet $\frac{5}{8}$ of a yard wide, at \$1.25 a yard; what will it cost her? *Ans.* \$62.

7. How many rolls of paper, 8 yards long and 20 inches wide, will be required to cover the walls and ceiling of a room 30 ft. long, $22\frac{1}{2}$ ft. wide and 10 ft. 8 in. high, deducting 142 sq. ft. for windows and doors? *Ans.* $41\frac{13}{40}$ rolls.

8. What will be the cost of papering the above room, at \$2.40 a roll, putting also a gilt moulding around the top of the walls, at 12 cents a foot? *Ans.* \$111.78.

9. A room contained 3 windows, which were curtained with brocatelle $\frac{1}{2}$ of a yard wide; 10 yards were required for each window @ \$1.50, and the curtains were lined with silk $\frac{5}{8}$ of a yard wide @ \$.87$\frac{1}{2}$; how many yards of silk were required, and what was the whole cost of the curtains?

Ans. 24 yd.; \$66.

MEASURES OF VOLUME.

349. A **Volume** is that which has length, breadth, and thickness or height. These three elements are called *dimensions*. A volume is also called a *solid*.

350. A **Rectangular Volume** or **Solid** is a volume bounded by six rectangles. The bounding rectangles are called *faces*. Cellars, boxes, rooms, etc., are examples of rectangular volumes.

3 feet high.

3 feet wide.

351. A **Cube** is a volume bounded by six equal squares. Or, a cube is a rectangular volume whose faces are all equal.

352. By the **Contents** or **Solidity** of a volume we mean the amount of space it contains. The contents are expressed by the number of times it contains a *cube* as a *unit of measure*.

Rule I.—*To find the contents of a cube or rectangular volume, take the product of its length, breadth, and height.*

For, in the volume above, the number of cubic units on the base equals the length multiplied by the breadth, or $3 \times 3 = 9$, and the whole number of cubic units equals the number on the base multiplied by the number of layers of these cubes, or $9\times 3=27$; hence the whole number of cubes, or the contents, equals the product of the length, breadth, and height.

Rule II.—*To find either dimension, divide the contents by the product of the other two dimensions.*

WRITTEN EXERCISES.

1. What are the contents of a room 18 ft. long, 14 ft. wide and 10 ft. high?

SOLUTION.—To find the contents, we multiply the length, breadth, and height together, and we have $18\times 14\times 10=2520$ cu. ft.; reducing this to cubic yards, we have 93 cu. yd. 9 cu. ft.

2. What are the solid contents of a cube whose edge measures 1 yd. 1 ft.? *Ans.* 2 cu. yd. 10 cu. ft.

3. A cistern 9 ft. square contains 405 cubic feet; what is its depth? *Ans.* 5 ft.

4. How many cubic inches in a rectangular block of marble 6 ft. long, 4 ft. wide, and $2\frac{1}{2}$ ft. thick? *Ans.* 103680.

5. How many cubic yards of air in a room 25 ft. long, 12 ft 6 in. wide, and $9\frac{1}{2}$ ft. high? *Ans.* $109\frac{103}{108}$ cu. yd.

6. A pile of bricks contains 125 cubic yards, and is 13 ft. 6 in. wide, and 8 ft. 4 in. high; what is its length? *Ans.* 30 ft.

7. How much earth will be dug out of a cellar 72 ft. long, 48 ft. wide, and 7 ft. 3 in. deep? *Ans.* 928 cu. yd.

THE CYLINDER.

353. A **Cylinder** is a round body of uniform size, with equal and parallel circles for its ends. The two circular ends are called *bases.*

354. The **Altitude** of a cylinder is the distance from the centre of one base to the centre of the other.

355. The **Convex Surface** of a cylinder is the surface of the curved part.

Rule I.—*To find the convex surface of a cylinder, multiply the circumference of the base by the altitude.*

Rule II.—*To find the contents of a cylinder, multiply the area of the base by the altitude.*

WRITTEN EXERCISES.

1. What is the convex surface of a cylinder, the diameter of whose base is 8 inches and whose altitude is 12 inches?

SOLUTION.—The circumference of the base equals 8×3.1416, which is 25.1328 inches; multiplying by the altitude, 12, we have 301.5936 square inches, the convex surface.

2. I have a log 18 ft. long and 20 inches in diameter; how many square feet of bark on the log? *Ans.* 94.248 sq. ft.

3. A well is 10 feet deep, and 3 feet in diameter; how many cubic feet does it contain? *Ans.* 70.686 cu. ft.

4. What is the cost of digging a well 15 ft. deep and 9 ft in circumference, at \$.62½ a cubic yard? *Ans.* \$2.24—.

5. How much zinc will it take to line the sides of a cistern 8 ft. in diameter and 8¼ feet deep? *Ans.* 23.0384 sq. yd.

6. Dr. Hiestand put in his house a cistern, 10 ft. in diameter and 4 ft. 6 in. high; how many cubic feet of water did it hold? *Ans.* 353.43 cu. ft.

WOOD MEASURE.

356. The **Measure of Wood** is the *cord*, which is divided into *cord feet*, etc.

357. A **Cord** of wood is a pile 8 feet long, 4 feet wide, and 4 feet high. It contains 8 cord feet, or 128 cubic feet.

358. A **Cord Foot** is a part of this pile 1 foot long. It is thus 1 foot long, 4 feet wide, and 4 feet high, and contains 16 cubic feet.

Rule.—*To find the number of cords in a pile of wood, find the number of cubic feet and reduce to cord feet and cords.*

WRITTEN EXERCISES.

1. How many cords in a pile of wood 28 ft. long, 10 ft high, and 10 ft. wide?

SOLUTION.—The number of cubic feet equals 28×10×10, which equals 2800; dividing by 16, to reduce this to cord feet, we have 175 cord feet; dividing by 8 to reduce this to cords, we have 21 cd. 7 cd. ft.

2. How many cords in a pile of wood 96 ft. long, 12 ft. wide, and 8 ft. high? *Ans.* 72 cords.

3. A load of wood containing exactly 1 cord, is 5 ft. 4 in. wide, and 3 ft. 9 in. high; what is its length? *Ans.* $6\frac{2}{5}$ ft.

4. What is the height of a pile of wood containing $27\frac{3}{4}$ cords, if it is 75 ft. long and 10 ft. wide? *Ans.* 4.736 ft.

5. What will be the cost of the wood that can be piled in a shed 20 ft. long, 10 ft. wide, and 8 ft. high, at $4.75 a cord?

Ans. $59.37½.

BOARDS AND TIMBER.

359. **Boards** and **Timber** are usually estimated in what are called *board feet*, instead of in cubic feet.

360. **A Board Foot** is 1 foot long, 1 foot wide, and 1 inch thick. A *cubic foot*, therefore, contains 12 *board feet*. Hence, board feet may be reduced to cubic feet by dividing by 12, and cubic feet to board feet by multiplying by 12.

A *standard board*, in commerce, is 1 inch thick, and its contents in board feet are the product of its length and breadth in feet. *Board feet* are usually known as *square feet*. Boards are quoted by the *hundred* or the *thousand*, meaning a *hundred square feet*, or a *thousand square feet*. *Round* timber, as masts, etc., is estimated in cubic feet; *hewn* timber, as beams, etc., either in board or cubic feet; *lumber* and *sawed* timber, as planks, scantling, joists, etc., in board feet.

Rule I.—*To find the contents of a board, multiply the length in feet by the width in inches, and divide the product by* 12.

Rule II.—*To find the contents of a plank, joist, etc., multiply the length in feet by the width and thickness in inches, and divide the product by* 12.

NOTES.—1. If one of the dimensions is inches and the other two are feet, the product will be *board feet*.

2. When a board tapers regularly, the length must be multiplied by the *mean width*, which is half the sum of the two ends.

WRITTEN EXERCISES.

1. What are the contents of a board 14 feet long and 9 inches wide?

SOLUTION.—Multiplying the length in feet by the width in inches, we have $14 \times 9 = 126$; and dividing by 12, we have $10\frac{1}{2}$ board feet, or *square feet*.

2. What are the contents of a board 16 feet long and $1\frac{1}{2}$ ft. wide? *Ans.* 24 sq. ft.

3. Required the contents of a board 20 ft. long, the ends being 18 and 14 inches respectively. *Ans.* $26\frac{2}{3}$ sq. ft.

4. How many square feet in 14 planks 16 ft. long, 18 inches wide, and 4 inches thick? *Ans.* 1344 sq. ft.

5. How many square feet in a stick of timber 40 feet long, 14 inches wide, and 9 inches thick? *Ans.* 420 sq. ft.

6. What must be the width of a board 6 ft. 4 in. long that it may contain $9\frac{1}{2}$ square feet? *Ans.* 18 inches.

7. How many square feet of inch boards will it require to make a box 3 ft. by 2 ft. 6 in. and 18 in. high on the outside? *Ans.* $29\frac{2}{9}$ sq. ft.

8. What is the cost of 40 boards 12 ft. long, 15 in. wide, at $2.75 per hundred square feet? *Ans.* $16.50.

9. What is the cost of 9 pieces of scantling 4 in. by 5 in. and 10 ft. long at $8.75 per thousand square feet? *Ans.* $$1.31\frac{1}{4}$.

10. What is the cost of flooring a three-story house, the floors being 56 ft. by 32 ft. and the plank $1\frac{1}{2}$ inches thick, at $33 per M? *Ans.* $266.112.

11. I wish to fence a field 36 rd. long and 18 rd. wide, the posts to be set 9 ft. apart, the boards to be 18 ft. long, and 11 inches wide, the fence being 3 boards high; the posts cost $30 per C, and the boards $15.50 per M, and it required 2 men 3 days at $3.50 each a day, to build the fence; required the number of posts, the amount of lumber, and the whole cost. *Ans.* 198 posts; $4900\frac{1}{2}$ sq. ft.; $156.357.

MASONRY, BRICKWORK, ETC.

361. Masonry is usually estimated by the *perch* and the *cubic foot;* sometimes by the *square foot* or the *square yard.*

362. A **Perch** *of stone* or of *masonry* is $16\frac{1}{2}$ ft. long, $1\frac{1}{2}$ ft. wide, and 1 ft. high; it contains $24\frac{3}{4}$ cubic feet, but when stone is built into a wall, 22 cubic feet make a perch, $2\frac{3}{4}$ cu. ft. being allowed for mortar and filling.

363. Excavations and **Embankments** are estimated by the *cubic yard.* A cubic yard of earth is called a *load.*

364. Brickwork is generally estimated by the *thousand bricks*, but sometimes in cubic feet.

In estimating *labor*, bricklayers and masons measure the length of the wall on the outside. The corners are thus measured twice, but this is considered an allowance for the greater difficulty of building them. No allowance is made for windows and doors, except by special contract, in which case it is customary to allow one-half the space actually required. In estimating *material*, allowance is made for doors, windows and corners.

The average size of bricks is 8 in.$\times 4 \times 2$, but Phila. and Baltimore bricks are $8\frac{1}{4}$ in.$\times 4\frac{1}{8} \times 2\frac{3}{8}$; Maine bricks, $7\frac{1}{2}$ in.$\times 3\frac{5}{8} \times 2\frac{3}{8}$; North River bricks, 8 in.$\times 3\frac{1}{2} \times 2\frac{1}{4}$; and Milwaukee bricks, $8\frac{1}{2}$ in.$\times 4\frac{1}{8} \times 2\frac{3}{8}$.

To build one *square foot* of wall 1 brick or 4 inches thick, requires 7 common bricks; 2 bricks, or 9 in. thick, 14 bricks; 3 bricks, or 13 in. thick, 21 bricks. In practice, the thickness of the wall is regarded as the same for each kind of brick.

Rule I.—*To find the number of perches in a piece of masonry, divide the number of cubic feet by* $24\frac{3}{4}$.

Rule II.—*To find the number of common bricks required for a wall or building, multiply the number of square feet in the wall by* 7, *if the wall is* 1 *brick thick; by* 14, *if* 2 *bricks thick; by* 21, *if* 3 *bricks thick.*

Rule III.—*To find the number of any kind of bricks required for a wall, or building, add* $\frac{1}{4}$ *of an inch to the length and the thickness of the brick, divide* 144 *by the product of these two sums to find the number of bricks in a square foot of wall* 1 *brick thick, and multiply by the number of bricks in the thickness, and this product by the number of square feet in the wall.*

NOTE.—An old rule was—*Deduct* $\frac{1}{10}$ *of the solid contents for the mortar and divide the remainder by the contents of one brick.* We may also *find the contents of a brick with the mortar surrounding it, and divide a cubic foot by this quantity, to find the number of bricks in a cubic foot.*

WRITTEN EXERCISES.

1. How many perches of masonry in a wall 60 ft. long, 4 ft. 6 in. high, and 15 inches thick?

SOLUTION.—Multiplying the length, breadth, and height together, we have $60 \times 4\frac{1}{2} \times 1\frac{1}{4}$, or $337\frac{1}{2}$ cu. ft., which, divided by $24\frac{3}{4}$, the number of cubic feet in a perch, equals $13\frac{7}{11}$ perches.

2. What will be the cost of digging a cellar 42 ft. long, 28 ft. wide, and 6 ft. 6 in. deep, at $.42 a load?

Ans. 118.90\frac{2}{3}$.

3. How many perches of stone, laid dry, will build a wall around a lot 20 rd. long and 18 rd. wide, the wall to be 5 ft. high and 2 ft. 6 in. thick? *Ans.* 628$\frac{28}{33}$ perches.

4. What will be the cost of filling in a street 600 ft. long and 65 ft. wide, averaging 4$\frac{1}{2}$ ft. below grade, at $.52 a cubic yard? *Ans.* $3380.

5. How many bricks of average size will it require to build the walls of a house 48 ft. long, 25 ft. wide, and 21 ft. high, the wall being 13 in. thick, allowing 240 sq. ft. for doors and windows? *Ans.* 57435 bricks.

6. What will be the cost of the bricks in a house 40 ft. square, 22 ft. high, the walls being three bricks thick, of Philadelphia brick, at 15\frac{1}{2}$ per M.? *Ans.* $1027.75.

7. Mr. Wilson had a well dug in his yard, 6 feet in diameter and 15 feet 9 inches deep; what did it cost at 50¢ a load?

Ans. $8.25—.

8. What will be the cost of digging and walling the cellar of a house 45 ft. by 24 ft., the cellar being 6 ft. deep and the wall 7$\frac{1}{2}$ ft. high and 1$\frac{1}{2}$ ft. thick, if the excavating cost 45¢ a load, and the masonry $4.25 a perch? *Ans.* 374.59\frac{1}{11}$.

MEASURES OF CAPACITY.

365. Measures of Capacity are volumes used to determine the quantity of fluids and many dry substances.

366. The **Principal Measures** of capacity are the *gallon* for liquid substances, and the *bushel* for dry substances.

CAPACITY OF CISTERNS, ETC.

367. The **Capacity of Cisterns,** etc., is usually expressed in *gallons* or *barrels*.

368. The **Standard Liquid Gallon** of the United States contains 231 cubic inches, and is equal to about 8$\frac{1}{3}$ lb Avoirdupois of pure water.

NOTE.—2150.42 is to 1728 as 5 to 4, nearly; hence a bushel is nearly equal to $1\frac{1}{4}$ cubic feet. Therefore, for practical purposes, *any number of cubic feet, diminished by $\frac{1}{5}$, will give their equivalent in bushels*, and *any number of bushels, increased by $\frac{1}{4}$, will give their equivalent in cubic feet.* The 5th example and those following will be solved by these rules.

WRITTEN EXERCISES.

1. How many bushels will be contained in a bin 8 ft. long, 6 ft. wide, and 3 ft. deep?

SOLUTION.—The contents equal $8 \times 6 \times 3$, or 144 cubic feet, which equals 248832 cubic inches; dividing by 2150.42, the number of cubic inches in a bushel, we have 115.71+ bushels.

2. A bin is 16 ft. long, 7 ft. wide, and $2\frac{1}{2}$ ft. deep; how many bushels will it hold? *Ans.* 224.99 bu.

3. What is the width of a bin 24 ft. long and 3 ft. 4 in. deep, to contain 640 bushels of wheat? *Ans.* 9.95+ ft.

4. One division of an elevator is 25 feet long, 15 feet wide, and contains 2000 bushels of grain; what is its depth? *Ans.* 6.63+ ft.

5. A bin is 10 ft. 5 in. long, 7 ft. wide, and 3 ft. 6 in. deep; how many bushels of shelled corn will it hold? how many of carrots? *Ans.* $204\frac{1}{6}$ bu. corn; $163\frac{1}{3}$ bu. carrots.

6. There is a rectangular box 4 ft. long, 3 ft. wide, and 2 ft. 4 in. deep; how many bushels of apples will it hold? how many of cranberries? *Ans.* $17\frac{23}{25}$ bu.; $22\frac{2}{5}$ bu.

7. A bin 8 ft. long, 6 ft. wide, and 3 ft. deep, is $\frac{3}{4}$ full of barley; what is its value at $1.25 a bushel? *Ans.* $108.

8. How many barrels of flour can be made from the contents of a bin 12 ft. long, $7\frac{1}{2}$ ft. wide, and 4 ft. deep, if one bushel of wheat makes 48 lb. of flour? *Ans.* $70\frac{26}{49}$ bbl.

9. A bin 5 ft. long, $4\frac{1}{2}$ ft. wide, and 3 ft. deep, is filled with Schuylkill white ash coal; what is its value at $6.50 a ton? *Ans.* $$12.53\frac{4}{7}$.

10. A shed 6 yd. long, $4\frac{1}{2}$ yd. wide, and 8 ft. high, is half full of Lehigh white ash coal; what is the value of the coal at $7.50 a ton? *Ans.* $211.304.

11. An ice-house 40 ft. long, 25 ft. wide, and 15 ft. high, is filled with ice; how many tons are there, if a cubic foot weighs $58\frac{1}{8}$ lb.? *Ans.* $435\frac{15}{16}$ tons.

12. A haymow is 24 ft. long by 18 ft. wide and 16 ft. high; what is the value of the hay when it is filled, valued at $12 a ton? *Ans.* $207.36.

13. Mr. Jenkins sold a rectangular stack of hay 8 ft. long, 7 ft. wide, and 6 ft. high, at $15 a ton; what was the value of the hay? *Ans.* 18.66\frac{2}{3}$.

14. A crib filled with corn in the ear measures on the inside 20 ft. in length, 12 ft. in width, and 7 ft. in height; what will be the value of the corn when shelled at $1.05 a bushel, if 2 bushels in the ear make 1 bushel when shelled?

Ans. $564.48.

COMPARISON OF MEASURES OF CAPACITY.

372. The **Dry Gallon,** or *half peck,* contains 268.8 cubic inches; hence 6 dry gallons equal nearly 7 liquid gallons.

NOTE.—The pupil will remember that the liquid gallon contains 231 cu. in., and the old beer gallon 282 cu. in.

WRITTEN EXERCISES.

1. How many more cubic inches in 495.3 dry gallons than in 495.3 liquid gallons? *Ans.* 18722.34 cu. in.

2. What part of 5 gal. 2 qt. of old beer measure is 3 gal. 2 qt. 1 pt. of liquid measure? *Ans.* .5399—.

3. How many bushels of oats can be put in a tank that holds 4550 gallons of water? *Ans.* 488.765— bu.

4. How many gallons of water can be poured into a bushel measure? *Ans.* 9.31—.

5. If 32 quarts of water are put into a vessel that holds exactly 32 quarts of strawberries, will they be more or less than the vessel will hold, and how much?

Ans. 302.4 cu. in. less.

6. A dishonest milkman charged for milk 7 cents a quart, beer measure, but wishing to cheat his customers, he measured it by the liquid quart; what was the loss to one taking $2\frac{1}{2}$ gallons? *Ans.* 12$\frac{31}{47}$ cents.

7. A grocer sold molasses at 15 cents a quart, but the clerk, by mistake, measured a day's sales by the dry quart; he sold 25 gallons; what did it cost a quart liquid measure?

Ans. 12$\frac{57}{64}$ cents.

COMPARISON OF WEIGHTS.

373. The **Troy Pound** and the **Apothecaries' Pound** each contains 5760 Troy grains; the *Avoirdupois* pound contains 7000 Troy grains. The Avoirdupois ounce contains 437½ grains, while the Troy ounce contains 480 grains.

Rule.—*Reduce the given denominations to grains, and divide by* 480 *to find Troy ounces, and* 437½ *to find ounces Avoirdupois.*

1. How many rings, each weighing $2\frac{3}{8}$ pwt., can be made from a bar of gold weighing 1 lb. Avoirdupois?

Ans. 101 rings, 31 gr. remaining.

2. What is the value of a silver pitcher weighing 2 lb. 10 oz. Avoirdupois, at $2.25 per ounce Troy?

Ans. $86.13.

3. Which is the heavier, $52\frac{5}{8}$ lb. of lead or 52.625 lb. of silver? *Ans.* The lead.

4. How many pounds of gold are actually as heavy as 10 lb. of iron? *Ans.* $12\frac{11}{72}$ lb.

5. How many times 3 lb. 10 oz. Troy is 5 lb. 12 oz. Avoirdupois? *Ans.* 1.8229+.

6. A young lady weighs 120 lb. Avoirdupois weight; how much would she weigh by Troy weight? *Ans.* $145\frac{5}{6}$ lb.

7. If a druggist buys 25 lb. Avoirdupois of drugs at $8⅓ a pound, and sells them in prescriptions at the rate of 75¢ an ℥, what is the gain? *Ans.* $65.10.

COMPARISON OF MONEY.

374. The **Pound Sterling** is valued at $4.8665; and the *Franc* at 19.3 cents; the *Mark* at 23.85 cents.

375. The gold **Dollar** weighs 25.8 gr.; the silver *Dollar*, 412½ gr.; the *Half-dollar* 192.9 gr.; the *Sovereign* (gold) 123.274 gr.; and the *Shilling* (silver) 87.27 gr.

1. How many dollars in £25? *Ans.* $121.66¼.

2. How many dollars in £14 12s.? *Ans.* $71.05.

3. How many pounds in $256.16? *Ans.* £52 12s. 9d—.

4. How many dollars in 240 guineas? *Ans.* $1226.358,

5. How many francs in £42? *Ans.* 1059.03 fr.

6. How many pounds in 875 francs?

Ans. £34 14.03s.

7. How many marks in £340? *Ans.* 6937.568 marks.

8. How many pounds in 4375 marks?

Ans. £214 8 s. 3 d.—.

9. How many pounds Avoirdupois in 1000 sovereigns?

Ans. 17.610$\frac{4}{7}$.

10. An ingot of pure gold was brought from California to be coined at the Philadelphia mint; if it weighed 15 lb. 8 oz. Troy, how many gold dollars will it make, $\frac{1}{10}$ of the coin being alloy? *Ans.* $3886; 7.08 gr. rem.

MISCELLANEOUS PROBLEMS.

1. How many planks laid crosswise, 1 ft. wide, will it take for a board walk 1 mi. 16 rd. long, and 4 ft. wide? *Ans.* 5544.

2. How many bushels of grain can a farmer store in a hogshead containing 122 gallons? *Ans.* 13.1+bu.

3. How much stair carpet will be required for a flight of 18 steps, each 10 in. wide and 7$\frac{1}{2}$ in. high? *Ans.* 8$\frac{3}{4}$ yd.

4. The cost of digging a sewer 1$\frac{1}{2}$ miles long, 5 feet wide, and 8 ft. deep, was $3716; what was the price per cubic yard? *Ans.* $.31$\frac{59}{88}$.

5. A farmer has a mow 20 ft. long, 12 ft. wide, and 10 ft. deep; how many tons of hay does it hold? *Ans.* 6 tons.

6. From a quartz rock yielding silver at the rate of $123.75 per ton, a miner obtained $75.64 worth; what was the weight of the rock? *Ans.* 12 cwt. 22$\frac{46}{99}$ lb.

7. If 4 persons can stand on one square yard of ground, how many people can be contained in a public square 32 rods on each side? *Ans.* 123904.

8. How many bunches of lath will be required for the walls and ceiling of a room 18 ft. long, 14 ft. wide, 10 ft. high, each bunch being estimated to cover 5 sq. yd.? *Ans.* 19$\frac{37}{45}$.

9. I wish to cover my parlor 25 ft. × 17 ft. 6 in. with Brussels carpet 26 inches wide; what will it cost me at 1.87\frac{1}{2}$ per yard, the pattern requiring an allowance of 2$\frac{9}{13}$ yd. for waste? *Ans.* $131.25.

10. A street 36 ft. wide, was paved with Nicholson pavement at \$3.25 per square yard; what did it cost to pave a square 32 rods long? *Ans.* \$6864.

11. A railroad tunnel is one-eighth of a mile long, averaging 24 ft. wide and 20 ft. high; what did the excavation cost, at \$1.50 a cubic yard? *Ans.* \$17600.

12. How many freight cars will be required to transport 12000 bu. wheat, 24000 lb. being the weight allowed for a single car? *Ans.* 30 cars.

13. A coal-dealer has a wagon which holds exactly one ton of Schuylkill red ash coal; if the wagon-bed was $7\frac{1}{2}$ ft. long and $4\frac{1}{2}$ ft. wide, what must be its depth? *Ans.* $12\frac{4}{5}$ in.

14. I used the earth taken from 4 cellars in grading a lot of ground; if the cellars were 30×21 ft., 28×18 ft., 24×16 ft., and 32×24 ft., respectively, and 5 ft. deep, how much earth did I use? *Ans.* $423\frac{1}{3}$ loads.

15. What costs the excavation for a cellar $5\frac{1}{2}$ ft. deep under the main building of a dwelling-house 30×25 ft. and an excavation for the walls of an L 16 ft. square, $1\frac{1}{2}$ ft. wide, and 2 ft. deep, at 50¢ per cu. yd.? *Ans.* $\$78.88\frac{8}{9}$.

PROBLEMS FOR REVIEW.

16. What costs the plastering of a house of 12 rooms, there being on each story 4 rooms 14×15 ft. and a hall 30×8 ft.; the first story being 10 ft. high, the second $9\frac{1}{2}$ ft., and the third $8\frac{1}{2}$ ft., allowance being also made for 24 doors $7 \times 3\frac{1}{2}$ ft., and 30 windows 6×3 ft., at 30¢ per sq. yd.? *Ans.* $\$357.86\frac{2}{3}$.

17. Required the cost of a cellar of a house 40×30 ft., the different items being as follows: excavating cellar, 4 ft. deep at 50¢ per cu. yd.; cellar wall, 7 ft. high and 18 in. thick, the lower 4 ft. common masonry, @ \$3.15 a perch, and the upper 3 ft. cut stone at 16¢ per sq. ft.

Ans. \$263 —.

18. Required the cost of the brick-work of the same house, the walls being 35 ft. high and 13 in. thick, and gable 10 ft. high, using common bricks at \$10 per M., bricklaying costing \$3 per M., allowing for 3 doors, each 7 ft. by $3\frac{1}{2}$ ft., and 30 windows each 6 ft. by 3 ft. *Ans.* \$1258.915.

INTRODUCTION TO PERCENTAGE.

MENTAL EXERCISES.

1. A gain of $2 on $5 is a gain of how many dollars on *the hundred?*

SOLUTION.—If the gain on $5 is $2, on $100, which is 20 times $5, the gain is 20 times $2, or $40.

2. A gain of $3 on $5 is a gain of how many dollars on *the hundred?*

3. What is the gain on a hundred when the gain is 4 on 20? 5 on 20? 4 on 25?

4. If the gain on $100 is $25, what is the gain on $4? on $12? on $20?

5. If the gain on $100 is $20, what is the gain on $5? on $15? on $25?

6. If the gain on $100 is $40, what is the gain on $1? on $12? on $36?

7. If the gain on $100 is $25, what part of the $100 equals the gain?

8. If the gain on $100 is $40, what part of the $100 equals the gain?

9. If the gain on $24 is at the *rate* of 25 on the 100, what is the gain?

10. If the gain on $25 is at the *rate* of 20 on the 100, what is the gain?

11. What is the gain on $50 at the *rate* of 10 on the *hundred?*

12. What is the gain on $250 at the *rate* of 20 on the *hundred?*

13. What is the gain on $360 at the *rate* of 15 on the *hundred?*

14. What is the *rate per hundred* at a gain of $6 on $30?

15. What is the *rate per hundred* at a gain of $15 on $60?

16. *Per cent.* means the same as *per hundred;* what then can we call the *rate per hundred?* *Ans. Rate per cent.*

17. A gain of $20 on $80 is a gain of what *per cent.?*

18. A loss of $15 on $75 is a loss of what *per cent.?*

19. What *per cent.* is a gain of 20 on 40? 5 on 25? 4 on 80? 3 on 60? 8 on 200?

20. What is 5 per cent. of 80? 4 per cent. of 24? 20 per cent of 40? 25 per cent. of 48?

SOLUTION.—5 per cent. is at the rate of 5 on the 100, and since 5 is $\frac{1}{20}$ of 100, 5 per cent. of 80 is $\frac{1}{20}$ of 80, which is 4.

21. What is 50 per cent. of 24? 30 per cent. of 60? 40 per cent. of 85? 60 per cent. of 45?

22. What per cent. is a gain of 15 on 60? 18 on 72? 12 on 48? 16 on 80? 20 on 60? 15 on 90?

SECTION VIII.

PERCENTAGE.

376. **Percentage** is the process of computation in which the basis of comparison is a *hundred.*

377. The **Term** *per cent.*—from *per*, by, and *centum, a hundred*—means *by* or *on the hundred;* thus, 6 per cent. of any quantity means 6 of every hundred of the quantity.

378. The **Symbol of Percentage** is %. The per cent. may also be indicated by a common fraction or a decimal; thus 6% $= \frac{6}{100} =$ .06.

379. The **Quantities** considered in percentage are the *Base*, the *Rate*, the *Percentage*, and the *Amount* or *Difference.*

380. The **Base** is the number on which the percentage is computed

381. The **Rate** is the number of hundredths of the base which are taken.

382. The **Percentage** is the result obtained by taking a certain per cent. of the base.

383. The **Amount** or **Difference** is the sum or difference of the base and percentage. They may both be embraced under the general term *Proceeds.*

NOTE.—In computation the rate is usually expressed as a decimal. For the difference between *Rate* and *rate per cent.*, see *Brooks's Philosophy of Arithmetic.*

EXPRESSION OF THE RATE.

1. Express 4% as a decimal and common fraction.

SOLUTION.—Since per cent. is so many on a hundred, 4% of a quantity is .04 of it; or, as a common fraction, $\frac{4}{100}$ or $\frac{1}{25}$ of it.

OPERATION.
4% $=.04=\frac{4}{100}=\frac{1}{25}$.

Express

2. 5%. *Ans.* .05 or $\frac{1}{20}$.
3. 6%. *Ans.* .06 or $\frac{3}{50}$.
4. 7%. *Ans.* .07 or $\frac{7}{100}$.
5. 8%. *Ans.* .08 or $\frac{2}{25}$.

6. 10%.	*Ans.* .10 or $\frac{1}{10}$.	10. $33\frac{1}{3}$%.	*Ans.* .$33\frac{1}{3}$ or $\frac{1}{3}$
7. $11\frac{1}{9}$%.	*Ans.* .$11\frac{1}{9}$ or $\frac{1}{9}$.	11. $37\frac{1}{2}$%.	*Ans.* .$37\frac{1}{2}$ or $\frac{3}{8}$.
8. $12\frac{1}{2}$%.	*Ans.* .$12\frac{1}{2}$ or $\frac{1}{8}$.	12. $\frac{1}{2}$%.	*Ans.* .005.
9. $16\frac{2}{3}$%.	*Ans.* .$16\frac{2}{3}$ or $\frac{1}{6}$.	13. $\frac{1}{4}$%.	*Ans.* .0025.

384. Cases. The subject of percentage is conveniently treated under three distinct cases:

1. Given the rate and base, to find the percentage or proceeds.

2. Given the rate and percentage or proceeds, to find the base.

3. Given the base and percentage or proceeds, to find the rate.

NOTE.—Authors usually present the subject in five or six cases, but it is thought that the method here adopted is to be preferred, on account of its logical accuracy and practical convenience.

CASE I.

385. ***Given, the base and the rate, to find the percentage or the proceeds.***

MENTAL EXERCISES.

1. What is 25% of 120 yards?

SOLUTION.—25% of anything is $\frac{25}{100}$ or $\frac{1}{4}$ of it; and $\frac{1}{4}$ of 120 yards is 30 yards. Therefore, etc.

What is

2. 20% of 75?	6. 15% of 60?
3. 25% of 80?	7. 35% of 120?
4. 50% of 132?	8. $12\frac{1}{2}$% of 144?
5. 75% of 96?	9. $16\frac{2}{3}$% of 108?

10. Out of a purchase of 120 dozen eggs, 20% turned out to be bad; how many were good?

11. From a hogshead of kerosene containing 108 gallons, $33\frac{1}{3}$% leaked out; how many gallons remained?

12. A train of cars running 20 miles an hour, increases its speed 15%; what is the rate of running after the increase?

13. A clerk's salary is $45 a month, but at the beginning of the year it was raised $11\frac{1}{9}$%; what did he then receive a month?

14. Mr. Smith paid a tax of $\frac{1}{2}$% on $3000; what was the amount of his tax?

15. In the 10th problem, which is the base, which the rate, and which the percentage?

WRITTEN EXERCISES.

1. What is 6% of \$275? What is the amount of \$275, increased by 6% of itself?

OPERATION.

$$\begin{array}{r} \$275 \\ .06 \\ \hline \$16.50 \end{array}$$

SOLUTION.—6% of \$275 equals .06 times \$275, which, by multiplying, we find to be \$16.50.

OPERATION.

$$\begin{array}{r} \$275 \\ 1.06 \\ \hline \$291.50 \end{array}$$

SOLUTION.—A number increased by 6%, or .06 times itself, equals 1.06 times itself; 1.06 times \$275 equals \$291.50.

Rule I.—*Multiply the base by the rate, to find the percentage.*

Rule II.—*Multiply the base by* 1 *plus the rate, to find the amount; or by* 1 *minus the rate, to find the difference.*

NOTES.—1. When the rate gives a small common fraction, take such a part of the base as is indicated by this fraction.

2. The *amount* equals the base *plus* the percentage; the *difference* equals the base *minus* the percentage.

What is

2. 12% of 475? *Ans.* 57.

3. 8% of 1875? *Ans.* 150.

4. 25% of 948 miles? *Ans.* 237.

5. 12½% of 1256 rd.? *Ans.* 157.

6. 35% of 1840 yd.? *Ans.* 644.

7. 66⅔% of \$124.65? *Ans.* \$83.10.

8. 33⅓% of \$234.54? *Ans.* \$78.18.

9. 45% of 18¾? *Ans.* 8.43¾.

10. ¾% of \$348? *Ans.* \$2.61.

11. ⅘% of $\frac{15}{16}$ lb.? *Ans.* .12 oz.

12. Find 25% of 46 lb. 12¾ oz. *Ans.* 11 lb. 11$\frac{3}{16}$ oz.

13. How much is 42½% of 6 lb. 8 oz. 12 pwt.

Ans. 2 lb. 10 oz. 5$\frac{1}{10}$ pwt.

14. A grain dealer bought 600 bar. of Western flour, and sold 16⅔% of it; how many barrels remained? *Ans.* 500.

15. A man's income is \$1800 a year, of which he pays 12% for house rent; what rent does he pay? *Ans.* \$216.

16. If the bread made from a barrel of flour weighs 33⅓ per cent. more than the flour, what is the weight of the *bread?* *Ans.* 261⅓ lb.

17. Mr. Hamlin had 360 acres of land, and sold $33\frac{1}{3}\%$ of it, how many acres remained? *Ans.* 240 acres.

18. The silver coin of the United States contains 10% of alloy; how much pure silver is there in $16\frac{2}{3}$ oz. of silver coin? *Ans.* 15 oz.

19. A land agent bought 1016 acres of land, and sold $12\frac{1}{2}\%$ to Mr. Chase and $37\frac{1}{2}\%$ of the remainder to Mr. Dunn; how much remained? *Ans.* $555\frac{5}{8}$ acres.

20. How much linseed oil can be extracted from 1 cwt. 27 lb. of flaxseed, if flaxseed contains 11% of oil, and a pint of oil weighs $\frac{3}{4}$ of a pound? *Ans.* 2 gal. $1.31\frac{1}{3}$ qt.

21. A clerk's salary is $2000 a year; he spends 10% of it the first quarter, 15% the second, 6% the third, and 4% the fourth; how much did he save? *Ans.* $1300.

22. Mr. Walton's income is $2500 a year, of which he spends 30% for board, $12\frac{1}{2}\%$ for clothes and books, and 10% for incidentals; what does he save in a year?

Ans. $1187.50.

23. A man owning $\frac{3}{4}$ of a machine shop worth $10,000, sold $16\frac{2}{3}\%$ of his share to his brother; what part of the whole shop did he still retain, and what was its value?

Ans. $\frac{5}{8}$; value, $6250.

CASE II.

386. ***Given, the rate and the percentage or proceeds, to find the base.***

MENTAL EXERCISES.

1. Twenty-four is 20% of what number?

SOLUTION.—If 24 is 20 per cent. of some number, it is $\frac{20}{100}$, or $\frac{1}{5}$ of that number; if 24 is $\frac{1}{5}$ of some number, $\frac{5}{5}$, or the number, equals 5 times 24 or 120.

Of what number is

2. 16 25%?
3. 19 $16\frac{2}{3}\%$?
4. 27 $33\frac{1}{3}\%$?
5. 1.5 50%?
6. 4 lb. 6%?
7. 2 bu. $12\frac{1}{2}\%$?
8. 8 A. $66\frac{2}{3}\%$?
9. £7.5 75%?

10. Thirty is 25% more than what number? 50% more than what number? 100% more than what number?

11. Sixty is 25% less than what number? 50% less than what number? 100% less than what number?

12. The population of a certain village is 2700, which is $12\frac{1}{2}$% more than it was 5 years ago; what was its yearly gain?

13. Mr. A. paid a tax of \$60, which was $\frac{2}{5}$% of the value of his property; what was the value of his property?

14. A gentleman gave his daughter \$50 as a Christmas present, which is 62% of what he gave to his wife; what did he give his wife?

15. On a certain day there were present at a graded school 160 pupils, which was $11\frac{1}{9}$% less than were registered; how many were registered?

WRITTEN EXERCISES.

1. 60 is 5% of what number? What number, increased by 20% of itself, equals 360?

SOLUTION.—If 60 is 5% of some number, then .05 times *some number* equals 60; if .05 times *some number* equals 60, the *number* equals 60 ÷ .05, which is 1200.

OPERATION.
$60 \div .05 = 1200$

SOLUTION.—A number increased by 20%, or .20 of itself, equals 1.20 times the number; and if 1.20 times *a number* equals 360, the *number* equals 360 ÷ 1.20, or 300.

OPERATION.
$360 \div 1.20 = 300$

Rule I.—*Divide the percentage by the rate, to find the base.*

Rule II.—*Divide the amount by 1 plus the rate, or the difference by 1 minus the rate, to find the base.*

Of what number is

2. 45 20%? *Ans.* 225.
3. 75 25%? *Ans.* 300.
4. 112 lb. 40%? *Ans.* 280 lb.
5. 456 A. 30%? *Ans.* 1520 A.
6. $\frac{5}{12}$ $33\frac{1}{3}$%? *Ans.* $\frac{5}{4}$.
7. $7\frac{3}{4}$ 75%? *Ans.* $10\frac{1}{3}$.
8. \$645 $62\frac{1}{2}$%? *Ans.* \$1032.
9. \$$450\frac{3}{4}$ $12\frac{1}{2}$%? *Ans.* \$3606.

10. What number increased by 40% of itself equals 1694?
Ans. 1210.

11. What number diminished by 20% of itself equals 468?
Ans. 585.

12. What fraction increased by 16% of itself equals $\frac{29}{30}$?
Ans. $\frac{5}{6}$.

13. What fraction diminished by 36% of itself equals $\frac{4}{5}$?
Ans. $\frac{5}{4}$.

14. 42 A. 112 P. is $16\frac{2}{3}$% of how much land?
Ans. 256 A. 32 P.

15. 14 lb. 10 oz. 16 pwt. is $33\frac{1}{3}\%$ more than what number?
Ans. 11 lb. 2 oz. 2 pwt.

16. A bookkeeper spends $600 per year, which is 24% of his salary; required his salary. *Ans.* $2500.

17. A young farmer owns 320 acres of land, which is 15% of what his father owns; how much has the father?
Ans. $2133\frac{1}{3}$ A.

18. A newsboy earned $15, which was 30% of what he then had in bank; how much had he in bank? *Ans.* $50.

19. A teacher spends 24% of his salary, and can thus save $760 a year; what was his salary? *Ans.* $1000.

20. Mr. Hays drew 35% of his bank deposit to pay a debt of $4788.56; what was his deposit? *Ans.* $13681.60.

21. A man bought some flour and sold 25% of it to A, and $33\frac{1}{3}\%$ of the remainder to C; how much did he buy if he sold C 640 barrels? *Ans.* 2560 bar.

22. Mr. Herr drew $62\frac{1}{2}\%$ of his money from the bank, and paid $33\frac{1}{3}\%$ of it for a house worth $4500; how much money had he remaining in bank? *Ans.* $8100.

23. A lady invested 90% of her money in bank stock, and some time after sold $33\frac{1}{3}\%$ of the stock, and still had $4500 invested; required the whole amount of her money.
Ans. $7500.

CASE III.

387. ***Given, the base and the percentage or the proceeds, to find the rate.***

MENTAL EXERCISES.

1. 12 is what per cent. of 48?

Solution.—48 is 100 per cent. of 48, and 12, which is $\frac{1}{4}$ of 48, is $\frac{1}{4}$ of 100 per cent., or 25 per cent. of 48.

What per cent.

2. Of 75 is 15?
3. Of $12 are $3?
4. Of 81 is 27?
5. Of 16 qt. are 5 qt.?
6. Of $\frac{3}{4}$ is $\frac{5}{8}$?
7. Of $\frac{4}{5}$ is $\frac{2}{7}$?
8. Of 75% is 25%?
9. Of 2.5% is 1.5%?

10. From a hogshead of wine containing 90 gallons, 12 gallons leaked out; what per cent. was lost?

11. If a man's income is $2000 a year and he saves $300; what per cent. does he spend?

12. The standard for gold and silver coin in the United States is 9 parts pure to 1 of alloy; what % of pure metal is there?

13. A merchant having put $5000 into a certain speculation, finds on settling up the business that he has received $5250; what % did he gain?

WRITTEN EXERCISES.

1. 20 is what per cent. of 80?

SOLUTION.—If 20 is some per cent. of 80, then 80 multiplied by *some rate* equals 20; if 80 multiplied by *some rate* equals 20, the *rate* equals 20 divided by 80, which is .25, or 25%.

OPERATION.

$20 \div 80 = .25$

2. 240 yd. being increased by a certain per cent. of itself equals 300 yd.; required the rate.

SOLUTION.—300 yd. minus 240 yd. equals 60 yd., which is the percentage. If 240 yd. multiplied by *some rate* equals 60 yd., the *rate* equals 60 divided by 240, which is .25, or 25%.

OPERATION.

$300 - 240 = 60$
$60 \div 240 = .25$

Rule I.—*Divide the percentage by the base, to find the rate.*

Rule II.—*Divide the difference between the proceeds and base by the base, to find the rate.*

NOTE.—The rate may also be found by dividing the proceeds by the base and taking the difference between 1 and the quotient.

What per cent. of

3. 360 is 90? *Ans.* 25%.
4. 675 is 135? *Ans.* 20%.
5. 900 is 360? *Ans.* 40%.
6. $880 is $528? *Ans.* 60%.
7. $\frac{5}{6}$ is $\frac{3}{4}$? *Ans.* 90%
8. $\frac{8}{9}$ is $\frac{2}{5}$? *Ans.* 45%.
9. 32% is $5\frac{1}{3}$%? *Ans.* $16\frac{2}{3}$%.
10. 4.5% is $3.37\frac{1}{2}$%? *Ans.* 75%.
11. 936 yd. is 312 yd.? *Ans.* $33\frac{1}{3}$%.
12. 18 lb. is 5 lb. 8 oz.? *Ans.* $30\frac{5}{9}$%.

13. The base is $14.10, the percentage $2.35; what is the rate? *Ans.* $16\frac{2}{3}$%.

14. If a miller takes 10 quarts of every bushel he grinds for toll, what per cent. does he take for toll? *Ans.* $31\frac{1}{4}$%.

15. My income last year was $1800 and my expenses $1356; what % of my income did I expend? *Ans.* $75\frac{1}{3}$%.

16. A regiment went into battle with 960 men, and came out with 600 men; what per cent. was lost? *Ans.* $37\frac{1}{2}\%$.

17. A merchant's liabilities are \$15760, and his assets \$7289; what % of his debts can he pay? *Ans.* $46\frac{1}{4}\%$.

18. A merchant bought 275 barrels of flour, and after losing 20% of it, he sold 25% of the remainder; what per cent. of the whole remained? *Ans.* 60%.

19. A put \$780 in a savings bank, which was 15% of all his money, and afterward deposited 25% of the rest of his money; what per cent. of all his money had he then in bank? *Ans.* $36\frac{1}{4}\%$.

20. A gold eagle of the United States weighs 258 gr. and the alloy in it weighs 25.8 gr.; what per cent. of the coin is alloy? *Ans.* 10%.

21. 35 per cent. of a regiment being sick, only 637 men were able to enter battle, of whom $\frac{1}{7}$ were killed; how many did the regiment number, and what per cent. of the whole number were killed? *Ans.* 980 men; $9\frac{2}{7}\%$.

GENERAL FORMULAS.

388. These methods and rules may all be represented in general formulas as follows:

CASE I.	CASE II.
1. Base × rate = Percentage.	1. Percentage ÷ rate = base.
2. Base × (1 + rate) = Amount.	2. Amount ÷ (1 + rate) = base.
3. Base × (1 − rate) = Difference.	3. Difference ÷ (1 − rate) = base.

CASE III.

Percentage ÷ base = rate.
Amount ÷ base = 1 + rate.
Difference ÷ base = 1 − rate.

NOTE.—These formulas apply to all the cases in the practical applications, and may be used instead of the rules, or with them, as the teacher prefers.

APPLICATIONS OF PERCENTAGE.

389. The **Applications of Percentage** are extensive, owing to the great convenience of reckoning by the hundred in business transactions.

390. These Applications of Percentage are of two

classes; those not involving time and those involving time The following are the most important of these applications

1ST CLASS.	2D CLASS.
1. Profit and Loss.	1. Simple Interest.
2. Commission.	2. Partial Payments.
3. Stocks, Dividends, etc.	3. True Discount.
4. Premium and Discount.	4. Discounting and Banking.
5. Brokerage.	5. Exchange.
6. Stock Investments.	6. Compound Interest.
7. Taxes.	7. Annuities.
8. Duties or Customs.	8. Insurance.

NOTES.—1. In the different cases of the application of percentage, care should be taken to see clearly the *base* upon which the percentage is reckoned.

2. A percentage deducted from the price of goods is called a *Discount*, and is treated under *Profit and Loss*. Successive *Discounts* called "Trade Discounts" are often taken off, as "10 and 5 per cent. off," meaning 10 per cent. off and 5 per cent. off of the remainder.

PROFIT AND LOSS.

391. **Profit** and **Loss** are terms which denote the gain or loss in business transactions.

392. The **Quantities** considered are as follows:

1. The **Cost,** which is the *base.*
2. The **Rate** of profit or loss.
3. The **Profit** or **Loss,** which is the *percentage.*
4. The **Selling Price,** which is the *amount* or *difference.*

NOTE.—In marking goods it is customary to take one or more words or a phrase or sentence, consisting of ten different letters, and let each letter in succession represent one of the Arabic figures. The prices marked thus can only be read by those who have the key.

CASE I.

393. ***Given, the cost and rate of profit or loss, to find the profit or loss, or the selling price.***

MENTAL EXERCISES.

1. A lady paid $10 for a shawl, and sold it at a loss of 20% required the loss.

SOLUTION.—At a loss of 20 per cent., $\frac{20}{100}$, or $\frac{1}{5}$ of the cost equals the loss. $\frac{1}{5}$ of $10 is $2. Therefore, etc.

2. A grocer bought tea at 80 cents a pound, and sold it so as to gain 25%; what was the gain?

3. Mr. Wilson paid $150 for a horse, and sold it at 16⅔% advance; what was his gain?

4. A dry goods merchant bought calico at 8 cents a yard, and sold it at a gain of 12½%; what was the selling price?

5. A farmer paid $48 for a cow; for what must he sell her to lose 6¼% by the transaction?

6. A furniture dealer bought a second hand sofa for $40, and having spent 10% on repairs, sold it at a gain of 25% on its whole cost: for what did he sell it?

WRITTEN EXERCISES.

1. A man bought a horse for $250, and sold it so as to gain 20%; what did he gain?

OPERATION.

$250
.20
$50.00

SOLUTION.—If the horse was bought for $250 and sold at a gain of 20%, the gain was .20 times $250, which is $50.

Rule I.—*Multiply the cost by the rate, to find the profit or loss.*

Rule II.—*Multiply the cost by* 1 *plus the rate of profit, or by* 1 *minus the rate of loss, to find the selling price.*

2. I bought $640 worth of English prints and sold them at a gain of 12%; what was the gain? *Ans.* $76.80.

3. Mr. Morgan sold his house, which cost $3680, at a loss of 5%; what did he receive for it? *Ans.* $3496.

4. A man weighing 162 lb., loses 33⅓% of his weight in a month; how much did he then weigh? *Ans.* 108 lb.

5. A dealer bought coal at $3.75 a ton, which he sells at 20% advance; what was his price? *Ans.* $4.50.

6. I sold a lot of envelopes marked $7.50 ℔ M, at 10 and 10% off; what was the price received? *Ans.* $6.075.

7. Some valentines marked $15 were sold at 25, 20, and 10% off; what was the selling price? *Ans.* $8.10.

8. I retail flour which cost $6.50 a barrel, at 15% advance; at what price per pound do I sell it? *Ans.* $.04—.

9. An agent gets pianos at 40% off, and sells at a gain of 25%; what does he get for a $650 piano? *Ans.* $487.50.

10. Mr. Bowman is obliged to mark down old-fashioned goods 20%; what does he deduct from those marked 6¼¢, 12½¢, 18¾¢, 25¢, and 40¢?

Ans. 1¼¢; 2½¢; 3¾¢; 5¢; 8¢.

11. What deduction from price is 10 and 15% off? what deduction is 5, 10, and 15% off? *Ans.* $23\frac{1}{2}$%; $27\frac{13}{40}$%.

12. What is the difference between 10% on and 5 and 5% on? between 20 and 10% on and 20 and 10% off?

Ans. $\frac{1}{4}$%; 60%.

13. A merchant bought 84 yd. of French chintz, at 20¢ per yard, and sold it a gain of $33\frac{1}{3}$%; what did he gain in the transaction? *Ans.* $5.60.

14. Henry bought a boat for $850, and sold it at a gain of 25%, and the buyer sold it at a loss of 20%; what did the latter receive for it? *Ans.* $850.

15. Mr. Warner's key for marking goods is, "now be quick;" if he buy a lot of calicoes at 10¢ a yard, how must he mark them to gain 35%? *Ans.* $nw\frac{n}{o}$.

CASE II.

394. ***Given, the rate and the profit or loss, or the selling price, to find the cost.***

MENTAL EXERCISES.

1. A boy sold a knife for 5 cents less than cost, and thereby lost 20%; what did it cost him?

SOLUTION.—If he lost 20 per cent., then $\frac{20}{100}$, or $\frac{1}{5}$ of the cost equals the loss, which equals 5 cents; if $\frac{1}{5}$ of the cost equals 5 cents, $\frac{5}{5}$, or the whole cost, equals 5 times 5 cents, or 25 cents.

2. A shoemaker charged 50 cents more for a pair of shoes than they cost him, and thereby gained 10%; what was the cost?

3. Edward sold his watch for $91, and thereby lost 35%; what did the watch cost him?

4. Mr. Pelton sold a set of maps for $15, and thereby lost 25%; for what should he have sold them to gain 30%?

5. A horse-jockey bought a horse for 15% less than its value, and sold it for 40% more than its value; what did he gain?

6. A picture-dealer sold two pictures for $120 each; on one he gained 25%, and on the other he lost 25%; what was his gain or loss?

7. A merchant sold a stove for $30, and thereby lost 25%; he then bought another stove for $30, and upon it gained 25%; what was the gain or loss?

WRITTEN EXERCISES.

1. A man gained $28 on a watch by selling it at a gain of 25%; what did the watch cost?

SOLUTION.—At a gain of 25%, .25 times the *cost* equals the gain, which is $28; if the *cost* multiplied by .25 equals $28, the *cost* equals $28 divided by .25, or $112.

OPERATION.
$28 ÷ .25 = $112

Rule I.—*Divide the profit or loss by the rate, to find the cost.*

Rule II.—*Divide the selling price by* 1 *plus the rate of profit, or by* 1 *minus the rate of loss, to find the cost.*

2. A merchant lost 15% by selling damaged delaines at 17¢ a yard; what did they cost him? *Ans.* 20¢.

3. Flour, sold at $7.54 a barrel, yields a profit of 16%; what did it cost per barrel? *Ans.* $6.50.

4. A furrier sold a set of sable furs at $12\frac{1}{2}$% less than cost, and lost $25; what did he get for them? *Ans.* $175.

5. What must I mark silks that cost $2.25 a yard so that I may fall 5% on my price and make 15%? *Ans.* 2.72\frac{7}{19}$.

6. A merchant's income is $5760 in a year, at a gain of $18\frac{3}{4}$% on his capital; how much would have been his income at a gain of 25% on his capital? *Ans.* $7680.

7. By selling my interest in a lead mine for $16,872, I gain 14%; how much would I have received for it if I had lost 14%? *Ans.* $12,728.

8. Prof. Winslow loses 16% by selling his library for $960 less than it cost; what must he have received for it if he had gained 16%? *Ans.* $6960.

9. On opening a case of goods that cost me $1.20 a yard, I find them slightly damaged; how shall I mark them that I may abate 25% and lose only 5%? *Ans.* $1.52.

10. A speculator sold two dwelling-houses for $6090 each; on one he gained 16%, and on the other he lost 16%; how much was gained or lost by the sale? *Ans.* Lost, $320.

11. A merchant bought a quantity of paper muslin @ 8¢, and marked it so that he could fall $9\frac{1}{11}$% on his marked price and gain 25% on cost; how must he mark it?

Ans. 11¢.

12. Mr. Smith's key for marking goods was "Republican;" if he marked some grenadines *e.bn* and gained at that rate $11\frac{1}{9}$% on the cost; what was the cost per yard? *Ans.* $2.25.

18. Mr. Baker sold Mr. Albert a farm for $85 an acre, and lost 15% on it; Mr. Albert sold the farm afterward to Mr. Hull, and made 15%; did Mr. Hull pay more or less an acre than Mr. Baker? *Ans.* $2.25 less.

CASE III.

395. ***Given, the cost and the profit or loss or the selling price, to find the rate.***

MENTAL EXERCISES.

1. A man bought a watch for $50, and sold it at a gain of $15; what was his gain per cent.?

SOLUTION.—If on $50 he gained $15, on $1 he gained $\frac{1}{50}$ of $15, which is $\$\frac{15}{50}$, or $\$\frac{3}{10}$; and on $100 he would gain 100 times $\$\frac{3}{10}$, which are $\$\frac{300}{10}$, or $30; hence the gain is 30 per cent.

2. If a coal dealer buys coal @ $5, and sells it at a gain of $2, what is his gain per cent.?

3. Sold a set of parlor furniture which cost $500 for $375; what was the loss per cent.?

4. What % is gained by selling goods at *double* their cost? What % is lost by selling them at *one-half* their cost?

5. A grocer buys a quantity of A No. 1 mackerel at $20 a barrel, and retails them at 15¢ a lb.; what is his gain %?

6. If a miller takes $3\frac{1}{5}$ quarts out of every bushel he grinds for toll, what % does he take for toll?

7. If goods are bought at 20% below market price and sold at 16% below that price, what is the gain %?

8. I bought a lot of goods at 5% below current prices, and sold them 5% above those prices; what was my gain %?

9. I imported a quantity of foreign goods, but some of them having got wet on the voyage, I sold them at a loss of $\frac{1}{16}$ of their cost; what was my loss %?

10. A lady sold some silk for $12, and thereby cleared $\frac{1}{4}$ of this money; what would she have lost %, if she had sold it for $6?

WRITTEN EXERCISES.

1. A man bought a horse for $200 and sold it at a loss of $20; what was the loss per cent.?

SOLUTION.—Since $200, the *base*, multiplied by the *rate*, equals $20, the rate must equal $20 divided by $200, which is .10, or 10%.

OPERATION.
$20 ÷ $200 = .10

Rule I.—*Divide the profit or loss by the cost, to find the rate.*

Rule II.—*Divide the difference between the cost and the selling price by the cost, to find the rate.*

2. Some muslin was bought for 8½¢ a yard and sold for 12¾¢; what was the gain %? *Ans.* 50%.

3. I sold a lot of damaged goods that cost me $.84 a yard for $.63; what was the loss per cent.? *Ans.* 25%.

4. If I buy paper at $3.50 a ream and sell it at 25¢ a quire, what is the gain %? *Ans.* 42$\frac{6}{7}$%.

5. If I buy at 20 and 10% off and sell for 20 and 10% on, what do I gain? *Ans.* 83⅓%.

6. Bought valentines at 25, 20 and 10% off and sold for 10 and 5% on; what per cent. did I gain? *Ans.* 113$\frac{8}{9}$%.

7. I bought a lot of goods for 15% below market price, and sold them for 15% above market price; what % did I clear? *Ans.* 35$\frac{5}{17}$%.

8. A man shipped 600 barrels of flour, and lost 16⅔% of it by a storm; he sold 75% of the remainder; what % of the whole remained? *Ans.* 20⅚%.

9. Mr. Jackson bought 500 shares of mining stock for $9000, and sold 400 shares for what they all cost; required the gain per cent. *Ans.* 25%.

10. Henry sold his horse and carriage for $450, and thereby cleared ⅕ of this money; what would he have gained % by selling them for $390? *Ans.* 8⅓%.

11. I bought a watch for $120, and set such a price on it that after falling $12, I still made 15% on the purchase; what % did I abate from the asking price? *Ans.* 8%.

12. A lady sold her piano for $350, and thereby cleared 20% of this money; what would she have gained %, if she had received $70 more than she did? *Ans.* 50%.

13. Mr. Marble bought a lot of cassimeres @ $3.75, and marked them *n.no*$\frac{j}{o}$, his key being "John Marble;" what was his gain % at the marked price? *Ans.* 18%.

14. I offered my house for sale at 35% advance on its cost, but finding no purchasers at that price it was finally sold at 35% less than was first asked; what was the gain or loss %? *Ans.* Loss 12¼%.

COMMISSION.

396. **Commission** is a percentage paid to an agent for the transaction of business.

397. An **Agent** is a person who transacts business for another; he is often called a *Commission Merchant*, a *Factor*, etc.

398. The **Base** in Commission is the actual *amount* of the *sale, purchase, collection*, or *exchange.*

399. The **Net Proceeds** is the sum left after the commission and charges have been deducted from the amount of a sale or collection.

400. The **Entire Cost** is the sum obtained by adding the commission and charges to the amount of a purchase.

401. The **Quantities** considered are: 1. The *Amount sold, bought*, etc.; 2. The *Rate of Commission;* 3. The *Commission;* 4. The *Entire Cost* or *Net Proceeds.*

The goods forwarded to be sold on commission are called a *consignment;* the person sending them is called the *consignor;* and the person to whom they are sent, the *consignee*, or *Factor*. An agent residing at a great distance from his employer, is often called a *correspondent;* the person for whom an agent does business is called the *Principal.*

CASE I.

402. ***Given, the base and rate to find the commission or net proceeds, or entire cost.***

1. An agent bought a house for $8650, his rate of commission being $3\frac{1}{2}\%$; what was his commission?

OPERATION.

SOLUTION.—The commission was $.03\frac{1}{2}$ times $8650, which equals $302.75.

$$\begin{array}{r} \$8650 \\ .03\frac{1}{2} \\ \hline \$302.75 \end{array}$$

Rule I.—*Multiply the base by the rate, to find the commission.*

Rule II.—*Multiply the base by* 1 *minus the rate, to find the net proceeds; or by* 1 *plus the rate, to find the entire cost.*

WRITTEN EXERCISES.

2. A factor sold goods to the amount of $7650, rate of commission being $3\frac{1}{3}\%$; required the commission and the amount paid over. Ans. $7395 paid over.

3. A sells $5472 worth of dry goods, charging $3\frac{1}{2}\%$ commission and $1\frac{1}{4}\%$ for insuring payment; what sum will he remit to his employer? *Ans.* $5212.08.

4. A lawyer having a debt of $1536 to collect, compromises for 95%; what is his commission at $4\frac{1}{2}\%$, and what does he remit to his employer? *Ans.* Com. $65.66.

5. My agent bought 40 horses for $150 each, and paid $25 for their keeping and $80 for transportation; his commission was $3\frac{1}{2}\%$; what did the horses cost me? *Ans.* $6315.

6. What would be the net proceeds of a sale of 450 bbl. of prime mess pork @ $17.12½, allowing $2\frac{1}{2}\%$ commission, and paying 5¢ a barrel storage for 30 days? *Ans.* $7491.09.

7. A tax collector had a warrant for $25,850, upon which he collected $12,500 at $1\frac{1}{2}\%$, and the balance at $2\frac{1}{2}\%$; required the amount of the collector's fees. *Ans.* $521.25.

8. An architect was employed to erect a city hall which cost $75,000, and was allowed $\frac{3}{8}\%$ for plans and specifications, and $1\frac{1}{2}\%$ for superintendence; but on settling accounts he claimed $1500; how much did he overcharge the city?

Ans. $93 75.

CASE II.

403. ***Given, the rate and the commission or the net proceeds or the entire cost, to find the base.***

1. An agent receives $84 commission for buying goods, at the rate of $1\frac{1}{3}\%$, what was the cost of the goods?

SOLUTION.—At a commission of $1\frac{1}{3}\%$, $.01\frac{1}{3}$ times the cost of the goods equals the commission, which is $84; hence, the *cost* equals 84 divided by $.01\frac{1}{3}$, which we find is $6300.

OPERATION.

$$\frac{84}{.01\frac{1}{3}}=6300$$

2. An agent receives $4920 to be invested in cotton after retaining his commission, $2\frac{1}{2}\%$; required the amount invested.

SOLUTION.—The sum to be invested, increased by $2\frac{1}{2}\%$ of itself, equals $1.02\frac{1}{2}$ times the sum, which equals $4920. If $1.02\frac{1}{2}$ times the *sum* equals $4920, the *sum* equals $4920 divided by $1.02\frac{1}{2}$, which we find is $4800.

OPERATION.

$$\frac{4920}{1.02\frac{1}{2}}=4800$$

Rule I.—*Divide the commission by the rate, to find the base.*

Rule II.—*Divide the net proceeds by 1 minus the rate, or the entire cost by 1 plus the rate, to find the base.*

WRITTEN EXERCISES.

3. A lawyer's commission for making collections for a firm at $2\frac{1}{2}\%$ is $1600; how much did he collect? *Ans.* $64,000.

4. A miller sent his Detroit agent $9270 to be invested in flour, after deducting his commission of 3%; what was the commission? *Ans.* $270.

5. An agent buys hides on commission, at $\frac{3}{4}\%$, and pays $25 for cartage; the entire bill was $4558.75; what was the commission? *Ans.* $33.75.

6. A commission merchant sells goods for a party at $1\frac{1}{4}\%$, and charges $2\frac{1}{2}\%$ for guaranteeing the payment of the money; his commission was $284.25; required the amount of goods sold. *Ans.* $7580.

7. A cotton factor received $1132.71 to invest in cotton at $.24 a pound, deducting $3\frac{1}{2}\%$ commission; how many pounds did he buy? *Ans.* 4560 lbs.

8. An agent bought 40 horses on commission, at $4\frac{1}{2}\%$; he paid $25 for keeping and $50 for transportation, which, with his commission, amounted to $345; what did the horses cost apiece? *Ans.* $150.

9. I sold some goods on commission at 5%, through an agent, who charged me 3%; my commission, after paying my agent, was $388; required the agent's commission, my commission, and the money paid to my employers.

Ans. My com., $970; agent's, $582; sum paid, $18,430.

CASE III.

404. ***Given, the base and the commission or the net proceeds or the entire cost, to find the rate.***

1. A commission merchant collects $7860, and his commission was $393; required the rate of commission.

SOLUTION.—The commission, $393, equals the base, $7860, multiplied by the *rate;* hence, the *rate* equals $393 divided by $7860, which we find is .05, or 5%.

OPERATION.

$$\frac{\$393}{\$7860} = .05$$

Rule I.—*Divide the commission by the base, to find the rate.*

Rule II.—*Divide the difference between the base and the net proceeds or the entire cost, by the base, to find the rate.*

WRITTEN EXERCISES.

2. A factor sold some land, and paid over $7742.10, retaining $117.90 as commission; required the rate.

Ans. $1\frac{1}{2}\%$.

3. An agent bought some flour, paid $$54\frac{1}{2}$ storage, and charged $180 commission; his entire bill was $$8234\frac{1}{2}$; what was the rate of commission? *Ans.* $2\frac{1}{4}\%$.

4. I sold a consignment of cotton goods through an agent for $2500; my commission was $112.50, and I paid the agent $37.50; what was the rate of commission of each?

Ans. Mine, $4\frac{1}{2}\%$; Agent's $1\frac{1}{2}\%$.

5. My factor sold a consignment of sugar for which he remitted a note for $8500; he charged $127.50 for guaranteeing payment and $191.25 for commission; what was his rate of commission and of guaranty?

Ans. Com. $2\frac{1}{4}\%$; guaranty, $1\frac{1}{2}\%$.

STOCKS AND DIVIDENDS.

405. A **Company** is an association of individuals for the transaction of business.

406. A **Corporation** is a company regulated in its operations by a general law or a special charter.

407. The **Stock** of a company is the capital invested in the business. The owners of stock are called *Stockholders*.

408. A **Share** is one of the equal parts into which the stock is divided. A share is usually $50 or $100.

409. An **Installment** is a sum required of stockholders as a payment on their subscription.

410. A **Dividend** is a sum paid to stockholders out of the gains of the company.

411. An **Assessment** is a sum required of stockholders to meet the expenditures or losses of the company.

412. The **Base** upon which dividends and assessments are estimated is the original or par value of the stock.

413. The **Quantities** considered are as follows: 1. The *Stock;* 2. The *Rate;* 3. The *Dividend* or *Assessment.*

CASE I.

414. ***Given, the stock and rate of dividend or assessment, to find the dividend or assessment.***

1. A owns $20,000 of the stock of a bank which declares a dividend of 8%; what is his dividend?

OPERATION.

$20000
.08
$1600.00

SOLUTION.—If A has $20,000 worth of stock, and the bank declares a dividend of 8%, his dividend is .08 times $20,000, which is $1600.

Rule.—*Multiply the par value of the stock by the rate, to find the dividend or assessment.*

NOTE.—It is often convenient to find the result by multiplying the dividend or assessment on one share by the number of shares.

WRITTEN EXERCISES.

2. Miss Atherton bought 78 shares of Reading R. R. stock, at $50; the company declares a dividend of 4%; what is her dividend? *Ans.* $156.00.

3. Miss Lyle owns 65 shares, at $50, in an insurance company, which on account of losses, requires an assessment of 2½ per cent.; what does she pay? *Ans.* $81.25.

4. The Union gas company, whose stock is $785,000, declares a semi-annual dividend of 3½ per cent.; required the amount of dividend. *Ans.* $27475.

5. A has 40 shares, $50 each, of stock in a bank, which declares a dividend of 5%; what is A's dividend, and how many shares of stock would it buy at par? *Ans.* 2 shares.

6. A man owns 50 shares of Salem turnpike stock ($100); the company declares a dividend of 8%, payable in stock; how many shares will he then own? *Ans.* 54 shares.

7. A company whose capital is $250,000, pays a dividend of $84 on 24 shares ($100), and reserves as a surplus, $5760; what were the net earnings? *Ans.* $14510.

CASE II.

415. ***Given, the rate and the dividend or assessment, or the result of increase or decrease of stock, to find the stock.***

1. A bank divides $8400 among the stockholders, being the amount of 7% dividend; required the whole amount of stock.

SOLUTION.—If $8400 is 7% of the stock, then .07 times the *stock* equals $8400; hence, the *stock* equals $8400 divided by .07, which is $120000.

OPERATION.

$$\frac{\$8400}{.07} = \$120000, \textit{ Ans.}$$

Rule I.—*Divide the dividend or assessment by the rate, to find the stock.*

Rule II.—*Divide the result of increase by 1 plus the rate, or the result of decrease by 1 minus the rate, to find the stock.*

WRITTEN EXERCISES.

2. I received $880 from a $5\frac{1}{2}$ per cent. dividend; how much stock do I own? *Ans.* $16000.

3. I receive $279 as my share of a 9% dividend; how many shares, at $50 each, do I own? *Ans.* 62 shares.

4. A company divides $72000 among its stockholders, as the result of an 8% dividend; what is B's stock, provided he owns $\frac{1}{8}$ of the entire stock? *Ans.* $112500.

5. A lady receives $1260 dividend at 7%; required the amount of stock she owns and the number of shares, valued at $50 each. *Ans.* 360 shares.

6. Mr. B receives $7800, payable in stock, as his share of a 12% dividend; how many shares had he at first, and how many has he now, shares at $50? *Ans.* 1456.

7. A gentleman received 7 shares and $25 in money, as his share of a 6% dividend; how many shares, valued at $50, did he then own? *Ans.* 132 shares.

8. In 1864 I received a stock dividend of 25% in the Camden and Amboy Railroad, and I then had 80 shares, at $100 each; how many shares had I at first?

Ans. 64 shares

9. I received a stock dividend of 10% in an oil company in March, 1865, and a similar dividend of 12% in November; I then owned 308 shares at $25; how many shares had I at the beginning of the year? *Ans.* 250 shares.

10. The expenses of an insurance company, capital $400,000, are 75% of the gross earnings; it reserves $10,000 and pays a dividend of $4\frac{1}{2}$%; what were the gross earnings? *Ans.* $112,000.

CASE III.

416. ***Given, the stock and dividend or assessment, or result of increase or decrease of stock, to find the rate.***

1. A company whose stock is $840000, clears $56000 in a year; what rate of dividend can it declare?

SOLUTION.—Since the dividend is some per cent. of the stock, the base, $840000, multiplied by the *rate* equals $56000; hence, the *rate* equals $56000 divided by $840000, which equals $.06\frac{2}{3}$.

OPERATION.

$$\frac{56000}{840000} = .06\tfrac{2}{3}, \textit{ Ans.}$$

Rule I.—*Divide the dividend or assessment by the stock, to find the rate.*

Rule II.—*Divide the difference between the stock and the result of increase or decrease, by the stock, to find the rate.*

WRITTEN EXERCISES.

2. A company whose stock is $125000, requires an assessment of $1875; what was the rate? *Ans.* $1\frac{1}{2}$%.

3. Mr. A owns 288 shares of stock, at $100, and draws a dividend of $1944; what was the rate? *Ans.* $6\frac{3}{4}$%.

4. The earnings of a canal company for 6 months are $70000, the stock is $2,330,000; if they declare a dividend whose rate is an integer, what is the largest rate, and what is the surplus? *Ans.* 3%; $100 surplus.

5. A owns 70 shares ($100) in a railroad company whose stock is $4000000, and his dividend is $402.50; required the rate of dividend, and the whole dividend. *Ans* $5\frac{3}{4}$%.

6. After receiving a stock dividend, I had 73 shares ($50) and $10 toward another share; what was the rate of dividend, if I had 61 shares at first? *Ans.* 20%.

7. I hold 350 shares in a Pittsburgh gas company ($50), and received two stock dividends, the first amounting to 42 shares, and the second to 58 shares and $40; what were the rates of dividend? *Ans.* 12% and 15%.

PAR, PREMIUM, AND DISCOUNT.

417. **Capital** is property consisting of *Money, Bonds, Stocks, Drafts,* etc.

418. **Drafts, Checks,** and **Bills of Exchange** are written orders for the payment of money at some definite place.

419. **Stocks** is a general name applied to the scrip or bonds of a corporation, and to government bonds and public securities.

420. **Scrip** or **Certificates of Stock** are the papers issued by a corporation to its stockholders, as evidence of the number of shares belonging to each respectively.

421. **Bonds** are written or printed obligations to pay certain sums of money at or before a specified time.

422. **State Stocks** or **United States Stocks** are bonds of a State, or of the United States, payable at some future time, with interest at a fixed rate.

423. The **Par Value** of capital is the value marked on its face, called the *nominal value* or *face.*

424. The **Real Value** or **Market Value** of capital is what it will sell for.

425. Capital is **Above Par,** or at a *premium* or *advance,* when it sells for more than its nominal value. Capital is *below par,* or at a *discount,* when it sells for less than its nominal value.

The stock of a company will generally be above par when the company is doing a lucrative business, and below par when it is doing a poor business. The stock of a town, city, etc., varies according to the confidence in its security, the fluctuations of the money market, etc.

Besides bonds, the U. S. Government issues *notes,* payable on demand without interest, which are a legal tender for all debts due the United States except duties. These notes, called "greenbacks," are, together with notes issued by the National Banks, the present circulating medium, and are called *currency.*

If the currency becomes depreciated in value, *gold* becomes an object of investment, the same as stocks. The value of gold being fixed, its fluctuations in price indicate the changes in the value of the currency. Thus, when gold is said to be at a *premium*, currency is really at a *discount*.

426. The **Base** upon which premium and discount are estimated is the *par value.*

427. The **Quantities** considered are four: 1. The *Par Value;* 2. The *Rate;* 3. The *Premium or Discount;* 4 The *Real Value.*

NOTE.—The problems under this subject are solved without brokerage—the sales and exchanges being regarded as direct without the aid of a broker.

CASE I.

428. ***Given, the par value and the rate of premium or discount, to find the premium or discount or real value.***

1. A broker bought 25 shares of stock ($50), at 5% premium; required the premium and cost, or real value.

SOLUTION.—The par value of 25 shares at $50 each is $50 × 25 = $1250; and the premium at 5% is .05 times $1250, which is $62.50; and this, added to the par value, equals $1312.50, the real value.

OPERATION.

$50 × 25 = $1250, par value.
.05
$62.50, premium.
1250
$1312.50, real value.

Rule I.—*Multiply the par value by the rate, to find the premium or discount.*

Rule II.—*Multiply the par value by 1 plus the rate of premium, or by 1 minus the rate of discount, to find the real value.*

WRITTEN EXERCISES.

2. B sold 46 shares of bank stock ($100), at 3% discount; required the discount and real value. *Ans.* $4462.

3. In 1858, I sold a $20 note on an Ohio bank, at $\frac{3}{5}$% discount; what did I receive for it? *Ans.* $19.88.

4. A broker paid currency for $560 in gold at $10\frac{1}{2}$% premium; how much currency did he pay, and what was the premium? *Ans.* He paid $618.80.

5. A speculator bought 35 shares of bank stock ($100), at $3\frac{1}{4}\%$ discount, and sold it at $1\frac{1}{2}\%$ premium; what was his gain? *Ans.* $166.25.

6. A banker bought 48 shares ($100) of canal stock, at 6% premium, and paid for them with $5000 in drafts, at $3\frac{1}{4}\%$ discount, and the balance in cash; how much cash did he pay? *Ans.* $250.50.

7. A lady exchanged 45 shares ($100) railroad stock, at $4\frac{1}{5}\%$ discount, for 70 shares of bank stock ($50), at 5% premium, receiving the difference in cash; what amount of cash did she receive? *Ans.* $636.

CASE II.

429. ***Given, the rate and the premium or discount or the real value, to find the par value.***

1. A man sold some securities at a discount of 5%, receiving $120 less than their face; what was their face value?

SOLUTION.—If the discount at 5% is $120, then .05 times the *par value* equals $120; hence, the *par value* equals $120 divided by .05, which we find is $2400.

OPERATION.

$120 ÷ .05 = $2400

Rule I.—*Divide the premium or discount by the rate, to find the par value.*

Rule II.—*Divide the real value by 1 plus the rate of premium, or by 1 minus the rate of discount, to find the par value.*

WRITTEN EXERCISES.

2. B sold some stocks at $3\frac{1}{2}\%$ premium and gained $210; what was their par value? *Ans.* $6000.

3. The premium on a draft at $\frac{3}{4}\%$ was $.90; required the face of the draft and its value. *Ans.* $120.90.

4. Mr. Allen paid $2587.50 for a bond, at $3\frac{1}{2}\%$ premium: required its face and the premium. *Ans.* Face, $2500.

5. Mr. Jones paid $5926.50 in currency for gold at $9\frac{3}{4}\%$ premium; how much did he purchase? *Ans.* $5400.

6. Sold stock bought at par, at an advance of $3\frac{1}{3}\%$, and gained $145; how many shares ($50) did I sell? *Ans.* 87

7. Miss Hartman sold 140 shares of Columbia National bank stock at $54 a share, premium 8%; required the par value and entire premium. *Ans.* $50; Prem. $560.

8. I gave a draft worth $\frac{3}{4}$% premium, for 75 shares of turnpike stock ($50) at 3% discount; what was the face of the draft? *Ans.* $3610.42+.

9. Mr. Dean sold 40 shares of stock at a premium of $4\frac{1}{2}$%, and received $180 advance; what was the par value of a share? *Ans.* $100.

10. A broker exchanged 700 shares of stock ($100), at 5% discount, for United States bonds ($100), at 5% premium, paying $70 in money; how many did he get?

Ans. 634 bonds.

11. Mr. Fish bought a number of shares of bank stock ($50), the discount at 5% being $200; $\frac{1}{4}$ of it he sold at par and the rest at 7% advance; what was the average gain on each share? *Ans.* 5.12\frac{1}{2}$.

CASE III.

430. ***Given, the par value and the real value or the premium or discount, to find the rate of premium or discount.***

1. I sold a note, drawn for $860, at a premium of $51.60; what was the rate of premium?

SOLUTION.—Since the premium equals the par value multiplied by the rate, $860 multiplied by *the rate* equals $51.60; hence, *the rate* equals $51.60 divided by $860, which we find is .06.

OPERATION.

$$\frac{51.60}{860} = .06$$

Rule I.—*Divide the premium or discount by the par value, to find the rate.*

Rule II.—*Divide the difference between the real value and the par value by the par value, to find the rate.*

WRITTEN EXERCISES.

2. I bought a draft, drawn for $1680, at a discount of $12.60; required the rate of discount. *Ans.* $\frac{3}{4}$%.

3. Mr. Peters bought 96 shares of railroad stock ($50) for $4476; what was the rate of discount? *Ans.* $6\frac{3}{4}$%.

4. If he sells these 96 shares for $4699.80, what is the rate of *discount*, and rate of *gain?* *Ans.* $2\frac{7}{80}$%; 5%.

5. Mr. Reed gave $7500 in notes, at 2% discount, for $6125 in gold; what was the rate of premium on the gold?

Ans. 20%.

6. A banker bought $500 in gold, premium $10\frac{1}{4}$%, and sold it for $600; what was the rate of premium on the sale, and what was the gain %? *Ans.* 20%; $8\frac{124}{147}$% gain.

BROKERAGE.

431. **Brokerage** is a percentage charged by brokers for the transaction of business.

432. A **Broker** is a person who buys or sells money, stocks, bills of exchange, real estate, etc., for others. A *Stock Broker* is one who deals in stocks, but is generally called simply a *broker*.

433. The **Base** upon which the commission for the purchase and sale of bonds and stocks is estimated is their *par value.*

434. The **Rate** is usually $\frac{1}{4}$%, and is so understood if no other rate is mentioned. In New York and Philadelphia custom has fixed the rate at $\frac{1}{4}$%.

435. The **Quantities** considered are: 1. The *Par value of the amount sold, bought,* etc.; 2. The *Rate of Brokerage;* 3. The *Brokerage;* 4. The *Market value of* $100, *or of* 1 *share;* 5. The *Entire Cost, or Net Proceeds.*

NOTES.—1. Stocks are quoted either at the price of one share, or at the price of $100 of par value of the stock, whatever be the par value of a share. The former method is used in Philadelphia; the latter in New York.

2. Stocks are often named from the rate of interest they draw; thus, we have 5's, 6's, 7-30's, etc. The time to run or date when due sometimes gives the name; as 5-20's, '81's, etc.

CASE I.

436. ***Given, the par value, the rate and the market value, to find the brokerage, net proceeds or entire cost.***

1. A broker bought for a party 15 shares Pennsylvania

R. R. ($50), rate of brokerage being $\frac{1}{4}\%$; required the brokerage.

SOLUTION.—The par value was $15 \times \$50$, or \$750. The brokerage was $.00\frac{1}{4}$ times \$750, which equals $\$1.87\frac{1}{2}$.

OPERATION.

$\$50 \times 15 = \$750$
$\$750 \times .00\frac{1}{4} = \$1.87\frac{1}{2}$

Rule I.—*Multiply the par value by the rate, to find the brokerage.*

Rule II.—*Multiply the par value by the market value minus the rate, to find the net proceeds; or by the market value plus the rate, to find the entire cost.*

NOTE.—It is often shorter to multiply the brokerage on one share, by the number of shares. When the par is \$50, one-half the rate should be used in applying the rule.

WRITTEN EXERCISES.

2. I bought through a broker 46 shares of bank stock (\$50) at par, brokerage being $\frac{1}{4}\%$; required the brokerage, and the cost of the stock. *Ans.* \$5.75.

3. A broker bought for me 76 shares of bank stock (\$50) at $47\frac{1}{2}$; what did the stock cost me, the brokerage being $\frac{1}{4}$ per cent.? *Ans.* \$3619.50.

4. Mr Lyte sold through his banker 72 shares New York Central (\$100) at 103; required the brokerage and net proceeds. *Ans.* Proceeds, \$7398.

5. My broker bought on my account 25 shares Bank of North America (\$100), at 150, and sold them at 161; what was his commission and my profit? *Ans.* Profit, \$262.50.

6. Shall I gain or lose if I buy 65 shares Northern Central (\$50) at $53\frac{1}{2}$, and after receiving two 4% dividends sell them for $51\frac{1}{4}$, brokerage $\frac{1}{4}\%$, interest on money not considered? *Ans.* \$97.50 gain.

CASE II.

437. ***Given, the rate, the brokerage, or the net proceeds, or entire cost, and the market value, to find the par value.***

1. A paid a broker \$150 for selling some drafts, at the rate of $2\frac{1}{2}\%$; what amount of drafts did he sell?

SOLUTION.—At a rate of $2\frac{1}{2}\%$, .025 times the par value of the drafts equals the brokerage, which is \$150; hence the *par value* equals \$150 divided by .025, which we find is \$6000.

OPERATION.

$$\frac{\$150}{.025}=\$6000$$

Rule I.—*Divide the brokerage by the rate, to find the par value.*

Rule II.—*Divide the net proceeds by the market value minus the rate, or the entire cost by the market value plus the rate, to find the par value.*

WRITTEN EXERCISES.

2. I paid a broker \$12.50 at $\frac{1}{8}\%$ for buying N. Y. Central (\$100); how many shares did he buy? *Ans.* 100 shares.

3. I paid my broker \$4712.50 for an investment in Missouri 6's (100), at 94, including brokerage at $\frac{1}{4}\%$; what was the par value of the bonds? *Ans.* \$5000.

4. I sent a New York broker a draft on Fisk & Hatch for \$4953, to cover an investment made by my order in Harlem Railroad at 95 (\$100), and his commission of $\frac{1}{4}\%$; how many shares shall I receive? *Ans.* 52 shares.

5. My broker sold \$3000 Philadelphia 6's at $101\frac{1}{4}$, and invested the proceeds in United Companies of New Jersey stock at $131\frac{5}{8}$ (\$100); how many shares did he buy, brokerage at $\frac{1}{4}\%$? *Ans.* 22 shares; \$128.75 surplus.

6. Mr. Westlake bought Pennsylvania R. R. stock (\$50) at $49\frac{7}{8}$, and sold it at $53\frac{1}{4}$; after paying brokerage, he found he had a profit of \$237.50; how many shares did he buy? *Ans.* 76.

7. A merchant wishing to meet a note for \$5000, directed his broker to sell sufficient West Philadelphia Pass. Railway stock (\$50) to cover the note and brokerage; if the stock was selling at $78\frac{3}{8}$, how many shares must he sell, and what would be the surplus? *Ans.* 64 shares; \$24 surplus.

8. I sold 25 shares of Philadelphia National Bank (\$100) at $156\frac{1}{2}$, and directed my broker to invest the proceeds in Norristown R. R. stock (\$50) at 99; what is the amount of investment, after deducting brokerage?

Ans. 39 shares; $\$40.37\frac{1}{2}$ surplus.

CASE III.

438. ***Given, the par value, and the brokerage, or the net proceeds, or entire cost, and the market value, to find the rate.***

1. A broker bought Reading convertible coupon 7's, par value $4000; his charge was $10; what was the rate of brokerage?

SOLUTION.—The brokerage, $10, equals the par value, $4000, multiplied by the *rate*; hence, the *rate* equals $10 divided by $4000, which we find is .00¼, or ¼%.

OPERATION.

$$\frac{\$10}{\$4000} = .00\tfrac{1}{4}$$

Rule I.—*Divide the brokerage by the par value, to find the rate.*

Rule II.—*Divide the difference between the real value of the stock, and the net proceeds or entire cost, by the par value, to find the rate.*

WRITTEN EXERCISES.

2. A broker buys 110 shares of gas stock, par value $25 a share; his charge was $6.87½; what was the rate of brokerage? *Ans.* ¼%.

3. A broker, having purchased, according to order, $5600 Rhode Island 6's at 110, informs me that the entire cost is $6188; what brokerage does he charge? *Ans.* ½%.

4. I sent a draft for $21250 to a Detroit broker, to invest in Michigan 6's at 106; he remitted me a balance of $25; what rate of brokerage did he charge? *Ans.* ⅛%.

INCOME FROM INVESTMENTS.

439. Investments in stocks, etc., may be made either for interest on the money or for the increase of capital.

440. There are **Several Classes** of stocks, viz.: those of *Corporations, States*, and the *General Government*.

441. Bonds are distinguished as *Registered* and *Coupon Bonds*. The *Registered* bonds are payable to order, and cannot be transferred without being indorsed.

442. The **Coupon** bonds have coupons or certificates of interest attached to them, which may be cut off and the interest collected when due.

443. The principal bonds of the United States are as follows:

U. S. 6's of '81 are payable in 1881, interest 6% in gold, due Jan. 1st and July 1st. U. S. 7-30's were bonds issued during the war and converted at maturity into 5-20's; interest $7\frac{3}{10}$% currency.

U. S. 5-20's, payable in not less than 5 or more than 20 years, at the option of the Government. There are several series of these bonds, called from the years in which they were issued, 5-20's of '62, '64, '65, '65 new issue (*n. i.*), '67, and '68. Interest 6% in gold, payable on the first three series May 1st and November 1st; on the last three, January 1st and July 1st.

U. S. 10-40's, payable in not less than 10 nor more than 40 years from date, at the option of the Government. Interest 5% in gold, payable on registered bonds and on $500 and $1000 coupon bonds March 1st and September 1st; on $50 and $100 coupon bonds once a year March 1st.

U. S. 5's of '81, payable in 1881. Interest 5% in gold, payable quarterly Feb. 1st, May 1st, Aug. 1st, Nov. 1st. U. S. Pacific Railroad 6's, payable in 1895, and thereafter, were issued to aid in constructing several railroads to the Pacific coast. Interest, 6% in currency, payable January and July. U. S. $4\frac{1}{2}$'s, redeemable after 1886; interest $4\frac{1}{2}$%; payable quarterly; also 4s due 1907, int. 4%, payable quarterly.

444. A **Mortgage** is a conditional conveyance of property as security for the payment of a debt.

Should the interest not be promptly paid, the mortgage may be *foreclosed*, and the property is then sold by the sheriff to the highest bidder, and the mortgage paid off from the proceeds. Property is usually not mortgaged beyond a certain part of its value, in order that the mortgagee may be secure from loss. A second mortgage is sometimes given, but this cannot be paid, in case of foreclosure, till the first is fully paid, and hence may not be a very good security.

445. A **Ground-Rent** is a fixed rent paid for ground, generally used for building purposes.

It is a common practice in some cities, when a person wishes to build one or more houses, instead of *buying* the ground required, to agree to pay the interest on its value as rent, the contract to continue in force as long as the rent is regularly paid. Ground-rents are *redeemable* or *irredeemable*. Some cities, as Philadelphia, prohibit the issue of any more irredeemable ground-rents.

Mortgages and *ground-rents* are not bought and sold at the Stock Exchange, but conveyancers are frequently employed in the transaction, as the title and condition of the property must be examined, and the necessary papers drawn up. Well-secured mortgages and ground-rents are in such high esteem as safe investments, that they are among the securities in which trust funds may be legally invested.

NOTE.—In changing from one investment to another, there is often a little more realized from the sale of the first than will procure an exact number of shares of the second. In such cases the income will be calculated on the number of shares, without noticing the surplus.

446. The **Quantities** considered are: 1. The *Amount Invested;* 2. The *Rate of Dividend or Interest;* 3. The *Income;* 4. The *Market Value of* $100; *or of one share;* 5. The *Rate of Income.*

CASE I.

447. ***Given, the amount of an investment, the market value, and the rate of dividend or interest, to find the income.***

1. If I invest $5100 in 7% bonds at 85, what will be my annual income from them?

SOLUTION.—Since for 85 cents you can buy $1 worth of stock, for $5100 you can buy as many dollars worth of stock as $.85 is contained times in $5100, or $6000. The annual income on this is $6000×.07 which equals $420.

OPERATION.

$5100÷.85=$6000
$6000×.07=$420

Rule.—I. *Divide the amount invested by the market value, to find the par value.*

II. *Multiply the par value by the rate to find the income.*

WRITTEN EXERCISES.

2. What annual income would I receive from $16050 invested in U. S. Pacific R. R. 6's at 107? *Ans.* $900.

3. If I invest $5631.25 in 5-20's at 112⅝, what is my annual income in currency, gold being worth 109¾? *Ans.* $329.25.

4. If I invest $5280 in United Companies of New Jersey, stock at 132, dividend 10%, what will be my income? *Ans.* $400.

5. Miss Brown has invested $8475 in 10-40's at 113; what will be her semi-annual income in currency, gold being 110? *Ans.* $206.25.

6. A conveyancer sold a lot 25 ft. front and 50 ft. deep on ground-rent, redeemable on payment of $1500; what is the ground-rent at 6%? *Ans.* $90.

7. Mr. Tompkins bought on ground-rent a lot 75 ft. front by 90 ft. deep, valued at $87.25 per foot front; what would be the ground-rent per foot front at 6%? *Ans.* $5.23½.

8. I made $5000 by a fortunate speculation, and wishing to invest it permanently, I bought $2000 6's of 81 at $117\frac{3}{8}$ and invested the remainder in the new $4\frac{1}{2}$'s at $110\frac{1}{4}$; what surplus remained after deducting brokerage, and what was my annual income in gold? *Ans.* Sur., $50.75; In., $225.75.

9. Mrs. Warner has $10,000 Philadelphia City 6's, at $103\frac{3}{4}$; would she increase or diminish her income for that year if she should exchange them for 5-20's at $110\frac{1}{8}$, gold being $108\frac{1}{2}$? *Ans.* Increase, $11.94.

10. Mr. Barton conveys a lot on a 6% ground-rent payable in gold and redeemable on payment of $4500; at what sum in currency must it be made redeemable to realize an equivalent rent, and what is the ground-rent in currency, gold being $112\frac{1}{2}$? *Ans.* $5062.50; Rent, $303.75.

CASE II.

448. ***Given, the income, the rate of dividend, and the market value, to find the amount invested.***

1. When U. S. 10-40's are selling at 110, how much must be invested to produce an income of $550?

SOLUTION.—Since $1 of stock gives an income of $.05, to give an income of $550 it will require $550÷.05, or $11000; $11000 of stock at 110% will cost $11000 × 1.10, or $12100.

OPERATION.

$550÷.05=$11000

$11000×1.10=$12100

Rule.—I. *Find the par value of the stock by dividing the income by the rate.*

II. *Multiply the par value by the market value of 1 share, to find the amount invested.*

WRITTEN EXERCISES.

2. A real estate dealer buys a 6% ground-rent of $300 per annum at par; what does it cost him? *Ans.* $5000.

3. A house subject to a ground-rent of $75 at 6% was sold for $5750; what was its value? *Ans.* $7000.

4. What sum must I invest in 5-20's at $119\frac{1}{2}$, to secure an annual income of $663 in currency, gold at $110\frac{1}{2}$, brokerage $\frac{1}{4}$%? *Ans.* $11975.

5. What sum must be invested in Kentucky 6's, at 103, to yield \$786 a year, brokerage $\frac{1}{4}$%? *Ans.* \$13525.75

6. When the new $4\frac{1}{2}$'s are selling at 105, what must I invest in them to secure an income of \$983.25 in currency, gold at $109\frac{1}{4}$, brokerage $\frac{1}{4}$%? *Ans.* \$21050.

7. If I sell \$8000 Ohio 6's at 118 and buy sufficient Georgia 7's at 103 to yield \$560 income, how much shall I have left, brokerage at the usual rate? *Ans.* \$1160.

8. What must I pay for Georgia 6's to realize 7% on the investment? What must I pay for Reading coupon 7's to give an income of $6\frac{3}{4}$%? *Ans.* $85\frac{5}{7}$; $103\frac{19}{27}$.

9. If gold is at 115, what must be paid for 5-20's to realize 7% on the investment? What must be the price of 10-40's to yield 6%, gold being 110? *Ans.* $98\frac{4}{7}$; $91\frac{2}{3}$.

10. What must be the price of gold so that 5-20's at 108, may realize 7%? What must be the price of gold, so that I may realize 5% from 10-40's bought at 105? *Ans.* 126; 105.

11. I bought a lot 50 ft. front and 85 ft. deep, at a ground-rent of \$5.40 per foot front; what would be the cost of the property, the ground-rent being 6% of it? *Ans.* \$4500.

12. How many shares of North Pennsylvania R. R., at 49, must be sold, that the proceeds, invested in Pennsylvania State 6's, at $115\frac{1}{2}$, may give an income of \$600, brokerage being deducted? *Ans.* 237 shares; $\$8.37\frac{1}{2}$ surplus.

13. Mr. Jackson sold \$15000 Union Pacific 7's at $101\frac{1}{8}$, and invested part of the proceeds in U. S. 6's of '81 at $117\frac{1}{2}$, sufficient to produce an income of \$750 in gold, and deposited the remainder, brokerage deducted, in bank: what was his bank deposit? *Ans.* \$412.50.

14. I have some California 7's, which bring me in an income of \$546, but preferring an investment nearer home, I decide to exchange them for Philadelphia 6's; if the California bonds are worth 117 and the Philadelphia 105, how much must I add to my investment to secure the same income, brokerage not considered? *Ans.* \$429.

CASE III.

449. ***Given, the market value, and the income or rate of dividend, to find the rate of interest on the investment.***

1. What per cent. of income will be realized by purchasing 7% bonds at 95?

SOLUTION.—$1 of stock will cost $.95, and pays $.07; if on $.95 the gain is $.07, on $1 it is as many per cent. as .07 ÷ .95, or $7\frac{7}{19}$%.

OPERATION.

$.07 \div .95 = .07\frac{7}{19}$

Rule.—*Divide the annual income or dividend of the stock by its market value, to find the rate of income.*

WRITTEN EXERCISES.

2. What is the rate of income of New York Central 6's bought at 106? *Ans.* $5\frac{35}{53}$%.

3. When U. S. 5-20's are at $112\frac{1}{2}$, and gold at 110, what per cent. will these bonds yield? *Ans.* $5\frac{13}{15}$%.

4. I bought an irredeemable ground-rent of $54 per annum for $850; what per cent do I realize? *Ans.* $6\frac{6}{17}$%.

5. Which is the better investment, 10-40's at $113\frac{3}{4}$, or the new $4\frac{1}{2}$'s at $110\frac{3}{4}$? *Ans.* The 10-40's.

6. Mr. Hull bought a ground-rent of $450 for $6575; what rate of income does it pay? *Ans.* $6\frac{222}{263}$%.

7. If I buy a $5000 mortgage at $2\frac{1}{2}$% discount, interest at 6%, what rate of income do I receive on it? *Ans.* $6\frac{2}{13}$%.

8. Which is the more profitable, Missouri 6's at 102, or 5-20's at 115, gold being 110? *Ans.* Mo. 6's.

9. Mr. Rogers bought Michigan 7's at $112\frac{1}{2}$, and afterwards exchanged them for 5-20's at $116\frac{1}{4}$; which was the better investment, gold selling at 112? *Ans.* Mich. 7's.

10. Wishing to make a permanent investment, I am recommended to take either 5's at 75, 6's at 85, or 7's at 90: which is the best investment? *Ans.* 7's at 90.

11. Gold being at $105\frac{1}{2}$, which pays the better interest on an investment, U. S. Pacific 6's at 108, or 5-20's at 110? *Ans.* 5-20's.

12

GENERAL TAXES.

450. A **Tax** is a sum of money assessed on persons or property for public purposes.

451. **Taxes** are assessed by the national government, a state, county, or town.

452. A **Property Tax** is a tax upon property. Property is of two kinds; *Real Estate* and *Personal Property.*

453. **Real Estate** is immovable property; as land, buildings, etc. *Personal Property* is movable property; as money, stock, furniture, etc.

454. A **Poll Tax** is a tax on the person. It is assessed in some states on each male citizen not exempt by law.

455. An **Assessment Roll** is a list or schedule containing the names of persons taxed, the valuation of their property, and the amount of their taxes.

456. An **Assessor** is an officer who appraises the property and prepares the assessment roll.

457. The **Quantities** to be considered are: 1. The *Taxable Property;* 2. The *Rate of Taxation;* 3. The *Amount of Tax.*

Real estate is usually assessed by the proper officer for not more than $\frac{1}{2}$ or $\frac{1}{3}$ of its real value. The value of personal property may be given in by the owner under oath, or if he neglects to do this, it is valued by the officer.

The term *poll* is from the German *polle*, the head. A poll tax is a *capitation tax*, from the Latin *caput*, the head. In some States the *income* from a person's occupation is assessed at a small sum and taxed. Money on interest secured by bond and mortgage is taxed in some States.

After the taxes have been assessed, each person receives a notice of his taxation, stating the day of appeal, when he may appear before the proper officers and show reasons for correcting any mistakes that have been made.

NOTE.—Government Taxes are taxes levied by the government, including Internal Revenue and Duties. They will be considered under the head of *Duties* and *Customs.*

CASE I.

458. ***Given, the taxable property and the rate of taxation, to find the amount of tax.***

1. The taxable property of a town is $794800, and the rate of taxation $.009 on a dollar; what is the tax?

SOLUTION.—If the tax is $.009 on $1, on $794800 it will be 794800 times $.009, or $7153.20.

OPERATION.

$794800 × .009 = $7153.20

Rule.—*Multiply the amount of taxable property by the rate, to find the tax.*

NOTE.—If there is a poll tax the sum produced by it should be added to the property tax to give the whole tax.

WRITTEN EXERCISES.

2. The real estate of a town is valued at $640876, and the personal estate at $750472; there are also 250 polls, at $1.50 each; what is the whole tax, the rate being 7 mills on a dollar? *Ans.* $10114.436.

459. Table.—In the assessment of taxes in a town, city, etc., a table is usually constructed by which the labor of calculation is greatly facilitated. The following table is based on the rate of $.015 to the dollar:

Prop.	Tax.	Prop.	Tax.	Prop.	Tax.	Prop.	Tax.	Prop.	Tax.
$1	.015	$10	.15	$100	$1.50	$1000	$15	$10000	$150
2	.030	20	.30	200	3.00	2000	30	20000	300
3	.045	30	.45	300	4.50	3000	45	30000	450
4	.060	40	.60	400	6.00	4000	60	40000	600
5	.075	50	.75	500	7.50	5000	75	50000	750
6	.09	60	.90	600	9.00	6000	90	60000	900
7	.105	70	1.05	700	10.50	7000	105	70000	1050
8	.12	80	1.20	800	12.00	8000	120	80000	1200
9	.135	90	1.35	900	13.50	9000	135	90000	1350

3. Find by the table A's tax, whose property is $7580, and who pays for 5 polls at the rate of $1.50 each.

SOLUTION.—We find from the table the tax on $7000, then on $500, then on $80, and then calculate the tax on 5 polls, and add the results together; the sum will be the entire tax.

OPERATION.

Tax on	$7000 =	$105
"	500 =	7.50
"	80 =	1.20
"	5 polls =	7.50
Whole tax	=	$121.20

4. Find B's tax, whose property is \$9750, and who pays for 2 polls. *Ans.* \$149.25.

5. A is worth \$7895, and his sister \$5634; what is the aggregate amount of their taxes? *Ans.* \$202.935.

6. Mr. Mark's property is assessed at \$8500; he pays for 1 poll and .1½ % on the income from his occupation, assessed at \$800; what was his entire tax? *Ans.* \$130.20.

7. Mr. Sidney's real estate is valued at \$75000 and his personal property at \$8670, both of which are assessed for ⅔ of their value; he pays for 4 polls at \$1.50 each, and also ¼% on an estimated income of \$1200; what is his entire tax? *Ans.* \$845.70.

8. Mr. Shank's property was assessed at \$3500 last year, and he paid .25% village tax, 1.025% county tax, .45% school tax, and \$1.25 poll tax; what amount of taxes did he pay? *Ans.* \$61.62½.

9. I find I have been assessed as follows: Real estate, \$50,000; personal property, \$3600; money at interest, \$15,000; income from occupation, \$1500; and 3 gold watches. I obtain an abatement of ⅘ on the real estate, ⅓ on personal property, \$5000 on money at interest, ⅘ for occupation, and 1 gold watch; how much does this lessen my tax, the rate being \$.004½, and one dollar for each watch? *Ans.* \$122.95.

CASE II.

460. ***Given, the rate of taxation and the tax or the amount left after payment of tax, to find the amount assessed.***

1. What is the assessed value of property taxed \$37.50, at the rate of 5 mills on a dollar?

SOLUTION.—At 5 mills on the dollar, .005 times the amount assessed equals the tax, which is \$37.50; hence the amount equals \$37.50 divided by .005, which we find is \$7500.

OPERATION.

$$\frac{\$37.50}{.005} = \$7500$$

Rule I.—*Divide the tax by the rate, to find the amount assessed.*

Rule II.—*Divide the amount left after payment of tax by 1 minus the rate.*

WRITTEN EXERCISES.

2. My tax is $37.80 at the rate of $4\frac{1}{2}$ mills on the dollar; required the property. *Ans.* $8400.

3. A's entire tax is $310.75; he pays for 3 polls at $2.25 each; the rate is 8 mills on the dollar; what is his property? *Ans.* $38000.

4. I have $12000 on interest, and my tax for money on interest is $$33.07\frac{1}{2}$, at $2\frac{1}{4}$ mills on the dollar; for how much money at interest am I overtaxed? *Ans.* $2700.

5. A bridge was built by a certain town at a cost of $7580, which was raised by a tax on the property-holders of $3\frac{1}{2}$ mills on the dollar; the collector's commission was $2\frac{1}{4}$%; what was the valuation of the property? *Ans.* $2215564.49.

NOTE.—The collector's commission is included in the tax.

6. Mr. Mills paid one year .45% township tax, .$3\frac{1}{2}$% county tax, .48% school tax, and 4 polls @ $1.25; his whole tax was $319.88; what was the value of his property? *Ans.* $24600.

CASE III.

461. ***Given, the assessed value and the tax, to find the rate.***

1. A tax of $6387.50 is to be assessed in a town; the real estate is valued at $345000, and the personal property at $477500; there are 420 polls, taxed @ $1.50; what is the rate of taxation?

SOLUTION.—The entire poll tax is $1.50 multiplied by 420, which is $630; subtracting this from the whole tax, we have remaining $5757.50, the property tax; dividing $5757.50 by $822500, the amount of property, we have 7 mills, the tax on $1.

OPERATION.

$1.50×420=$630
$6387.50−$630=$5757.50
$5757.50÷$822500=.007

Rule.—*Divide the property tax by the amount of taxable property, the quotient will be the rate of taxation.*

NOTE.—If there is a poll tax, subtract it from the whole tax before dividing.

WRITTEN EXERCISES.

2. A's property is valued at $7580, his tax is $35: required the rate of taxation. *Ans.* .0046+.

3. A tax of $17250 is to be assessed on a town; the real estate is valued at $850000 and the personal property at $250000; there are 500 polls, each of which is taxed $1.50; what is the rate of taxation? *Ans.* $.015.

4. A market-house costing $7650, was built in a certain town, in which the taxable property was estimated at $804796, the collector's commission being $2\frac{1}{2}\%$; what was the rate of taxation? *Ans.* .0097+.

5. In a certain school the expenses are as follows: salary of teacher, $500; fuel, $42.75; apparatus, $32.50. The school fund amounted to $125.25, and the rest of the expenses was paid by a rate bill; if the entire attendance was 7280 days, what was C's bill, who sent 4 pupils 90 days each? *Ans.* $22.25.

MENTAL EXERCISES.

MISCELLANEOUS PROBLEMS.

1. How much currency will $50 in gold buy, when gold is at a premium of 8%?

2. How much gold can be bought for $220 in currency, when gold is at a premium of 10%?

3. If I take 10% off and then another 10% off, what is the rate off?

4. What is the actual rate off, when I take 20% off and then another 20% off?

5. What is the difference between 10% off and 5 and 5% off? between 15% off and 10 and 5% off?

6. What is the difference between 20% off and two 10%'s off? between 10% on and 5 and 5% on?

7. How much is 10 and 10% on? 20 and 10% on? 20 and 5% on? 20 and 20% off? 20 and 20% on?

8. What is the difference between 15% on, and 10 and 5% on? between three 10%'s on, and three 10%'s off?

9. If an article is sold so as to gain $\frac{3}{8}$ on the cost, what is the gain per cent.?

10. I paid an agent $200 for selling a house for me, what did I receive for it, the rate of commission being $1\frac{1}{4}\%$?

11. What cost 5 shares of bank stock at 102, brokerage $\frac{1}{8}\%$?

12. Required the cost of 4000 U. S. $4\frac{1}{2}$'s, at $108\frac{1}{4}$, brokerage $\frac{1}{8}\%$.

13. A boy bought oranges at $3 a hundred, and sold them for 5 cents apiece; what per cent. did he gain?

14. A milliner sold some old-fashioned hats at \$3.60, which was 25% below marked price; at what price were they marked?

15. A lady bought some canal stock at 80, and sold it at par, gaining \$500; how many shares at \$50 each?

16. I bought R. R. stock at 87½, and sold at 102½, and gained \$3000; how many shares at \$100 each?

17. What sum must I invest in stock at 106, paying 8% yearly dividends, to realize an income of \$1200?

18. What per cent. does stock, paying 6% dividends, yield, when bought at 90? 8% stocks at 120?

19. How much must I pay for stock that pays 6% dividends, in order to realize 8% on my investment?

20. I bought Phila. 6's at a rate by which I realized 5½%; what did I pay for them?

21. What will be the annual premium of insurance, at 1½%, on goods valued at \$2000?

22. I paid \$8 for insuring \$400; what was the rate? Paid \$16 for insuring \$1000; what was the rate?

REVIEW PROBLEMS.

1. The list price of a lot of ginghams is 20¢ a yard; if I buy them at 20% off and sell at 25% on, what is my gain per yard? *Ans.* 9¢.

2. Bought French merinos at \$1.15 a yard, and marked them *cln*, my key being "Charleston;" what was my gain % if I sell the goods at the marked price? *Ans.* $30\frac{10}{23}$%.

3. Mr. Behmer bought mining stock (\$50) at 9% premium, and sold it at a loss of \$5 on a share; at what rate was it sold? *Ans.* 1% disc.

4. An agent sold 150 barrels of flour, charging 2½% commission, and 2¼% guaranty; the net proceeds due the consignor were \$1143; for how much was the flour sold per barrel? *Ans.* \$8.

5. I exchanged \$5000 U. S. 6's at 108 for 30 shares of stock of Bank of North America (\$100); what did I pay for the latter? *Ans.* 180%.

6. Mr. Albert received in April a stock dividend of 10 shares in a mining company and another of 21 shares in November; he then owned 231 shares; what were the rates of dividend? *Ans.* 5% and 10%.

SIMPLE INTEREST.

462. Interest is a percentage on a sum of money for its use a certain time.

463. The **Principal** is the sum on which the interest is computed.

464. The **Rate** of interest is the rate per cent. on $1 for a certain time. The usual time is one year.

465. The **Time** is the period during which the money is on interest.

466. The **Amount** is the sum of the principal and interest.

467. Simple Interest is the interest on the principal only. **Compound Interest** is the interest on both the principal and interest.

468. Legal Interest is interest at the rate fixed by law. The rate varies in different states, as follows:

	Legal Rate.		Rate agreed on.
5%	Louisiana,	7%	Pennsylvania in certain cases.
7%	New York, Michigan, Wis., Minnesota, Kansas, S. Carolina, Georgia, Dakota, and Connecticut.	8%	Louisiana, N. Carolina, and Ohio.
8%	Alabama, Florida, and Texas.	10%	Mississippi, Missouri, Tennessee, Wisconsin, Michigan, Kentucky, Iowa, Indiana, Illinois, Georgia, District of Columbia.
10%	Nebraska, Nevada, California, Colorado, Oregon, Arizona, Montana, Idaho, and Washington Territory.	12%	Virginia, Texas, Oregon, Minnesota, Kansas.
12%	Wyoming,	15%	Nebraska.
6%	Debts due the U. S. in District of Columbia, and all States not mentioned.	Any rate agreed upon	Arkansas, Arizona, California, Colorado, Dakota, Florida, Maine, Massachusetts, Nevada, Utah, Rhode Island, S. Carolina, Washington Territory.

469. Usury is a rate of interest greater than the law allows. Various penalties are attached to the taking of usury in different States.

The legal rate in England and France is 5%; and in Ireland, Canada, and Nova Scotia is 6%.

In notes, contracts, accounts, mortgages, etc., when no rate is specified, the legal rate is understood.

Notes draw interest after they become due, though interest is not mentioned in them; and interest is reckoned on book accounts after the expiration of the term of credit.

470. The **Quantities** are five: 1. The *Principal;* 2. The *Interest;* 3. The *Rate;* 4. The *Time;* 5. The *Amount.*

NOTE.—In computing interest it is customary to reckon a month as $\frac{1}{12}$ of a year, and a day as $\frac{1}{30}$ of a month. In dealing with the U. S. Government, each day is $\frac{1}{365}$ of a year.

CASE I.

471. ***Given, the principal, the rate per cent., and the time, to find the interest or the amount.***

MENTAL EXERCISES.

1. What is the interest of $80 for 2 yr. 6 mo. at 6%?

SOLUTION.—6 months equal $\frac{6}{12}$ or $\frac{1}{2}$ of a year, which with 2 yr. equals $2\frac{1}{2}$, or $\frac{5}{2}$ years. At 6 per cent. for 1 yr., $\frac{6}{100}$ of the principal equals the interest, and for $2\frac{1}{2}$ or $\frac{5}{2}$ yr., $\frac{5}{2}$ times $\frac{6}{100}$ or $\frac{15}{100}$ or $\frac{3}{20}$ of the principal equals the interest, $\frac{3}{20}$ of $80 equals $12.

What is the interest of

2. $60 for 6 yr. at 5%?

3. $40 for 4 yr. at 5%?

4. $30 for 5 yr. at 4%?

5. $600 for 2 yr. 3 mo. at 8%?

6. $300 for 4 yr. 6 mo. at 6%?

7. $240 for 3 yr. 9 mo. at 8%?

8. $330 for 7 yr. 6 mo. at 4%?

9. $500 for 3 yr. 7 mo. 6 da. at 5%?

10. What is the interest of $300 for 5 yr. 3 mo. 18 da. at 10%?

11. What is the interest of $500 for 2 yr. 2 mo. 12 da. at 5%?

12. What is the amount of $50 for 2 yr. 8 mo. at 6 per cent.?

REMARK.—We find that $\frac{16}{100}$ or $\frac{4}{25}$ of the principal equals the interest, hence $\frac{29}{25}$ of the principal equals the amount; $\frac{29}{25}$ of $50=$58.

13. What is the amount of $600 for 7 yr. 6 mo. at 6 per cent.?

14. What is the amount of $300 for 8 yr. 10 mo. at 6 per cent?

COMMON METHOD.

1. What is the interest of $2400 for 6 yr. 7 mo. 15 da., at 7%?

SOLUTION.—By reduction, we find that 6 yr. 7 mo. 15 da. equals $6\frac{5}{8}$ yr. At 7%, .07 times $2400 equals the interest for 1 year, which is $168; if the interest for 1 year is $168, for $6\frac{5}{8}$ yr. it is $6\frac{5}{8}$ times $168, which by multiplying we find is $1113. Hence the following

OPERATION.

$$\begin{array}{r} \$2400 \\ .07 \\ \hline \$168.00 \\ 6\frac{5}{8} \\ \hline \$1113.00 \end{array}$$ *Ans.*

Rule.—I. *Multiply the principal by the rate, and that product by the time expressed in years, to find the interest*

II. *Add the interest to the principal to find the amount.*

WRITTEN EXERCISES.

Required the interest

2. Of $360 for 3 yr. 6 mo at 7%? *Ans.* $88.20.

3. Of $940 for 7 yr. 8 mo. at 6%? *Ans.* $432.40.

4. Of $860 for 5 yr. 9 mo. at 5%? *Ans.* $247.25.

5. Of $780 for 8 yr. 4 mo. at 7%? *Ans.* $455.

6. Of $590 for 3 yr. 10 mo. at 8%? *Ans.* $180.93.

7. Of $1296 for 5 yr. 10 mo. 15 da. at 6%? *Ans.* $456.84.

8. Of $4080 for 3 yr 3 mo. 9 da. at 5%? *Ans.* $668.10.

9. Of $7856 for 2 yr. 4 mo. 5 da. at 7%? *Ans.* $1290.78.

SIX PER CENT. METHOD.

472. The **Six Per Cent. Method** is so called because the process is based upon that rate.

1. What is the interest of $240 for 6 yr. 8 mo. 18 da. at 6%?

OPERATION.

$1 for			
1 yr. - .06	× 6 yr.	-	0.36
1 mo. - .005	× 8 mo.	-	.04
1 da. - .000$\frac{1}{6}$	×18 da.	-	.003
			.403
			240
			$96.72

SOLUTION.—The interest on $1 for 1 year is 6¢, and for 6 yr. it is 6 times 6¢, or 36¢; for 1 month, or $\frac{1}{12}$ of a year, it is $\frac{1}{12}$ of 6¢, or $\frac{1}{2}$ of a cent, hence for 8 months it is 8 times $\frac{1}{2}$ of a cent, or $.04; for 1 month or 30 days, the interest on $1 is $\frac{1}{2}$ of a cent, or 5 mills, and for 1 day it is $\frac{5}{30}$ or $\frac{1}{6}$ of a mill; hence for 18 days it is $\frac{18}{6}$ of a mill, or $.003. Adding, we have $.403, the interest on $1 for 6 yr. 8 mo. 18 da., and on $240 the interest will be 240 times $.403, or $96.72.

Rule.—I. *Multiply the number of years by the rate, take $\frac{1}{2}$ of the number of months as cents, and $\frac{1}{6}$ of the number of days as mills; their sum will be the interest of* $1 *for the given time at 6%.*

II. *Multiply this sum by the principal, the product will be the interest at* 6 *per cent. For any other rate take as many sixths of it as that rate is of six.*

NOTES.—1. Another "6 per cent. method" is to reduce the years to months, and take *half* the number of *months* for *cents*, etc., as before.

2. Another "6 per cent. method" is to take the *number* of *months as cents*, and *one-third* of the number of *days as mills*, and multiply the sum by half the principal.

WRITTEN EXERCISES.

Required the interest.

2. Of $380 for 3 yr. 4 mo. 12 da. at 6%. *Ans.* $76.76.

3. Of \$975 for 5 yr. 6 mo. 6 da. at 6%. *Ans.* \$322.72½.
4. Of \$834 for 9 yr. 10 mo. 15 da. at 6%. *Ans.* \$494.14½.
5. Of \$45.95 for 8 yr. 6 mo. 24 da. at 7%. *Ans.* \$27.55+.
6. Of \$23.75 for 7 yr. 7 mo. 21 da. at 5% *Ans.* \$9.07+.
7. Of \$.325 for 9 yr. 5 mo. 14 da. at 8%. *Ans.* \$.24½+.
8. \$147 37½, 4 yr. 11 mo. 13 da. 7%. *Ans.* \$51.094.
9. \$635.32½, 9 yr. 9 mo. 11 da. 9%. *Ans.* \$559.51—.
10. \$387.13¾, 10 yr. 7 mo. 7 da., 10% *Ans.* \$410.53—.
11. \$570.05, 3 yr. 5 mo. 5 da. 6½%. *Ans.* \$127.11+.

EASY BUSINESS METHOD.

473. The following six per cent. method for *Months* and *Days* will be found concise and practical.

1. What is the int. of \$2400 for 3 yr. 2 mo. at 6%? at 7%?

SOLUTION.—Since the interest for 1 yr., or 12 mo., at 6% is $\frac{6}{100}$ of the principal, the interest for 2 mo., or $\frac{1}{6}$ of a year, is $\frac{1}{6}$ of $\frac{6}{100}$, or $\frac{1}{100}$ of the principal, and for 1 mo. it is $\frac{1}{2}$ of $\frac{1}{100}$ of the principal; hence the interest on \$2400 for 1 mo. is $\frac{1}{2}$ of $\frac{1}{100}$ of \$2400 or $\frac{\$24.00}{2}$, and for 3 yr. 2 mo., or 38 mo. it is $\frac{\$24.00\times38}{2}$ which reduced by cancellation, is \$456.

OPERATION.

$$\frac{\overset{12}{\$\cancel{24}.00}\times38}{\cancel{2}}=\$456.\ \textit{Ans.}$$

SOL.—Since the int. at 6% is $\frac{\$24.00\times38}{2}$ the int. at 1% is $\frac{\$24.00\times38}{2\times6}$, and the int. at 7% is $\frac{\$24.00\times38\times7}{2\times6}=\$532.$

OPERATION.

$$\frac{\overset{\overset{2}{\cancel{4}}}{\$\cancel{24}.00}\times38\times7}{\cancel{2}\times\cancel{6}}=\$532.$$

2. What is the int. of \$4800 for 63 days at 6%? at 4½%?

SOLUTION.—The interest for 2 mo., or 60 da., at 6%, by the previous analysis, is $\frac{1}{100}$ of \$4800; and for 6 da., or $\frac{1}{10}$ of 60 da., it is $\frac{1}{10}$ of $\frac{1}{100}$, or $\frac{1}{1000}$ of \$4800; and for 1 day it is $\frac{1}{6}$ of $\frac{1}{1000}$ of \$4800, and for 63 da. it is 63 times $\frac{1}{6}$ of $\frac{1}{1000}$ of \$4800, or $\frac{\$4.800\times63}{6}$, which reduced is \$50.40.

OPERATION.

$$\frac{\overset{.800}{\$\cancel{4.800}}\times63}{\cancel{6}}=\$50.40.\ \textit{Ans.}$$

Rule for Months.—*Multiply the Principal by the number of months, and divide by* 200; *or, Point off two places in the principal to express its division by* 100, *divide by* **2**, *and multiply by the number of months.*

Rule for Days.—*Multiply the principal by the numb r*

of days, and divide by 6000 ; or, *Point off three places in the principal to express its division by* 1000, *divide by* 6, *and multiply by the number of days.*

NOTE.—For any other rate than 6%, divide by 6 and multiply by the rate. The method for days is popularly expressed thus: "*Multiply dollars by days and divide by* 6000."

WRITTEN EXERCISES.

Required the interest

3. Of \$1500 for 4 mo. at 6%. *Ans.* \$30.
4. Of \$780 for 10 mo. at 6%. *Ans.* \$39.
5. Of \$960 for 2 yr. 9 mo. at 7%. *Ans.* \$184.80.
6. Of \$1260 for 33 days at 6%. *Ans.* \$6.93.
7. Of \$800 for 63 days at 6%. *Ans.* \$8.40.
8. Of \$87.50 for 3 yr. 3 mo. at $4\frac{1}{2}$%. *Ans.* \$12.80—.
9. Of \$18.28 for 93 days at 7%. *Ans.* \$0.33+.
10. Of \$2480 for 124 days at $7\frac{1}{2}$%. *Ans.* \$64.07—.
11. On a 3 mo. note for \$3600, dated April 15, 1876, at 6%, allowing 3 days of grace (exact time). *Ans.* \$65.40.
12. Of \$1200, from Jan. 9, 1880, to July 16, 1880, at $7\frac{3}{10}$% (exact number of days). *Ans.* \$45.99.

METHOD OF EXACT INTEREST.

474. Exact Interest is that which is obtained by reckoning 365 days to the year.

475. Exact Interest is reckoned by the United States Government, and is growing in favor with business men.

Bankers and business men often use *Interest Tables*, which are sometimes calculated to exact interest.

1. What is the exact interest of \$785 from July 20 to December 1st, at 7%?

SOLUTION.—From July 20 to December 1 there are 134 days; the interest of \$785 for 1 year of 365 days, at 7%, is \$54.95, and for 134 days it is $\frac{134}{365}$ of \$54.95, which is \$20.17+.

OPERATION.

$$\begin{array}{r} \$785 \\ .07 \\ \hline 54.95 \\ 134 \\ \hline 365)7363.30 \\ \hline \$20.17\frac{1}{73} \end{array}$$

Rule.—*Multiply the principal by the rate, and that product by the integral number of years; then multiply the in* *terest for one year by the exact number of days, and divide by 365; and take the sum of the two results.*

NOTE.—The exact interest for any number of days less than 1 year, may also be found by deducting from the common interest $\frac{1}{73}$ of itself.

WRITTEN EXERCISES.

2. What is the interest, at 7%, of $327.25 from January 5th, 1860, to July 12th, 1862? *Ans.* $57.613.

3. What is the amount of $480, on interest at 6%, from Apr. 7th, 1851, to Aug. 25th, 1860? *Ans.* $750.25—.

4. A had $1200 on interest from May 20th, 1856, to Sept. 5th, 1861; what was the int. at $5\frac{1}{2}$%? *Ans.* $349.53—.

5. Required the amount of 1900\frac{3}{4}$ on int. at 5% from June 9th, 1850, to Jan. 14th, 1860. *Ans.* $2813.11.

6. B gives his note, August 6th, 1857, for $670, interest at 7%; he pays the note and interest May 17th, 1861; how much did he pay? *Ans.* $847.19.

7. Required the amount of $875.48, on interest at 6%, from Dec. 19th, 1845, to Feb. 29th, 1860. *Ans.* $1621.24.

8. Which is the greater, exact interest or common interest, and why? Prove that $\frac{1}{73}$ off from common interest will give exact interest.

CASE II.

476. ***Given, the time, the rate, and the interest or the amount, to find the principal.***

MENTAL EXERCISES.

1. What principal will in 3 yr. 4 mo., at 6%, give $80 interest?

SOLUTION.—We find that $\frac{1}{5}$ of the principal equals the interest, which is $80; if $\frac{1}{5}$ of the principal is $80, $\frac{5}{5}$, or the principal, equals 5 times $80, or $400.

What principal will give an interest of

2. $60 in 8 yr. at 5%?
3. $70 in 7 yr. at 4%?
4. $12 in 8 yr. at 6%?
5. $18 in 7 yr. 6 mo. at 6%?
6. $54 in 2 yr. 3 mo. at 8%?
7. $106 in 8 yr. 10 mo. at 6%?

8. How much money must a person borrow that he must pay an annual interest of $180 at 6%?

9. How much money has Howard on interest, supposing he receives $320 for 5 yr. 4 mo. at 6 per cent.?

10. The interest on A's money for 4 years at 5% is $200, and A's money is $\frac{2}{3}$ of B's; how much money has each?

WRITTEN EXERCISES.

1. What principal will in 4 yr. 8 mo., at 6%, give $151.20 interest?

OPERATION.

4 yr. 8 mo.=56 mo.

56 ÷ 2 = .28.

$$\frac{\$151.20}{.28} = \$540$$

SOLUTION.—We find the interest of $1 for 4 yr. 8 mo., at 6%, is $.28. If $1 gives an interest of $.28, to give $151.20 interest it will require as many dollars as $.28 are contained times in $151.20, which is $540. Hence the following

Rule.—*Divide the given interest by the interest of* $1 *for the given rate and time; or divide the amount by the amount of* $1.

2. What principal will in 3 yr. 8 mo., at 6%, give $462 interest? *Ans.* $2100.

3. What principal will in 12 yr. 9 mo., at 7%, give $64.26 interest? *Ans.* $72.

4. What principal will in 7 yr. 4 mo., at 8%, amount to $749.70? *Ans.* $472.50.

5. What principal will in 5 yr. 8 mo. 15 da., at 5%, give $575.40 interest? *Ans.* $2016.

6. What principal will in 7 yr. 7 mo. 13 da., at 7%, amount to $2400? *Ans.* $1565.19.

7. What principal will in 4 yr. 11 mo. 17 da., at 7%, amount to $3363.79? *Ans.* $2496.37+.

8. The sum of A's and B's money on interest for 4 yr. 6 mo., at 6%, gives $5400 interest; how much money has each, if 3 times B's equals A's? *Ans.* $15000; $5000.

CASE III.

477. ***Given, the principal, the rate, and the interest or the amount, to find the time.***

MENTAL EXERCISES.

1. In what time will $200 at 5% give $60 interest?

SOLUTION.—At 5 per cent. for *one* year, $\frac{1}{20}$ of the principal equals the interest; $\frac{1}{20}$ of $200 is $10; if it require *one* year for $200 to gain $10, to gain $60 it will require as many years as $10 are contained times in $60, which are 6.

In what time will

2. $250 at 4% give $80 int.?
3. $150 at 6% give $45 int.?
4. $100 at 8% give $32 int.?
5. $60 at 5% give $21 int?
6. $25 at 6% am't to $43?
7. $50 at 9% am't to $86?

8. In what time will a principal gain 2 times, 3 times, or 4 times itself at 10%?

9. In what time will a principal double itself at 5%? at 6%? at 8%? at $12\frac{1}{2}$%?

10. In what time will a principal treble itself at 5%? at 10%? at 20%? at 25%?

11. The amount of a principal for a certain time at 5% is $250, and for the same time at 8% is $280; required the principal and the time.

WRITTEN EXERCISES.

1. In what time will $234 give $49.14 interest, at 6%?

OPERATION.

$$\begin{array}{r} \$234 \\ .06 \\ \hline 14.04 \text{ Int. 1 yr.} \end{array}$$

$$\frac{49.14}{14.04} = 3\tfrac{1}{2} \text{ yr.}$$

$$= 3 \text{ yr. } 6 \text{ mo.}$$

SOLUTION.—The interest of $234, at 6%, for *one* year, is $14.04. If in *one* year the principal gives $14.04 interest, to give $49.14 interest it will require as many times 1 yr. as $14.04 is contained times in $49.14, which is $3\frac{1}{2}$ yr., or 3 yr. 6 mo. Hence we have the following

Rule.—*Divide the given interest by the interest of the principal at the given rate for* ONE *year.*

NOTE.—When the amount is given, subtract the principal from the amount to find the interest, and then proceed as before.

2. In what time will $750, at 6 per cent., give $105 interest? *Ans.* 2 yr. 4 mo.

3. In what time will $720, at 6 per cent., give $957.60 amount? *Ans.* 5 yr. 6 mo.

4. In what time will $960, at 5 per cent., give $54.40 interest? *Ans.* 1 yr. 1 mo. 18 da.

5. In what time will $1800, at $4\frac{1}{2}$ per cent., give $3047.40 amount? *Ans.* 15 yr. 4 mo. 24 da.

6. In what time will $26.50, at $7\frac{1}{2}$ per cent., give $17.46 interest? *Ans.* 8 yr. 9 mo. 12 da.

7. In what time will $18.20, at $5\frac{3}{4}$ per cent., give $10.23 interest? *Ans.* 9 yr. 9 mo. 9 da.

8. The amount of a certain principal, in a certain time, at 5 per cent., is $833, and the amount for the same time at 12 per cent. is $1047.20; required the principal and time.

Ans. Prin. $680; Time, 4 yr. 6 mo.

SUG.—The difference of the amounts equals the interest at 7%.

9. A certain sum of money on interest amounts at 4 per cent., for a certain time, to $1216, and at 10 per cent., for the same time, to $1600; required the principal and time.

Ans. Prin. $960; Time, 6 yr. 8 mo.

CASE IV.

478. ***Given, the principal, the time, and the interest or the amount, to find the rate.***

MENTAL EXERCISES.

1. At what rate will $60 in 5 years give $21 interest?

SOLUTION.—For 5 years at *one* per cent. $\frac{5}{100}$ or $\frac{1}{20}$ of the principal equals the interest; $\frac{1}{20}$ of $60 equals $3; if $60 in 5 years at *one* per cent. gains $3, to gain $21 it will require as many times 1 per cent. as $3 is contained times in $21, which are 7.

At what per cent. will

2. $40 in 5 yr. give $20 int.?

3. $50 in 6 yr. give $15 int.?

4. $60 in 7 yr. give $21 int.?

5. $10 in 4 yr. give $12 am't?

6. $90 in 5 yr. give $117 am't?

7. $60 in 3 yr. 4 mo. give $70 am't?

8. At what rate will a principal gain 2 times itself in 30 years? 4 times itself? 5 times itself?

9. At what rate will a principal double itself in 10 years? in 12 yr.? in 20 yr.? in 25 yr.?

10. At what rate will a principal treble itself in 20 years? in 25 yr.? in 40 yr.? in 80 yr.?

11. The amount of a certain principal for 7 years at a certain rate per cent. is $540, and for 10 yr. is $600; required the principal and the rate per cent.

WRITTEN EXERCISES.

1. At what rate will $234 give $49.14 interest in 3 yr. 6 mo.?

OPERATION.

$$\begin{array}{r} \$234 \\ .01 \\ \hline 2.34 \\ 3\frac{1}{2} \\ \hline \$8.19 \text{ int. at } 1\% \end{array}$$

$$49.14 \div 8.19 = 6\%.$$

SOLUTION.—We find the interest of $234 for 3 yr. 6 mo., at *one* per cent., is $8.19. If the principal in the given time, at *one* per cent., gives $8.19 interest, to give $49.14 interest, it will require as many times one per cent. as $8.19 is contained times in $49.14, which is 6 per cent. Hence we have the following

Rule.—*Divide the given interest by the interest of the principal for the given time, at* ONE *per cent.*

NOTE.—When the amount is given, subtract the principal from the amount to find the interest, and proceed as before.

2. At what rate will $240, in 5 yr. 4 mo., give $64 interest? *Ans.* 5%.

3. At what rate will $654, in 7 yr. 8 mo., give $350.98 interest? *Ans.* 7%.

4. At what rate will $72.50, in 3 yr. 4 mo. 15 da., give $14.681¼ interest? *Ans.* 6%.

5. At what rate will $3975, in 6 yr. 7 mo. 20 da., give $2375.06¼ interest? *Ans.* 9%.

6. At what rate will $13.25, in 8 yr. 10 mo. 18 da., give $7.062¼ interest? *Ans.* 6%.

7. The amount of a certain principal for 5 yr. at a certain rate is $2430, and for 12 yr., at the same rate, it is $3312; required the principal and the rate. *Ans.* $1800; 7%.

8. The amount of a certain principal for 4 yr., at a certain rate per cent., is $3551, and for 19 yr., $6929¾; required the principal and rate. *Ans.* Prin., $2650; Rate, 8½%.

MENTAL EXERCISES.

1. A's fortune is $200, which is ¼ of B's; what interest will each receive on his money in 4 years, at 5%?

2. C's money is $300, which is ¾ of D's money; what is the amount of the money of each on interest, for 5 years at 6 per cent.?

3. A's money is $400, which is ⅔ of B's money; how much more interest will B receive than A in 8 years, at 5 per cent.?

4. The amount of $250 for 6 years at 10% is to be divided between C and D, so that C shall have 3 times as much as D; what does each receive?

5. A, B, and C, together, have $1200, of which A has twice, and B 3 times, as much as C; what is the interest of each for 5 years, at 6%?

6. If the interest of $2500 for 4 years, at 10 per cent., be divided into two parts, which are as 2 to 3, it will respectively give ⅓ of B's and ⅓ of A's money: how much has each?

7. The interest on ½ of A's and ⅓ of B's fortune, for 5 years, at 6 per cent., is $240; what is the fortune of each, provided ⅓ of A's equals ⅓ of B's?

8. A's money is 4 times B's, and the sum of the interest received by both for 3 years, at 8 per cent., is $600; how much money has each?

9. The interest for 4 years, at 5 per cent., on the money Martin owes, is $40; and the interest for the same time and rate per cent. on

the money due him, is $70; how much more has he due than he owes?

10. The interest on the money A paid for a farm, house, and store, for 8 years, at 5 per cent., equals $18000; what was the cost of each, provided the farm cost 3 times as much as the house, and the house twice as much as the store?

11 A man wishes to place such a sum of money on interest at 6 per cent., that it will give an annual interest of $360 for a poor sister; required the amount invested.

12. Two-thirds of A's fortune, plus ¾ of B's, being on interest for 6 years, at 5 per cent, amounts to $7800; what is the fortune of each, supposing ⅔ of A's equals ¾ of B's?

13. ½ of the cost of Bowman's house, plus ⅔ of the cost of his farm, being on interest for 5 years, at 8 per cent., amounts to $2100; what is the cost of each, provided the house cost ⅓ as much as the farm?

14. Two times the value of a horse, plus 3 times the value of a cow, which is ⅓ of the value of the horse, in 8 years, at 5%, gives $84 interest; required the value of each.

15. The money Henry paid for a horse, carriage, and harness, in 10 years, at 5 per cent., would give such an interest that if put on interest for the same time and rate, it would amount to $270; how much did he pay for each, if the horse cost twice as much as the carriage, and the carriage 3 times as much as the harness?

INTEREST ON PROMISSORY NOTES.

479. **A Promissory Note** is a written promise to pay some one a certain sum of money on demand, or at a specified time.

480. The **Face** of a note is the sum whose payment is promised. It is written in *words* in the body of the note, and in *figures* at the top or bottom.

481. The **Maker** of a note is the party who signs it. The *Payee* is the party to whom it is made payable. The *Holder* is the one who owns it.

If a note reads "with interest," it draws interest from date; otherwise it draws interest from the time of maturity until paid. A note may draw interest from a particular time after date, if so specified in the note. When no rate is mentioned the legal rate of the State is understood. In business language a note is said to be "made in favor of" the payee.

A note should contain the words, "value received," otherwise the holder may be required to prove that value was received.

482. A Negotiable Note is a note that can be transferred from one party to another. A note is negotiable when it is made payable to the "bearer," or to the "order" of the payee.

A note payable "to order" becomes negotiable by the payee writing his name on the back of it, which is called *indorsing* the note. A note payable "to bearer" is negotiable without indorsement. A note payable to a particular person only, is not negotiable.

The words "without defalcation" are required in Pennsylvania to make a note negotiable; in New Jersey, "without defalcation or discount."

483. The **Indorser** of a note is the party who puts his name on the back as security for its payment.

It is customary in raising money on notes, to have one or more responsible persons write their names on the back of the note as security for its payment. In case of the refusal of the maker to pay the note when due, each indorser is liable for the whole amount of the note in the order of signing, unless he writes above his name the words "without recourse," or unless there is an agreement between two or more indorsers to share the loss between them.

When the maker fails to pay a note, it is usual for the holder to make his demand on the last *liable* indorser, who pays the note and then gets the amount from the preceding indorser, and so on, up to the first indorser. The holder, however, has the option of collecting the amount from *any* liable indorser, and when so collected, all *subsequent* indorsers are released, the indorser who pays becomes the holder, and may collect from any *prior* liable indorser, and so on up to the first.

484. The **Maturity** of a note is its becoming legally due at the expiration of the time. In most of the States a note matures *three days* after the time specified, unless the words "without grace" are inserted.

485. Days of Grace are the three days usually allowed by law for the payment of a note after the expiration of the time specified in the note.

When grace is allowed the note matures on the *last day of grace*. When no grace is allowed, it matures at the expiration of the time specified. If a note is payable *on demand*, it is legally due when presented.

If a note becomes legally due on Sunday or a legal holiday, it must be paid in most States on the day preceding. In Connecticut, three days' grace is allowed on notes for $35 or more, but not on notes for a less amount; if the last day is a legal holiday falling on Sunday, the note is due on Monday. In Maine and Nebraska, if the third day is a legal holiday falling on Monday, the note is payable on Tuesday; and in New York a note maturing on a legal holiday, or Monday observed as such holiday, is payable the following day. The following notation indicates when a note is nominally and legally due: July 4|7, 1876.

When the time of a note is stated in months, calendar months are meant. A note for 4 months, dated Oct. 15, would mature Feb. 15 | 18; but if dated Oct. 29th, 30th, or 31st, it would expire on the last day of February, and be legally due on the 3d of March.

486. A **Protest** is a written declaration made by a *notary public*, that the maker of a note has failed to pay it.

A protest must be made out on the day the note matures, and sent to the indorser immediately, to *hold him responsible*. The neglect to protest a note on maturity releases an indorser from all obligation to pay it, unless the words "waiving demand and notice" appear above the indorser's signature.

There are two methods of estimating the time between different dates. The first is by compound subtraction, which is still generally used in partial payments. The second is by determining the number of entire years, if any, and then reckoning the number of days left, either by adding the number in the different months between the dates, or from the table, Art. 296. This latter method is now generally adopted by merchants in finding interest on items in an account, and for calculations for short periods, and will be used in the following examples.

487. The **Principal Kinds** of notes will now be given, and the calculation of the interest upon them required.

A *Time Note* is one made payable at a specified time; when no time of payment is specified, the note is due on *demand*. A *Joint Note* is a note signed by two or more persons who are jointly liable for its payment. A *Joint* and *Several Note* is a note signed by several persons who are both jointly and singly liable for its payment.

A *Principal* and *Surety Note* is one in which another person becomes security for the payment of the note by the maker. A surety note should be made payable to the order of the surety, who should indorse it on the back to the order of the creditor. It is held that a note made in favor of the creditor and indorsed by the surety, does not bind the latter to the payment of the debt. In reckoning the interest on notes, 3 days of grace are to be allowed.

DEMAND NOTE.

$315 MILLERSVILLE, PA., *July* 7, 1876.

For value received, I promise to pay Jacob M. Frantz, or order, on demand, Three Hundred and Fifteen Dollars, without defalcation. P. W. HIESTAND.

1. TIME NOTE.

$225 INDIANA, PA., *Oct.* 15, 1876.

Sixty days after date, I promise to pay Joseph H. Landis, or order, Two Hundred and Twenty-five Dollars, with interest, for value received, without defalcation.

R. W. FAIR.

What is due on this note at maturity? *Ans.* $227.36.

2. PRINCIPAL AND SURETY NOTE.

$779.25 TRENTON, N. J., *Nov.* 15, 1876.

Two months after date, I promise to pay Philip Dunn, or order, Seven Hundred and Seventy-nine $\frac{25}{100}$ Dollars, with interest, for value received, without defalcation or discount. HENRY WOOD.

Surety, PHILIP DUNN.

What will be due on this note at maturity?

Ans. $787.56.

3. JOINT NOTE.

$650 JEFFERSON CITY, MO., *Aug.* 21, 1876.

On demand, for value received, we promise to pay James Mackay, or order, Six Hundred and Fifty Dollars, with interest, negotiable and payable without defalcation or discount. JOHN TOMLINSON,
CHARLES LEROY.

What will be due on this note, Jan. 1, 1877?

Ans. $664.41.

4. JOINT AND SEVERAL NOTE.

$727.75 NEW YORK, *Sept.* 25, 1876.

Six months after date, we jointly and severally promise to pay Matthew Wilcox, or order, Seven Hundred and Twenty-seven $\frac{75}{100}$ Dollars with interest, value received.
SAMUEL MORGAN,
RICHARD J. MENDENHALL.

What will be due on this note at maturity?

Ans. $753.79.

5. COMPANY NOTE PAYABLE AT A BANK.

$480 PHILADELPHIA, *April* 1, 1876.

Ninety days after date we promise to pay Claxton & Co., or order, at the National Bank of Northern Liberties, Four Hundred and Eighty Dollars, for value received, without defalcation. WILLIAMS, FRENCH & Co.

What was the value of this note, August 12, 1876?

Ans. $483.20.

6. A 30-day note for $750, without interest, was paid in 50 days; what was the amount due? *Ans.* 752.12\frac{1}{2}$.

7. A 60-day note for $630, with interest from date, was paid in 105 days; what sum was due? *Ans.* $641.025.

8. What is the difference of the interest on a note for $700 given June 1, 1876, due 3 months after date, and on one given at the same date for the same amount, due 90 days after date? *Ans.* $.23$\frac{1}{8}$.

ANNUAL INTEREST.

488. Annual Interest is the simple interest of the principal, and of each year's interest from the time of its accruing until settlement.

489. Annual Interest is sanctioned by some States when the note is written "with interest payable annually."

1. *Simple Interest* is not due, and cannot be collected until the principal is due, unless the note reads, "with interest payable annually." *Annual Interest* allows interest on the *unpaid interest* of a debt as well as upon the debt itself.

2. In *Compound Interest*, each year's interest is added to the principal, and the sum forms a new principal for the succeeding year.

3. The neglect to collect the annual interest on a note drawn "with interest payable annually," is in some States, regarded as a waiving of the contract requiring it.

1. What is the amount due on a note of $300, at 6% for 3 yr. 3 mo., interest payable annually?

OPERATION.

$300 × .06 = $18 int. for 1 yr.
$18 × 3$\frac{1}{4}$ = 58.50, int. for 3$\frac{1}{4}$ yr.
$18 × .135 = 2.43, int. on 1st int.
$18 × .075 = 1.35, int. on 2d int.
$18 × .015 = .27, int. on 3d int.
300.00 principal.
$362.55

SOLUTION.—The interest on $300 for one year is $18, and for 3 yr. 3 mo. is $58.50; the first year's interest is on interest 2 yr. 3 mo., giving $2.43 interest; the second year's is on interest for 1 yr. 3 mo., amounting to $1.35; the third year's interest is on interest 3 mo., amounting to $.27; adding the interest on the principal, the interest on each year's interest, and the principal, we have $362.55 as the amount due.

Rule.—I. *Find the interest on the principal for the given time and rate: also find the interest on each year's interest for the time it has remained unpaid.*

II. *The sum of these interests will be the annual interest, and this, added to the principal, will be the amount due.*

NOTE.—The work may be shortened by calculating the interest for the sum of the times during which the different interests remain unpaid.

WRITTEN EXERCISES.

2. What is the interest due on a note for $840, dated March 2, 1872, interest payable annually, if no payments are made till Sept. 9, 1876? *Ans.* $252.90.

3. How much is due Jan. 1, 1877, on a note for $1000, dated June 16, 1873, interest payable annually at 7%, if the yearly interest has been regularly paid? *Ans.* $1038.69.

4. $1250. CONCORD, N. H., *Feb.* 10, 1871.

For value received, I promise to pay to the order of Jacob Clark, on demand, One Thousand Two Hundred and Fifty Dollars, with interest annually. THOMAS MAYNARD.

What was due on this note, June 11, 1875, if the annual interest was paid up for the first two years?

Ans. $1432.73.

PARTIAL PAYMENTS.

490. Partial Payments are payments in part of notes or other obligations bearing interest.

491. An **Indorsement** is an acknowledgment of a payment written on the back of the obligation, stating the time and amount of the payment.

The term *Indorsement* is used in different business papers, in each case, however, meaning a *writing on the back*, from the Latin *dorsum*, the back.

1. The writing of the name on the back of a check, draft, note, etc., is called a *General Indorsement*, or an *indorsement in blank*.

2. A *Special Indorsement* directs the obligation to be paid to some particular person, or to his order.

3. An acknowledgment of the payments on a note, written on the back of it, is also an indorsement. The person holding the obligation signs his name to this statement as a receipt.

492. The Supreme Court of the United States, and nearly all the States, adopt the following rule for partial payments, called

THE UNITED STATES RULE.

I. *Find the amount of the principal to the time of the first payment; if the payment equals or exceeds the interest, subtract the payment from the amount and treat the remainder as a new principal.*

II. *If the payment is less than the interest, find the amount of the same principal to the time when the sum of the payments shall equal or exceed the interest due, and subtract the sum of the payments from the amount.*

III. *Proceed in the same manner with the remaining payments until the time of settlement.*

NOTE.—This rule is founded upon the decision of Chancellor Kent. The principle is, that neither interest nor payment shall draw interest. It has been adopted by nearly all the States—New Hampshire, Vermont, and Connecticut being the principal exceptions.

1. $600 MILLERSVILLE, PA., *July* 12, 1870.

Four years after date, I promise to pay Henry Wilson, or order, Six Hundred Dollars, with interest, for value received. CHARLES HARDING.

On this note were the following indorsements:

May	24, 1871,	received	$131.20
Dec.	18, 1872,	"	40.00
Sept.	12, 1873,	"	175.00

How much remained due July 12, 1874?

OPERATION.

Principal, or face of note	$600.00
Interest to first payment	31.20
Amount due May 24, 1871	631.20
First payment to be deducted	131.20
Balance due after first payment	500.00
Interest on balance to second payment is $47.00. The payment being less, is not deducted.	
Interest from first payment to third payment	69.00
Amount due Sept. 12, 1873	569.00
Sum of second and third payments to be deducted . . .	215.00
Balance due after third payment	354.00
Interest from Sept. 12, 1873, to July 12, 1874	17.70
Balance due on settlement, July 12, 1874	371.70

2. $4000 LANCASTER, PA., *May* 10, 1870.

Five years after date, for value received, I promise to pay Robert Turner, or order, Four Thousand Dollars, with interest from date. MORTON BLACK, JUN.

Indorsements: May 10, 1871, $300; May 22, 1872, $250; June 16, 1873, $70; July 30, 1874, $175.

How much was due May 10, 1875? *Ans.* $4389.58.

3. $800. COLUMBIA, PA., *March* 10, 1870.

For value received, on demand, I promise to pay W. H. Fisher, Eight Hundred Dollars, with interest.

W. H. CROTHERS.

Indorsements: Feb. 16, 1871, $10.00; Oct. 20, 1871, $75.00; Jan. 14, 1872, $15.00; April 26, 1872, $10.00.

The note was settled Sept. 1, 1872; what was then due?

Ans. $808.25.

4. A note of $7000 was given Jan. 1, 1872.

Indorsements: May 3, 1873, $400; Aug. 8, 1874, $70; Sept. 9, 1875, $120; Oct. 7, 1876, $950.

What was due Jan. 1, 1877, Int. 7%? *Ans.* $7910.

5. A note of $5860 was given Sept. 10, 1874.

Indorsements: Aug. 16, 1875, $150; May 18, 1876, $350; Dec. 28, 1877, $95; Nov. 17, 1878, $112.

What was due Jan. 1, 1879, Int. 5%? *Ans.* $6414.66.

6. A note of $3500 was given May 12, 1870.

Indorsements: Jan. 16, 1871, $50; July 10, 1871, $25; Dec. 18, 1871, $250; June 20, 1872, $475; Aug. 20, 1873, $75; Sept. 30, 1873, $35.

What was due Jan. 1, 1874, Int. 6%? *Ans.* $3320.72.

493. Business men generally settle notes and interest accounts, payable within a year, by the following rule, called the

MERCHANTS' RULE.

I. *Find the amount of the principal to the time of settlement, and also the amount of each payment to the time of settlement.*

II. *Subtract the sum of the amounts of the payments from the amount of the principal, for the balance due.*

NOTES.—1. In some States merchants apply this rule to notes for longer periods by reckoning the interest for 1 year, and subtracting from the amount the amounts of the payments made during the year, and taking this balance for a new principal.

2. As the periods in these notes are all short, the interest should be calculated for the number of days.

1. $5480. PHILADELPHIA, *Jan.* 1, 1875.

Sixty days after date, for value received, I promise to

pay Joseph Trotter, or order, Five Thousand Four Hundred and Eighty Dollars, without defalcation.

JAMES TAYLOR.

Indorsements: March 15, $200; June 12, $300; Aug. 9, $500; Oct. 1, $700.

What is due Dec. 3, 1875? *Ans.* $3994.86.

2. A note of $4774.25 was given Nov. 9, 1875.

Indorsements: Jan. 1, 1876, $500; Feb. 12, $600; April 17, $450; June 10, $247.50; Aug. 1, $250.

What is due Oct. 1, 1876, at 7%? *Ans.* $2953.592.

3. A note was given for $1250, April 20, 1874.

Indorsements: May 10, $200; July 17, $50; Sept. 25, $140; Oct. 19 $150; Dec. 12, $350.

What was due April 20, 1875, at 5%? *Ans.* $396.89.

NOTE.—For Connecticut, Vermont, and New Hampshire rules, see *Brooks's Higher Arithmetic.*

DISCOUNT AND PRESENT WORTH.

494. Discount is an allowance made for the payment of money before it becomes due.

495. The **Present Worth** of a debt payable at a future time without interest is such a sum as, being on interest for the time at a certain rate, will amount to the debt.

496. The **True Discount** is the difference between the amount of the debt and the present worth.

NOTES.—1. The *true discount* is the *interest* on the present worth for the time between the payment of the debt and the time it becomes due.

2. The present worth corresponds to the principal, the discount to the interest, and the debt to the amount; hence the different cases may be solved as in Interest.

1. What is the present worth of $585, due 5 years hence without interest, money being worth 6%?

SOLUTION.—The amount of $1 for 5 years, at 6%, is $1.30, hence the present worth of $1.30 is $1, and the present worth of $585 is as many times $1 as $1.30 is contained times in $585, which is $450. Hence

OPERATION.

$0.06×5=$0.30

Amount=$1.30

$585÷1.30=$450, *Ans.*

Rule.—I. *Divide the given sum by the amount of* $1 *for the given rate and time, to find the present worth*

II *Subtract the present worth from the given sum to find the discount.*

NOTE.—When several payments are made without interest, find the present worth of each separately, and take their sum.

WRITTEN EXERCISES.

2. What is the present worth of $1206, due 5 yr. 8 mo. hence without interest, money worth 6%? *Ans.* $900.

3. What is the discount of $6460, due 4 yr. 10 mo. 12 da. hence without interest, money worth 6%? *Ans.* $1460.

4. A owes $2178 payable in 3 yr. 9 mo. without interest, but wishing to pay it immediately, what should he in equity pay, money worth 7 per cent.? *Ans.* $1725.14+.

5. B bought $2500 worth of goods on 6 mo. 18 da. credit; what allowance should be made, if the bill be paid immediately, money being worth 6%? *Ans.* $79.87—.

6. I can sell my horse for $280 cash or $300 on 1 yr. 6 mo. credit; I choose the latter; how much did I lose, money being worth 6 per cent.? *Ans.* $4.77+.

7. A gives his note for $850 in 2 yr. 8 mo. without interest; at the end of 8 mo. he wishes to pay the note; what should the holder of the note receive? *Ans.* $758.93—.

8. A man owes $600, of which one-third is to be paid in one year and the remainder in two years; what is the present value, money worth 6 per cent.? *Ans.* $545.82.

9. What is the present worth of $2400, one-fourth due in 8 mo., one-third in 1 year, and the remainder in 18 mo., money being worth 6 per cent.? *Ans.* $2249.07.

BANK DISCOUNT AND BANKING.

497. A **Bank** is an incorporated institution which receives and loans money, or furnishes a paper circulation.

498. A **Bank of Deposit** is one which receives money or its equivalent on deposit, to be drawn at the order of the depositor.

499. A **Bank of Discount** is one that lends money, discounts notes, drafts, etc. A *Bank of Issue* is one that makes and issues notes to circulate as money.

Some banks unite two and some all of these offices. A *Savings Bank* is one that receives small sums on deposit, and pays interest to its depositors.

500. A **Check** is an order on a bank, given by one of its depositors, to pay a certain amount to some person or his order, or to bearer.

501. **Bank Discount** is the interest on the face of the note for the time from the day of discount to the day of payment.

502. The **Proceeds** or **Avails** of a note is the sum received for it when discounted, and equals the face less the discount.

503. The **Term of Discount** is the number of days from the time of discounting to the time of maturity of the note.

When a person wishes to borrow money at a bank, he presents a note, either made or indorsed by himself, payable at a certain time, and receives for it a sum equal to the face *less* the interest for the time the note has to run. This amount is withheld by the bank in consideration of advancing money on the note prior to its maturity.

In Pennsylvania, Delaware, Maryland, Missouri, and the District of Columbia, the *day of discount* and *day of payment* are both reckoned, which, with the three days of grace, make 4 days. A 60-day note, in these States, would be discounted for 64 days.

Business men often discount notes by deducting the interest for a given time, with or without grace, as may be agreed upon. The rate is fixed by agreement, and may be other than the legal rate.

504. The difference between bank discount and true discount may be shown as follows:

If I take my note to the bank promising to pay $106 at the end of 1 year, to get it cashed, by the method of true discount I would receive $100; but by the method of bank discount, not counting days of grace, I would receive $106 minus the interest of $106 for 1 year, that is, $106 −$6.36=$99.64.

CASE I.

505. ***Given, the face of the note, the rate, and the time, to find the discount and the proceeds.***

1. What is the present worth or proceeds of a note for $600, due in 21 days, discounted at a bank at 6 per cent.?

SOLUTION.—We find the interest of $600 for 21 da. plus 3 da., or 24 da., is $2.40, which is the discount. Subtracting this from $600, we have the proceeds, equal to $597.60.

OPERATION.

$600	21+3=24
.004	24÷6 = .004
$2.400	
$597.60, *Ans.*	

Rule.—I. *Find the interest on the face of the note for three days more than the specified time, for the discount.*

II. *Subtract the discount from the face, to find the present worth.*

Note.—The discount of an *interest-bearing* note is computed on the *amount* of the note at its maturity. Banks compute interest for the *actual number of days* a note has to run, whether a note is drawn for *months* or *days*.

WRITTEN EXERCISES.

2. What is the discount of a note for $275, due in 60 days, discounted at a bank at 7%? *Ans.* $3.37—.

3. What are the proceeds of a note for $965, at 90 days, discounted by a broker at 7%? *Ans.* $947.55.

4. Required the proceeds of a note for $876.50, due in 60 days, discounted by a bank at 6%. *Ans.* $867.30.

5. Required the difference between the true discount and the bank discount of $690, due in 2 yr. 6 mo., money worth 6%, not reckoning days of grace. *Ans.* $13.50.

Find the time when due, the time to run, the discount, and the proceeds of the following notes:

6. 650\frac{25}{100}$. PHILADELPHIA, *March* 16, 1870.

Four months after date I promise to pay Thomas Newman, or order, Six Hundred and Fifty $\frac{25}{100}$ *Dollars, at the Girard Bank, value received, without defalcation.*

HENRY OSBORN.

Discounted, April 1st, 1870, at 6%. *Ans.* Dis., $11.92.

7. 135\frac{50}{100}$. WASHINGTON, *Aug.* 20, 1872.

Three months after date, for value received, I promise to pay W. H. Seal, or order, One Hundred Thirty-five $\frac{50}{100}$ *Dollars, without defalcation.* D. NEWLIN FELL.

Discounted, Sept. 7, 1872, at 6%. *Ans.* Discount, $1.76.

8. $750. CHICAGO, *June* 16, 1876.

Nine months after date, for value received, I promise to pay Mary Smith, or order, Seven Hundred Fifty Dollars, with interest, at 6 per cent. FANNIE E. WILLARD.

Discounted, at 6%, Oct. 24, 1876. *Ans.* Dis., $19.08.

CASE II.

506. ***Given, the rate, the time, and the proceeds or the discount, to find the face.***

1. I wish to borrow $800 from a bank; for what must I give my note at 30 days, discounting at 6 per cent.?

SOLUTION.—We find the interest of $1 for 33 days and subtract it from $1, which gives the proceeds of $1. If for every $1 in the face of the note the proceeds are $0.9945, to give $800 proceeds will require as many times one dollar as $0.9945 is contained times in $800, which are $804.42.

OPERATION.

$$\begin{array}{r} \$1.0000 \\ .0055 \\ \hline .9945 \end{array} \text{, Proceeds of \$1.}$$

$$\frac{800}{.9945} = \$804.42+.$$

Rule.—*Divide the given proceeds by the proceeds of* $1 *for the given time and rate; or divide the discount by the discount of* $1.

WRITTEN EXERCISES.

2. A wishes to borrow $1000 from a bank for 60 days; for what sum must he give his note, discounting at 6 per cent.? *Ans.* $1010.61.

3. What is the face of a note at 90 days, the proceeds of which, discounted at 6%, are $2000? *Ans.* $2031.50.

4. For what sum must a note be drawn at 60 days to net $5000, when discounted at 6 per cent.? *Ans.* $5053.06.

5. A broker buys a 60 day note for $20 less than the face; what was the face, discount 6 per cent.? *Ans.* $1904.76.

6. Find the face of a 6 mo. note which, when discounted at 1 per cent. a month, yields $685.50. *Ans.* $730.03.

7. Mr. Brown, owing $1000, gave a 90 day note, which was discounted at 1¼ per cent. a month; required the face of the note to pay the exact debt. *Ans.* $1040.31.

8. Mr. Schofield presented a note for 30 days at a Baltimore bank for discount; the proceeds being $954.56, what was the face of the note? *Ans* $960.

CASE III.

507. ***Given, the face, the rate, and the proceeds or the discount, to find the time.***

1. The proceeds of a note for $600, discounted at 6%, are $593.70; what was the time?

SOLUTION.—Subtracting $593.70 from $600, we find the discount is $6.30. The discount on $1.00 for *one* day, at 6%, is $\frac{1}{6}$ of a mill; and on $600 it is 600×$.000$\frac{1}{6}$, or $0.10. Hence the note was discounted for as many days as $0.10 is contained time in $6.30, or 63 days. Therefore, the time was 63—3, or 60 days.

OPERATION.

$600
593.70
6.30 discount.
600×.000$\frac{1}{6}$=.10
6.30÷.10=63 days.
63—3=60 days.

Rule.—*Divide the discount by the interest on the face for one day, and subtract 3 days of grace from the quotient.*

NOTE.—When the time a note has to run after being discounted is required, we wish to know the actual time, and therefore do not subtract the days of grace.

WRITTEN EXERCISES.

2. A merchant discounts a note for $2000 at a bank, and receives $1969; what is the time? *Ans.* 90 days.

3. A commission merchant sold a consignment of cotton for $4500, receiving in payment a note, which yielded, on being discounted, $4475.25; what was the time of the note? *Ans.* 30 days.

4. A note dated June 21st, 1875, was discounted July 1st at 7%; the face of the note was $6540 and the proceeds $6472.60; how long had it to run after it was discounted? *Ans.* 53 days.

5. A note dated Jan. 15th, 1876, at 6 months, was discounted at the First National Bank, St. Louis; the proceeds were $8402.25, and the face $8500; what was the date of discount? *Ans.* May 11.

6. An interest-bearing note, dated Aug. 1, 1872, at 90 days was discounted at 8%; the face was $750, and the proceeds $759.982; what was the date of discount? *Ans.* Oct. 1.

CASE IV.

508. ***Given, the face, the time, and the proceeds or the discount, to find the rate.***

1. The proceeds of a note for $300, at 30 days, are $298.35; what is the rate?

D

SOLUTION.—We find the discount on $300 is $1.65; and the discount on $300 at *one* per cent. for 33 days is $0.27½. Hence the required rate is as many times 1% as 27½ is contained times in $1.65, which is 6%.

OPERATION.

$300
298.35
1.65 discount.
$300 \times .00\frac{11}{120} = 27\frac{1}{2}$.
$1.65 \div 27\frac{1}{2} = 6$.

Rule.—*Divide the discount by the interest on the face, at 1% for the given time.*

WRITTEN EXERCISES.

2. Mr. Herr buys goods to the amount of $4000, and to pay for them gets his note for 60 days discounted at a bank; if the face is $4042.45, what is the rate? *Ans.* 6%.

3. A note dated July 1st, 1875, at 3 months, was discounted at bank on Aug. 10, 1875; the face was $2500, and the proceeds $2473.264; what was the rate? *Ans.* 7%.

4. A note dated September 12th, 1875, at 6 months, was discounted at Wilmington, Del., December 9th, 1875; the face of the note was $5750 and the proceeds $5624.77⅞; what was the rate of discount? *Ans.* 8%.

STOCK INVESTMENTS WITH INTEREST.

509. In **Stock Investments** operators take into consideration the interest on the money invested.

Since money is worth its interest while invested, to know the actual gain or loss of an investment, we should reckon the interest on the money invested.

Stock speculators frequently, instead of paying for stock, deposit a sum called a "margin," to secure the broker against loss, should the stock fall in price before delivery or sale.

NOTE.—As the following examples are worked principally by a combination of methods previously given, it has been thought unnecessary to divide them into cases. Brokerage at ⅛ per cent. is to be reckoned on all purchases and sales. Money is considered worth 6%.

1. What is the annual rate of interest of an investment which pays 5% semi-annually, if reinvested at 6%?

SOLUTION.—The interest for the first half year may be on interest during the second half year at 6%; hence, at the end of the year the interest for the first half year will amount to $.05 × 1.03, or $.0515, which, added to the interest of *the second* half year, $.05, gives $.1015 as the yearly interest on $1.

OPERATION.

$\$.05 \times 1.03 = \$.0515$
$\$.05 + \$.0515 = \$.1015$
$= 10\frac{3}{20}\%$

WRITTEN EXERCISES.

2. When the Penn. Railroad pays 2% quarterly, what yearly dividend will be equal to this? *Ans.* $8\frac{9}{50}$%.

3. If I buy Michigan 6's at 108, interest payable semi-annually, what annual rate % do I receive? *Ans.* $5\frac{271}{433}$%.

4. If I buy 15 shares United Companies of New Jersey at $137\frac{1}{2}$ ($100), and receive $$37\frac{1}{2}$ dividend quarterly, what annual rate of interest do I receive? *Ans.* $7\frac{233}{551}$%.

5. Mr. Westlake sold $4000 Illinois 6's at 106, interest payable quarterly, and bought Kentucky 6's at 105, interest payable semi-annually; did he increase or diminish his yearly income if each dividend was put out at interest as soon as received? *Ans.* Diminished $1.80; surplus, $20.

6. I buy in August 20 shares Second and Third Sts. Pass. Railway ($50) at 83, and receive in October, January, April, and July, a 3% dividend; what % of income do I receive during the year, and what will be my entire dividend if each dividend is invested for the remainder of the year, interest at 8%? *Ans.* Div., $123.60; rate, $7\frac{289}{665}$%.

COMPOUND INTEREST.

510. Compound Interest is interest on both principal and interest, when the interest is not paid when due.

Compound interest assumes that if the borrower does not pay the interest when due, it is proper that he should pay interest for it until paid. Some regard it as just, but it has not the sanction of law.

1. What is the compound interest of $400 for 2 yr. at 6%?

SOLUTION.—Multiplying by the rate per cent., we find the interest for 1 year to be $24; adding this to the principal, we find the amount to be $424, which is the principal for the 2d year. Multiplying the new principal by the rate, we find the interest for the 2d year to be $25.44, and adding this to the 2d principal, we find the amount for 2 years to be $449.44, from which subtract the 1st principal, and the remainder, $49.44, is the compound interest. Hence the following

OPERATION.

$400	
.06	
24.00	= Int. 1st yr.
400	
424.00	= Amt. 1st yr.
.06	
25.44	= Int. 2d yr.
424	
449.44	= Amt. 2d yr.
400	
49.44	= C. Int. for 2 yr.

Rule.—I. *Find the amount of the principal for the first period of time for which interest is reckoned, and make this the principal for the second period.*

II. *Find the amount of this principal for the next period; and thus continue till the end of the given time.*

III. *Subtract the given principal from the last amount, and the result will be the compound interest.*

NOTES.—1. When the interest is due semi-annually or quarterly, we find the interest for such time and proceed as above directed.

2. When the time is for years, months, and days, find the amount for the years, then compute the interest on this for the months and days, and add to the last amount before subtracting.

WRITTEN EXERCISES.

2. What is the compound interest of $568, for 3 yr., at 6 per cent.? *Ans.* $108.50.

3. What is the amount, at compound interest, of $90, for 6 yr., at 7 per cent.? *Ans.* $135.06½.

4. What is the compound interest of $347.50, for 4 yr. 8 mo., at 6 per cent.? *Ans.* $108.76.

5. What is the compound interest of $1728¼, for 2 yr. 6 mo., at 6 %, payable semi-annually? *Ans.* $275.27.

6. What is the amount of $240, for 2 yr. 3 mo., at 8 per cent., payable quarterly? *Ans.* $286.82.

7. What is the amount of $450, for 8 yr., at 6 per cent., compound interest?

SOLUTION.—We look in the table under 6 per cent., and opposite 8 yr. we find the amount of $1 to be $1.5938481; multiplying this amount by 450, we have the amount of $450, which is $717.23.

OPERATION.

$1.5938481
450
$717.2302950

8. What is the amount of $780, for 9 yr., at 8 per cent., compound interest? *Ans.* $1559.22.

9. What is the amount of $300, for 16 yr., at 7 per cent., compound interest? *Ans.* $885.65.

10. Required the compound interest of $950, for 20 yr., at 4 per cent. *Ans.* $1131.57.

11. What is the difference between the simple and compound interest of $600 for 6 yr. 6 mo. 6 da. at 6 %? *Ans.* $42.90.

511. The calculation of compound interest is facilitated by the use of the following table. Similar tables are also used for simple interest.

TABLE.

Amount of $1 at Compound Interest in any number of years not exceeding 25.

Yr.	2 per cent.	2½ per cent.	3 per cent.	3½ per cent.	4 per cent.	4½ per cent.
1	1.0200 0000	1.0250 0000	1.0300 0000	1.0350 0000	1.0400 0000	1.0450 0000
2	1.0404 0000	1.0506 2500	1.0609 0000	1.0712 2500	1.0816 0000	1.0920 2500
3	1.0612 0800	1.0768 9062	1.0927 2700	1.1087 1787	1.1248 6400	1.1411 6612
4	1.0824 3216	1.1038 1289	1.1255 0881	1.1475 2300	1.1698 5856	1.1925 1860
5	1.1040 8080	1.1314 0821	1.1592 7407	1.1876 8631	1.2166 5290	1.2461 8194
6	1.1261 6242	1.1596 9342	1.1940 5230	1.2292 5533	1.2653 1902	1.3022 6012
7	1.1486 8567	1.1886 8575	1.2298 7387	1.2722 7926	1.3159 3178	1.3608 6183
8	1.1716 5938	1.2184 0290	1.2667 7008	1.3168 0904	1.3685 6905	1.4221 0061
9	1.1950 9257	1.2488 6297	1.3047 7318	1.3628 9735	1.4233 1181	1.4860 9514
10	1.2189 9442	1.2800 8454	1.3439 1638	1.4105 9876	1.4802 4428	1.5529 6942
11	1.2433 7431	1.3120 8666	1.3842 3387	1.4599 6972	1.5394 5406	1.6228 5305
12	1.2682 4179	1.3448 8882	1.4257 6089	1.5110 6866	1.6010 3222	1.6958 8143
13	1.2936 0663	1.3785 1104	1.4685 3371	1.5639 5606	1.6650 7351	1.7721 9610
14	1.3194 7876	1.4129 7382	1.5125 8972	1.6186 9452	1.7316 7645	1.8519 4492
15	1.3458 6834	1.4482 9817	1.5579 6742	1.6753 4883	1.8009 4351	1.9352 8244
16	1.3727 8570	1.4845 0562	1.6047 0644	1.7339 8604	1.8729 8125	2.0223 7015
17	1.4002 4142	1.5216 1826	1.6528 4763	1.7946 7555	1.9479 0050	2.1133 7681
18	1.4282 4625	1.5596 5872	1.7024 3306	1.8574 8920	2.0258 1652	2.2084 7877
19	1.4568 1117	1.5986 5019	1.7535 0605	1.9225 0132	2.1068 4918	2.3078 6031
20	1.4859 4740	1.6386 1644	1.8061 1123	1.9897 8886	2.1911 2314	2.4117 1402
21	1.5156 6634	1.6795 8185	1.8602 9457	2.0594 3147	2.2787 6807	2.5202 4116
22	1.5459 7967	1.7215 7140	1.9161 0341	2.1315 1158	2.3699 1879	2.6336 5201
23	1.5768 9926	1.7646 1068	1.9735 8651	2.2061 1448	2.4647 1555	2.7521 7635
24	1.6084 3725	1.8087 2595	2.0327 9411	2.2833 2849	2.5633 0417	2.8760 1383
25	1.6406 0599	1.8539 4410	2.0937 7793	2.3632 4498	2.6658 3633	3.0054 3446

Yr.	5 per cent.	6 per cent.	7 per cent.	8 per cent.	9 per cent.	10 per cent.
1	1.0500 000	1.0600 000	1.0700 000	1.0800 000	1.0900 000	1.1000 000
2	1.1025 000	1.1236 000	1.1449 000	1.1664 000	1.1881 000	1.2100 000
3	1.1576 250	1.1910 160	1.2250 430	1.2597 120	1.2950 290	1.3310 000
4	1.2155 063	1.2624 770	1.3107 960	1.3604 890	1.4115 816	1.4641 000
5	1.2762 816	1.3382 256	1.4025 517	1.4693 281	1.5386 240	1.6105 100
6	1.3400 956	1.4185 191	1.5007 304	1.5868 743	1.6771 001	1.7715 610
7	1.4071 004	1.5036 303	1.6057 815	1.7138 243	1.8280 391	1.9487 171
8	1.4774 554	1.5938 481	1.7181 862	1.8509 302	1.9925 626	2.1435 888
9	1.5513 282	1.6894 790	1.8384 592	1.9990 046	2.1718 933	2.3579 477
10	1.6288 946	1.7908 477	1.9671 514	2.1589 250	2.3673 637	2.5937 425
11	1.7103 394	1.8982 986	2.1048 520	2.3316 390	2.5804 264	2.8531 167
12	1.7958 563	2.0121 965	2.2521 916	2.5181 701	2.8126 648	3.1384 284
13	1.8856 491	2.1329 283	2.4098 450	2.7196 237	3.0658 046	3.4522 712
14	1.9799 316	2.2609 040	2.5785 342	2.9371 936	3.3417 270	3.7974 983
15	2.0789 282	2.3965 582	2.7590 315	3.1721 691	3.6424 825	4.1772 482
16	2.1828 746	2.5403 517	2.9521 638	3.4259 426	3.9703 059	4.5949 730
17	2.2920 183	2.6927 728	3.1588 152	3.7000 181	4.3276 334	5.0544 703
18	2.4066 192	2.8543 392	3.3799 323	3.9960 195	4.7171 204	5.5599 173
19	2.5269 502	3.0255 995	3.6165 275	4.3157 011	5.1416 613	6.1159 390
20	2.6532 977	3.2071 355	3.8696 845	4.6609 571	5.6044 108	6.7275 000
21	2.7859 626	3.3995 636	4.1405 624	5.0338 337	6.1088 077	7.4002 499
22	2.9252 607	3.6035 374	4.4304 017	5.4365 404	6.6586 004	8.1402 749
23	3.0715 238	3.8197 497	4.7405 299	5.8714 637	7.2578 745	8.9543 024
24	3.2250 999	4.0489 346	5.0723 670	6.3411 807	7.9110 832	9.8497 327
25	3.3863 549	4.2918 707	5.4274 326	6.8484 752	8.6230 807	10.8347 059

EXCHANGE.

512. Exchange is the method of making payments in distant places by means of *Drafts* or *Bills of Exchange.*

513. Exchange is of two kinds, *Domestic* and *Foreign.* Exchange between two places in the same country is called *Domestic* or *Inland Exchange;* that between different countries is called *Foreign Exchange.*

514. A **Draft** or **Bill of Exchange** is a written order for the payment of money. In domestic exchange a bill is usually called a *Draft.*

515. A **Sight Bill** is one payable "at sight" or on its presentation. A *Time Bill* is one payable at a specified time after sight or after date.

516. The **Indorsement** of a bill is the writing upon the back of it, by which the payee transfers the payment to another.

A *special* indorsement is an order to pay the bill to some particular person, who is then called the *Indorsee,* and he alone can collect the bill. An indorsement *in blank* is the writing of the holder's name upon the back, which makes the bill payable to the bearer.

The person who signs the bill is called the *Maker* or *Drawer;* the person requested to pay is called the *Drawee;* the person to whom the money is to be paid, is the *Payee;* the person who has possession of the bill is called the *Owner* or *Holder.*

517. The **Acceptance** of a bill is the promise of the Drawee, when presented, to pay it at maturity. The Drawee *accepts* by writing across the face of the bill, "Accepted," with the date and his signature; the bill is then called an *Acceptance,* and is of the character of a promissory note.

If a bill is protested for non-acceptance, the maker is under obligations to pay it immediately, although the time specified in it has not expired. Bills of exchange are entitled to "days of grace," unless a particular day is named. In New York, Pennsylvania, and a few other States, no grace is allowed to bills of sight. If a note is payable on demand, it is legally due when presented, as bank-notes, etc. If a particular time is specified in a note, it is legally due on that day. If a draft is drawn at *usance,* the time is regulated by custom or the law of the place where it is payable.

When a bill is drawn "acceptance waived," it is not subject to protest until maturity. When an indorser writes over his name, "demand and notice waived," he is liable even if the bill is not protested. If the indorser writes "without recourse" over his indorsement, he is not liable for the payment of the bill

In reckoning the time of maturity of a bill payable after date, the day on which it is dated is not included, and in the case of a bill payable after sight, the day of presentment is not included.

518. The **Rate of Exchange** is the rate per cent. which is reckoned upon a draft. The *Course of Exchange* is the current price paid in one place for bills of exchange upon another.

The *brokerage* is usually included in the quotation of exchange.

519. The **Par of Exchange** is the established value of the monetary unit of one country in the monetary unit of another; it is either *intrinsic* or *commercial.*

520. Exchange is at *par* when a draft or bill sells for its face; at a *premium* when it sells for more than its face; and at a *discount* when it sells for less than its face.

The rate of exchange between two places or countries depends upon the course of trade. If the trade between New York and Chicago is equal, exchange is at par. If New York owes Chicago, the demand in New York for drafts on Chicago is greater than the demand in Chicago for drafts on New York, hence the drafts are at a *premium* in New York. But if Chicago owes New York, the demand for drafts is less in New York than in Chicago; hence drafts in New York on Chicago are at a *discount.*

The reason why the banks in New York should charge a premium, is that they must be at the expense of actually sending money to the Chicago banks, or be charged with interest on their unpaid balance; the reason why the Chicago banks will sell at a discount is that they are willing to sell for less than the face of a draft in order to get the money owed them in New York immediately.

A check, draft, or certificate of deposit *on a bank in the place where drafts are selling at a premium,* is often sent to pay a debt in the place where drafts are *selling at a discount,* and such a check or draft will command a premium.

If the course of exchange is unfavorable in drawing, the discount is sometimes avoided by means of a circuitous exchange through several intermediate places between which the course is favorable.

DOMESTIC EXCHANGE.

521. Domestic or **Inland Exchange** is the exchange between two places in the same country.

522. The **Base** of an inland bill is the *face;* the *Rate* is the rate of premium or discount.

523. The **Forms** and **Use** of drafts may be seen by the following examples and explanations:

FIRST NATIONAL BANK OF NEW ORLEANS,

$8000. NEW ORLEANS, *Jan.* 16, 1877.

At sight, pay to the order of John Smith, Eight Thousand Dollars. THOMAS HASKINS,

To the MERCHANTS' NATIONAL BANK, *Cashier.*

PHILADELPHIA, PA.

EXPLANATION.—Suppose John Smith, of New Orleans, owes James Thomson & Co., of Philadelphia, $8000; he goes into a bank in New Orleans and gets the above draft. He then writes on the back of the note, "Pay to the order of James Thomson & Co.," signing his name, and forwards it to James Thomson & Co., in Philadelphia, who take it to the Merchants' National Bank, and writing the name of their firm on the back, receive the money.

THIRD NATIONAL BANK,

$5600. ST. LOUIS, MO., *Jan.* 11, 1877.

At ten days sight, pay to the order of H. B. Claflin & Co., Five Thousand Six Hundred Dollars, and charge the same to the account of JAMES SIMPSON,

To the FIFTH NATIONAL BANK, *Cashier.*

NEW YORK.

EXPLANATION.—Suppose that Harvey Williams, of St. Louis, wishing to pay a debt of $5600 to H. B. Claflin & Co., of New York, buys the above draft on the Fifth National Bank of New York. He forwards it to H. B. Claflin & Co., who, having indorsed it, will present it at the bank. The "ten days after sight" means after acceptance. It should be presented to the bank upon which it is drawn as soon as received, when the cashier writes upon it "accepted," with the date of acceptance, and signs his name as cashier. This makes the bank liable for its payment, and is an agreement to pay it after ten days.

NOTE.—If Harvey Williams has an account with the Fifth National Bank, he may draw on it directly as one bank draws on another. A person sometimes draws on a party who owes him in order to collect the bill.

CASE I.

524. ***To find the cost of a bill of exchange at sight, or on time.***

1. What must I pay in Philadelphia for a draft of $800 on New Orleans, exchange being $1\frac{1}{2}$ per cent. premium?

SOLUTION.—At a premium of $1\frac{1}{2}$% the cost of exchange of $1 is $1 + $1\frac{1}{2}$ ct. = $1.015, and the cost of $800 is 800 times $1.015, which are $812. Hence for sight exchange we have the following

OPERATION.

$$\begin{array}{rl} \$1.000 & \\ .015 & = \text{rate of exchange.} \\ \hline 1.015 & = \text{cost of } \$1. \\ 800 & \\ \hline \$812.000 & \textit{Ans.} \end{array}$$

Rule.—*Find the cost of* $1 *by adding the rate to* $1, *when at a premium, or subtracting it, when at a discount; and multiply the result by the face of the draft.*

2. What must be paid in New York for a draft of $2000 on St. Louis at 30 days, exchange being 2% premium?

SOLUTION.—The draft being on time should be purchased at a discount. The discount of $1, at the rate in St. Louis, for 30 + 3 or 33 days, is $.0055, which subtracted from $1, equals $.9945, the cost of $1 of the draft if the exchange was at par, but there is a premium of 2 per cent., hence adding $.02 we find the actual cost of $1 of the draft to be $1.0145, and multiplying this by 2000, we have $2029, the entire cost. Hence, for time exchange, the following

OPERATION.

$1.0000	
.0055	= discount for 33 da.
$.9945	= cost of $1 at par.
.02	= rate of exchange.
$1.0145	= cost of $1 of draft.
2000	
$2029	= whole cost.

Rule.—*From* $1 *subtract the bank discount of* $1 *for the time and rate, where the draft is purchased; to this result add the rate of exchange when at a premium, and subtract it when at a discount, and multiply the result by the face of the draft.*

WRITTEN EXERCISES.

3. Manson & Co., of Harrisburg, owe a party in Cleveland $4750; what must they pay at a Harrisburg bank for a draft on Cleveland, exchange $\frac{3}{4}$% discount? *Ans.* $4714.375.

4. What will a draft of $3500 cost, payable 30 days after sight, at 6%, exchange $1\frac{1}{2}$% premium? *Ans.* $3533.25.

5. A New York firm received a shipment of flour from Milwaukee, amounting to $7500, and remitted the money by a 15 day draft, at 7%, exchange being at a discount of $1\frac{1}{2}$%; what did they pay for the draft? *Ans.* $7361.25.

6. A Detroit merchant bought an assortment of spring goods in New York at a cost of $1500; what will be the cost of a 2 mo. draft on New York, at $1\frac{1}{4}$% premium, which will discharge the debt? *Ans.* $$1500\frac{5}{8}$.

7. A Philadelphia firm send their check for $4500 to their agent in Des Moines, where drafts on Philadelphia are selling at $1\frac{1}{4}$% premium; what will the Des Moines bankers pay for it? *Ans.* $4556.25.

CASE II.

525. ***Given, the cost of a bill of exchange, to find its face.***

1. A Boston merchant paid $2029 for a draft on Pittsburgh at 30 days, exchange 2% premium; required the face of the draft.

SOLUTION.—We find by Case I. that a draft for $1 will cost $1.0145, therefore a draft that costs $2029 must be for as many dollars as $1.0145 is contained times in $2029, which are $2000. From the above solution we derive the following

OPERATION.

$1.000
.0055 = discount for 30 da.
$.9945 = cost of $1 at par.
.02 = rate of exchange.
$1.0145 = cost of $1 of draft.

$$\frac{2029}{1.0145} = \$2000 \textit{ Ans.}$$

Rule.—*Find the cost of a draft of* $1, *and divide the given cost by it; the quotient will be the face of the draft.*

WRITTEN EXERCISES.

2. Jones & Bro., of St. Paul, purchased a sight draft for $2587 on Cincinnati, at a discount of $\frac{1}{2}$%; required the face of the draft. *Ans.* $2600.

3. A merchant in Maine buys a draft on New York at 45 days for $601.95 at a premium of $1\frac{1}{8}$%; what is the face of the draft? *Ans.* $600.

4. I received from Philadelphia a check for $40.20, which cost $\frac{3}{4}$% to have cashed; what should have been the face of the check that I might have realized $40.20? *Ans.* $40.50.

5. My agent sold $5000 worth of goods on commission, at $2\frac{1}{2}$%, and remits the proceeds in a draft bought at $1\frac{1}{4}$% premium; what did I receive for the sale? *Ans.* $4814.81.

6. A Baltimore merchant wishes to pay a debt of $1500 in Detroit by a sight draft on the First National Bank, Baltimore; if exchange on Baltimore is $\frac{1}{8}$% premium at Detroit, what must be the face of the draft? *Ans.* $1498.13—.

NOTE.—Since the draft is at a premium of $\frac{1}{8}$ per cent. in Detroit, it must be drawn for such a sum as, with the premium, will amount to $1500; hence the face will equal $1500 ÷ 1.00\frac{1}{8}$.

7. If the Baltimore merchant in the previous problem buy, instead of a sight draft, a draft at 90 days, what will be the cost of the draft? *Ans.* $1474.91.

8. A Boston merchant sends to a creditor in Savannah a sight draft on Boston for $1498.13; what was the debt, exchange on Boston being at a premium of $\frac{1}{8}$%? *Ans.* $1500.

CASE III.

526. ***Given, the face and the cost of a draft, to find the rate of exchange.***

1. A draft on Baltimore for $2000 at 30 days cost me $2029; what was the rate of exchange?

SOLUTION.—We find that the cost of $1 of the draft, if exchange was at par, is $.9945, and of $2000 is 2000 times $.9945, or $1989; the difference between $1989 and $2029, the actual cost, is $40, which is the premium; dividing the premium, $40, by the face, $2000, we have the rate, 2%.

OPERATION.

$1.0000
.0055 = discount for 33 da.
$.9945 = cost of $1 at par.
$.9945 × 2000 = $1989
$2029 — $1989 = $40
40 ÷ 2000 = .02, or 2%

Rule.—*Find the premium or the discount, and divide it by the face, to find the rate.*

WRITTEN EXERCISES.

2. A Savannah cotton broker bought a 30 day draft on Philadelphia for $3530.04, the face being $3500; what was the rate of exchange? *Ans.* $1\frac{1}{2}$% premium.

3. Sold grain on commission to the amount of $5000; having reserved $2\frac{1}{2}$%, I bought with the proceeds a draft for $4814.81, which I remitted to the consignor; what rate of exchange did I pay? *Ans.* $1\frac{1}{4}$% premium.

4. Mr. Bair, of Cincinnati, buys of Hood & Co., Phila., a lot of woollen goods amounting to $750, and forwards in payment a draft at 3 mo., which costs him 727.12\frac{1}{2}$; what was the rate of exchange? *Ans.* $1\frac{1}{2}$% discount.

FOREIGN EXCHANGE.

527. Foreign Exchange is the exchange that takes place between different countries.

A *Set of Exchange* consists of three bills of the same tenor and date, each containing a condition that it shall continue payable only while the others are unpaid.

To prevent loss, or delay, each bill of a set is remitted in a different manner, and when one bill of the set has been paid the others are worthless.

528. The **Money of Account** of any country consists of the denominations of the money of that country in which accounts are kept.

529. The Act of March 3, 1873, provides that "the value of the standard coins . . . of the world shall be estimated annually by the Director of the Mint, and be proclaimed on the first day of January by the Secretary of the Treasury."

530. In accordance with this law, the following table was published by the Secretary of the Treasury, Jan. 1, 1877:

COUNTRY.	MONETARY UNIT.	STANDARD.	VALUE IN U. S. MONEY.
Austria,	Florin,	Silver,	.45,3
Belgium,	Franc,	G. and S.,	.19,3
Bolivia,	Dollar,	G. and S.,	.96,5
Brazil,	Milreis of 1000 reis,	Gold,	.54,5
British America,	Dollar,	Gold,	$1.00
Bogota,	Peso,	Gold,	.96,5
Central America,	Dollar,	Silver,	.91,8
Chili,	Peso,	Gold,	.91,2
Denmark,	Crown,	Gold,	.26,8
Ecuador,	Dollar,	Silver,	.91,8
Egypt,	Pound of 100 piasters,	Gold,	4.97,4
France,	Franc,	G. and S.,	.19,3
Great Britain,	Pound Sterling,	Gold,	4.86,6½
Greece,	Drachma,	G. and S.,	.19,3
German Empire,	Mark,	Gold,	.23,8
Japan,	Yen,	Gold,	.99,7
India,	Rupee of 16 annas,	Silver,	.43,6
Italy,	Lira,	G. and S.,	.19,3
Liberia,	Dollar,	Gold,	1.00
Mexico,	Dollar,	Silver,	.99,8
Netherlands,	Florin,	G. and S.,	.38,5
Norway,	Crown,	Gold,	.26,8
Peru,	Dollar,	Silver,	.91,8
Portugal,	Milreis of 1000 reis,	Gold,	1.08
Russia,	Rouble of 100 copecks,	Silver,	.73,4
Sandwich Islands,	Dollar,	Gold,	1.00
Spain,	Peseta of 100 centimes,	G. and S.,	.19,3
Sweden,	Crown,	Gold,	.26,8
Switzerland,	Franc,	G. and S.,	.19,3
Tripoli,	Mahbub of 20 piasters,	Silver,	.82,9
Tunis,	Piaster of 16 caroubs,	Silver,	.11,8
Turkey,	Piaster,	Gold,	.04,3
U. S. of Colombia,	Peso,	Silver,	.91,8

531. Bills of Exchange are usually made payable either 3 days after sight or 60 days after sight. The latter are quoted at a lower rate, on account of the discount.

532. Most of the dealings in foreign exchange are with the commercial centres mentioned in the following table, taken from a recent New York paper:

	60 days.	3 days.
Prime banking sterling bills on London,	4 82½@4 83	4 84 @4 84½
Good bankers' and prime com'l,	4 81½@4 82½	4 83 @4 84
Good commercial,	4 80 @4 81	4 81½@4 82½
Paris (francs),	5 24⅜@5 24¼	5 21⅞@5 19⅜
Antwerp (francs),	5 24⅜@5 24¼	5 21⅞@5 19⅜
Swiss (francs)	5 23⅛@5 20	5 20⅝@5 18⅛
Amsterdam (guilders),	. 39⅞@ 40	40⅛@ 40¼
Hamburg (reichmarks,)	93¾@ 94	94½@ 94¾
Frankfort (reichmarks),	93¾@ 94	94½@ 94¾
Bremen (reichmarks),	93¾@ 94	94½@ 94¾
Berlin (reichmarks),	93¾@ 94	94½@ 94¾

Remittances to and from other places are frequently made in bills on these leading ones, especially London.

In the London quotations "prime" bills are those on the best banking houses, "good" are those on houses in good credit, but in less demand than the prime. "Commercial" signifies merchants' drafts, which generally rate below bankers'. In the quotations on Paris, Antwerp, and Switzerland the franc is the unit, and the quotation gives the number of francs and centimes to the dollar. The exchange on Amsterdam is the number of cents to the guilder; while on Hamburg, Frankfort, Bremen, and Berlin, the quotation gives the number of cents in 4 reichsmarks. 4.82½@4.83 indicates the highest and lowest prices on the day on which the quotations were made.

REMARK.—United States securities are quoted in London on a gold basis instead of a greenback one, of 4 shillings to the dollar, hence they usually appear lower than with us.

533. A **Letter of Credit** is a letter from a banking house in one country to one or more of their correspondents in another, directing them to pay to the person in whose favor the letter is written, any sum not exceeding a certain amount specified in the letter.

1. What must be paid in New York in gold for a bill of exchange on London for £450, at 3 days sight, at $4.88 to the pound sterling?

OPERATION.

$4.88
450
$2196.00

SOLUTION.—If £1 cost $4.88, £450 cost 450 times $4.88, which is $2196. Hence the following

Rule.—*Find the cost of a unit of the currency in which the bill is given, and multiply the face by it for the cost, or divide the cost by it for the face.*

WRITTEN EXERCISES.

2. What will be the cost in Philadelphia of the following draft, exchange at 60 days being $4.84, and gold at $110\frac{1}{2}$? *Ans.* $2674.10.

Exchange for £500. PHILADELPHIA, *July* 1, 1875.

Sixty days after sight of this First of Exchange (second and third unpaid) Pay to the order of Chas. Smith, Five Hundred Pounds Sterling, for value received, and charge the same to account of PETER WRIGHT & SONS.

TO MESSRS. BROWN BROTHERS, LIVERPOOL.

3. What must a merchant in Canton pay for a draft of $1134, if 1 tael=$1.63? *Ans.* 695.71— taels.

4. I wish to remit 5400 francs to Paris; what will a draft cost me in New York, if 1 franc =19.6¢? *Ans.* $1058.40.

5. A merchant in London sold a consignment of wheat from Odessa for £420; what will be the face of a draft on Odessa for the amount, if £1=6 roubles 70 copecks? *Ans* 2814 roubles.

6. By the first quotation in the table, Art. 532, what amount of exchange on Geneva at 3 days sight will $1200 in gold buy? *Ans.* 6247.50 francs.

7. At the first quotation, how much exchange on Berlin at 60 days sight, will $1000 currency buy, gold selling at 115? *Ans.* 3710.14 + reichsmarks.

8. How much in currency will a bill on Amsterdam for 4000 guilders cost, at the first quotation, 60 days sight, gold $113\frac{1}{4}$? *Ans.* $$1806.33\frac{3}{4}$.

9. Mecke & Co., of Bremen, wish to remit 4870 reichsmarks to their correspondent in New York; what will be the face of a draft for 3 days at the second quotation? *Ans.* $$1153.58\frac{1}{8}$.

10. A commission merchant in Cadiz having sent to his correspondent in Philadelphia, an invoice of sherry, valued at 8400 pesetas, draws on him for the cost, exchange being 1 peseta=20.1¢; what would have been the advantage to the consignee of remitting a draft on Cadiz, 1 peseta being worth 19.3¢ in Philadelphia? *Ans.* $67.20.

11. A gentleman about to visit Europe obtains a letter of credit from Fisk, Hatch & Co., depositing bonds as security; he draws in Paris 2000 francs July 1, the bill of exchange at 60 days reaching New York, July 17; on his return, Sept. 15, he settled the account; what must he pay, commission 1%, exchange 5.24¼, gold 112½?

Ans. $433.48—.

ARBITRATION OF EXCHANGE.

534. Arbitration of Exchange, also called *Circular Exchange,* is the method of making exchange between two places by means of one or more intermediate exchanges.

535. Simple Arbitration is that in which there is only one intermediate exchange; *Compound Arbitration* is that in which there are two or more intermediate exchanges.

As rates of exchange constantly vary, it is often more advantageous to make the exchange by passing through several intermediate places than by a direct remittance, and the object of arbitration is to enable a person to ascertain which will be most profitable.

1. A merchant wishes to pay a debt in Paris of 4680 francs, remitting through London, exchange between Paris and London being at £1 = 26 francs, and between London and New York £1 = $4.90; what will it cost in New York?

SOLUTION.—If we represent the required number of dollars by x, we have x = 4680 francs, 26 francs = £1, and £1 = $4.90. Now, the product of the first set of values will equal the product of the second set; hence the product of the second set, divided by the product of all the first set except x, will equal x, from which we have x = $882.

OPERATION.

$x = 4680 francs
26 f. = £1
£1 = $4.90

x = $882

2. A merchant in Savannah wishes to remit $2000 to Cincinnati; exchange on New Orleans is ¼% premium; between New Orleans and St. Louis 1% discount; between St. Louis and Cincinnati ½% discount; what was the value of the remittance in Cincinnati if sent through these cities?

SOLUTION.—According to the given rates $1.00¼ in Savannah = $1 in New Orleans; and $0.99 in New Orleans = $1 in St. Louis; and $0.99½ in St. Louis = $1 in Cincinnati; hence, as above explained, x = $2025.29.

OPERATION.

x Cin. = $2000 S.
$1.00¼ S. = $1 N. O.
$0.99 N. O. = $1 St. L.
$0.99½ St. L. = $1 Cin.

x = $2025.29

Rule.—I. *Represent the sum required by* X, *affixing the proper unit of currency, place it equal to the given sum, and arrange the given rates of exchange so that in any two consecutive equations the same unit of currency shall stand on opposite sides.*

II. *If commission is charged for* DRAWING, *place* 1 *minus the rate on the* LEFT *if the* COST OF EXCHANGE *is required, and on the* RIGHT *if* PROCEEDS *are required; but if commission is charged for* REMITTING, *place* 1 *plus the rate on the* RIGHT *if* COST *is required and on the* LEFT *if* PROCEEDS *are required.*

III. *Divide the product of the numbers on the right by the product of the numbers on the left, cancelling equal factors; the result will be the required sum.*

WRITTEN EXERCISES.

3. If exchange between New York and Amsterdam is 40¢ per florin, and between Amsterdam and St. Petersburg is 15 florins to 8 roubles, what must be paid in St. Petersburg for a bill on New York for $1200? *Ans.* 1600 roubles.

4. When exchange between Boston and London is £9=$46, between London and Paris is £2=54 francs, and between Paris and Stockholm is 7 francs=5 crowns; how much must be paid in Boston for a bill on Stockholm, for 2400 crowns? required the difference between it and the direct exchange at 1 crown=27¢. *Ans.* $11.95.

5. A merchant in Baltimore having purchased goods in Berlin to the value of 5000 marks, remits through London, Antwerp, and Amsterdam, at the following rates: £1=$4.85; £1=25.15 francs; 1 guilder=2.5 francs; 1 guilder=1.875 reichsmarks; what will the remittance cost him in Baltimore, allowing $\frac{1}{4}$% brokerage in London? *Ans.* $1288.83.

6. A man in San Francisco wishes to pay a debt of $5200 in Philadelphia; the direct exchange is 2% in favor of Philadelphia, but the exchange on Baltimore is $1\frac{1}{2}$% in favor of Baltimore, and between Baltimore and Philadelphia $\frac{3}{4}$% in favor of Philadelphia; required the difference between the direct and circular exchange. *Ans.* 13.58\frac{1}{2}$.

7. A gentleman spending the winter in Berlin, wishing to obtain some funds from his agent in Cincinnati, directs his agent in London to draw on Cincinnati through New York, for $1000, and remit to him through London and Amsterdam, the rates being as follows; $\frac{3}{4}\%$ in favor of Cincinnati; £1=$4.855; £1=11.75 guilders; 1.12 guilders=2 marks, and the London agent charges $\frac{1}{4}\%$ brokerage both for drawing and remitting; what does he receive and which is best, the circular exchange, or the direct at 1 mark=24¢ in Cincinnati?

DUTIES OR CUSTOMS.

536. Duties, or **Customs,** are taxes levied by government upon imported goods; they are of two kinds, *ad valorem* and *specific*.

537. An **Ad Valorem** duty is a certain percentage assessed on the cost of the goods in the country from which they were imported.

538. A **Specific Duty** is a certain sum assessed on goods without regard to their cost.

539. A **Tariff** is a schedule showing the rate of duty fixed by law on all kinds of imported merchandise.

540. Tare is an allowance for the weight of the box, cask, or covering containing the goods.

For some articles certain rates of tare are fixed by law: in other cases the *real* tare only, ascertained under regulations prescribed by the Secretary of the Treasury, is allowed. If the tare is specified in the original invoice, the collector may, if he chooses, with the consent of the consignee, accept it as the correct tare.

541. Breakage is an allowance for the loss of liquors imported in bottles.

The allowance for breakage is 5% on ale, beer, porter, liquors, and sparkling wine in bottles; no allowance is now made on still wines.

542. Gross Weight or **Value** is the weight or value of the goods before any deductions have been made.

543. Net Weight or **Value** is the weight or value of the goods after all allowances have been deducted.

By the present tariff, most duties of the United States are *ad valorem*, but some duties are specific, and some articles are charged both a specific and an *ad valorem* duty. The duty is reckoned on the actual cost at the place of purchase or manufacture, increased by all charges for transportation previous to final shipment.

Seaport towns, where customs are collected, are called *ports of entry*. The offices in which they are collected are called *custom-houses;* and the officer who superintends the collection of duties and other business of the custom-house is called *collector of the port*.

A vessel is *entered* at a port by lodging at the custom-house a *manifest* or statement of its cargo, and also a list of passengers, if it have any, these papers verified by the oath of the master. The *clearance* from its port of departure and papers proving its nationality must also be deposited, before it is permitted to discharge its cargo.

A vessel is *cleared* from a port by lodging at the custom-house a manifest of its outward cargo, verified by oath, and agreeing with the shippers' manifests of parts of cargo. All government charges must be paid also, and everything connected with the discharging of the inward cargo settled, after which a "general clearance" is issued, and the vessel is at liberty to leave the port, having received its papers of nationality again.

The illegal introduction of goods into a country otherwise than through the regular ports of entry is called *smuggling*.

544. All merchandise imported from foreign ports or places must be consigned in the manifest, invoice, or bill of lading, to some person or firm at the port of importation, by whom it must be duly entered—either for immediate consumption or for warehouse

Merchandise not intended for immediate consumption may be deposited in U. S. Bonded Warehouses, and remain there not longer than three years, the owner being at liberty to withdraw it at any time upon payment of the duties and charges for storage.

In custom-house business the *long* ton, cwt., and qr. are used. Foreign money is reduced by the table given in Art. 530, unless the invoice is accompanied by a consular certificate stating that a different rate of exchange is ruling at the time the invoice is made out.

545. The Quantities considered are: 1. The *Cost* of the goods, or the *Quantity;* 2. The *Duty;* 3. The *Rate;* 4. The *Allowances*.

NOTE.—Duties may be treated under three cases, as other applications of Percentage; but as the second and third cases are merely theoretical, they are not given here.

CASE I.

546. ***Given, the base and rate, to find the duty.***

1. What is the specific duty on 75 drums of figs, each weighing 57 lb., tare 21 lb. to the cwt. at $11 a cwt.?

SOLUTION.—We find the number of cwt. to be $38\frac{19}{112}$; multiplying 21 lb., the tare on 1 cwt., by $38\frac{19}{112}$, we have the whole tare equal to $7\frac{281}{1792}$ cwt., which, subtracted from $38\frac{19}{112}$ cwt., leaves the net weight, $31\frac{23}{1792}$ cwt. On 1 cwt. the duty is \$11, and on 31 cwt. (the fraction being less than $\frac{1}{2}$ is not reckoned) it is 31 times \$11, or \$341.

OPERATION.

$$\frac{75 \times 57}{112} = 38\tfrac{19}{112} \text{ cwt.}$$

$$21 \times 38\tfrac{19}{112} = 7\tfrac{281}{1792} \text{ cwt. tare.}$$

$$38\tfrac{19}{112} - 7\tfrac{281}{1792} = 31\tfrac{23}{1792}.$$

$$31 \times \$11 = \$341.$$

Rule I.—*For ad valorem duties, multiply the cost of the goods by the rate of duty.*

Rule II.—*For specific duties, deduct first the allowances, and compute the duty on the remainder.*

NOTE.—In reckoning duties, whole dollars, pounds, gallons, etc., are used as the base, fractions less than $\frac{1}{2}$ being rejected and more than $\frac{1}{2}$ being reckoned as 1. Duties are payable in gold.

WRITTEN EXERCISES.

2. I received from Havre an invoice of 50 dozen bottles of champagne, costing \$12.50 per dozen; what is the duty at \$6 a dozen, breakage 5%? *Ans.* \$288.

3. H. B. Claflin & Co. received an invoice of Brussels laces costing 2800 francs in Brussels, charges 101.56 francs; what was the duty at 30%? *Ans.* \$168.

4. What is the duty at 20% ad valorem on 350 boxes of Naples oranges, invoiced at 20 lire per box, charges 475 lire, commission $2\frac{1}{2}$%? *Ans.* \$295.80.

5. What is the duty in currency on 25 hhd. of sugar, each weighing 5 cwt. 1 qr. 3 lb., tare 21 lb. per hhd., duty \$$1\frac{1}{4}$ per cwt., and 35 hhd. of molasses, 120 gallons each, duty 5¢ a gallon, gold at $111\frac{1}{2}$? *Ans.* \$411.16.

6. Sharpless Brothers imported from London 10 cases of woollen goods, net weight 1350 lb., value £756 15 s., commission $2\frac{1}{2}$%; what did the goods cost in store, duty 50¢ per lb. and 35% ad valorem, gold $113\frac{1}{8}$? *Ans.* \$6528.49—.

7. Chas. Ford & Son, linen merchants, Belfast, shipped to John Ford & Co., April 2, 1876, 40 pcs. $\frac{3}{4}$ Ducks marked W. R., No. 14, containing $2063\frac{3}{4}$ yd. @ $7\frac{1}{8}$ d.; commission, $2\frac{1}{2}$%; ship charges 5 s.; what will be the duty at 35%? *Ans.* \$107.45.

8. Invoice of 75 Drums Caustic Soda, shipped per S. S. Germania for New York, by the undersigned on acc't and risk of Messrs. Smith & Bro.

W A 1 / R 75 — 75 Drums.

	Cwt. qr. lb.		
Gr.	421 3 3		Cwt. qr. lb.
Tr.	14 3 23	Net	406 3 8@13/3
			45564 lb. discount 2½% _____

Liverpool, April, 1874. Com. 2½% _____

McKeon Bros.,

Duty,—45,564 lb.@1½¢=$ £

What was the cost of this invoice, including duty, when delivered to Messrs. Smith & Bro.? *Ans.* $1994.25 gold.

9. Invoice of Mdse., purchased, paid for in London, and shipped by David Taylor & Sons, per steamer Ohio to Philadelphia, for % and risk of Messrs. Snow, Gilbert & Co.

F B No. 21 *One Case Pallet Knives.*

24 doz. 3 in. Plain.	Net 2/8	£
12 " 4 " "	2/8	
4 " 5 " "	3/5	_____
Duty as Manuf. Steel,	disc't, 2½%	_____
		£
$ @45%=$12.15	case	0-2-0
	com. 2½%	_____
		£

Find the duty on the above invoice. *Ans.* $12.15.

NOTE.—The 8th and 9th problems are taken from actual custom-house transactions, and indicate the exact forms used. In problem 8, the letters at the left are the trade mark on the drums; $\frac{1}{75}$ indicates numbers of the drums from 1 up to 75; gross weight is 421 cwt. 3 qr. 3 lb., tare, 14 cwt. 3 qr. 23 lb.; net weight is 406 cwt. 3 qr. 8 lb. or 45564 lb., at 13s. 3d. per cwt.; from which a discount of 2½ per cent. must be subtracted, giving the amount paid, and adding a commission of 2½ per cent, we have the cost up to shipment, which must be reduced to United States currency, reckoning $4.8665 = £1. The duty being specific is, however, reckoned only on the weight. In the 9th example, having deducted discount and added price of case and the commission, and reduced the amount of the invoice to United States money, 45 per cent. of the nearest exact number of dollars will be the duty.

INTRODUCTION TO RATIO AND PROPORTION.

MENTAL EXERCISES.

1. Eight is how many times 4?

2. What is the relation of 8 to 4? *Ans.* 8 is *two times* 4.

3. What is the relation of 12 to 3? Of 16 to 4? Of 18 to 6? Of 20 to 5? Of 24 to 6? Of 30 to 5?

4. What is the relation of 3 to 6? Of 4 to 12? Of 6 to 24? Of 7 to 35? Of 8 to 57? Of 9 to 62?

5. The measure of the relation of two numbers is called their *ratio.*

6. What is the ratio of 12 to 4?

Ans. The ratio of 12 to 4 is *three.*

7. What is the ratio of 18 to 9? Of 25 to 5? Of 48 to 8? Of 63 to 7? Of 64 to 4? Of 70 to 10? Of 80 to 8?

8. What is the ratio of 3 to 6?

Ans. The ratio of 3 to 6 is *one-half.*

9. What is the ratio of 4 to 12? Of 3 to 18? Of 5 to 30? Of 9 to 108? Of 11 to 132? Of 12 to 144?

10. What is the ratio of $\frac{1}{2}$ to $\frac{1}{3}$? Of $\frac{1}{5}$ to $\frac{1}{6}$? Of $\frac{2}{3}$ to $\frac{3}{4}$? Of .5 to .25? Of .2 to .04? Of .03 to .12?

11. The ratio of two numbers may be expressed by writing the colon between them; thus 8 : 4 denotes the ratio of 8 to 4.

12. Required the value of 12 : 6; of 28 : 7; of 42 : 6; of 24 : 12; of 12 : 24.

13. How does the ratio of 8 to 4 compare with the ratio of 12 to 6?

Ans. They are equal.

14. What number has the same ratio to 12 that 18 has to 6?

15. What number has the same ratio to 20 that 40 has to 10?

16. The ratio of 9 to 36 is the same as the ratio of 15 to what number?

17. 25 is to 5 as 40 to what number? 24 is to 12 as 15 is to what number?

18. When we express the ratio of two numbers equal to the ratio of two other numbers, as, 24 is to 4 as 36 is to 6, we have a *proportion.*

19. What proportion can we derive from the two ratios 40 to 8 and 60 to 12?

20. How many numbers do we have in a proportion? How many ratios? Are the ratios equal or unequal?

21. The equality of two ratios may be expressed by writing the symbol = between them; thus 8 : 4 = 12 : 6.

22. Write the proportion 16 is to 8 as 24 is to 12; also 15 is to 45 as 18 is to 54.

SECTION IX.

RATIO AND PROPORTION.

RATIO.

547. Ratio is the measure of the relation of two similar quantities; thus, the ratio of 8 to 4 is 2.

548. The **Symbol** of ratio is the colon (:); thus, 8 : 4 signifies the ratio of 8 to 4. Ratio is also expressed by writing the numbers in the form of a fraction; thus, $\frac{8}{4}$.

549. The **Terms** of a ratio are the two numbers compared, called respectively the *antecedent* and the *consequent.*

550. The **Antecedent** is the number compared with the consequent; thus, in the ratio 8 : 4, 8 is the antecedent.

551. The **Consequent** is the number with which the antecedent is compared; thus, in 8 : 4, 4 is the consequent.

552. A **Ratio** is found by dividing the antecedent by the consequent; thus, in 8 : 4, the ratio is $\frac{8}{4}$, or 2.

553. A **Simple Ratio** is the ratio of two numbers, as 6 : 3. A *Compound Ratio* is the product of two or more simple ratios; as $(3:4)\times(5:6)$, or $\frac{3}{4}\times\frac{5}{6}$.

554. A **Compound Ratio** is usually expressed by writing the simple ratios one under another; thus, $\left\{\begin{matrix}3:4\\5:6\end{matrix}\right\}$.

555. Ratio exists only between similar quantities, and is always an abstract number.

NOTES.—1. The symbol of ratio (:) is supposed to be a modification of the symbol of division.

2. Ratio is usually defined as the *relation* of two numbers. This is indefinite, for the ratio is the *measure* of the relation.

3. A few authors divide the *second* term by the *first*, calling it the *French Method.* The method and name are both founded in error; nearly all the French mathematicians, like the German, English, etc., divide the first term by the second.

PRINCIPLES.

1. *The ratio equals the quotient of the antecedent divided by the consequent.*

Thus, if the antecedent is represented by a, and the consequent by c, and the ratio by r, we have $a \div c = r$, or $\frac{a}{c} = r$.

2. *The antecedent is equal to the product of the consequent and ratio.*

For, since $\frac{a}{c} = r$, multiplying by c, we have $a = c \times r$.

3. *The consequent is equal to the quotient of the antecedent divided by the ratio.*

For, since $\frac{a}{c} = r$, $a = c \times r$, from which we see that $c = \frac{a}{r}$.

MENTAL AND WRITTEN EXERCISES.

What is the ratio of

1. 12 to 3? *Ans.* 4.
2. 24 to 4? *Ans.* 6.
3. 90 to 16? *Ans.* $5\frac{5}{8}$.
4. 488 to 61? *Ans.* 8.
5. \$256 to \$856? *Ans.* $\frac{32}{107}$.
6. £144 : £256? *Ans.* $\frac{9}{16}$.
7. $\frac{3}{4} : \frac{2}{3}$? $\frac{9}{10} : \frac{4}{5}$? *Ans.* $\frac{9}{8}$.
8. $\frac{25}{16} : \frac{15}{12}$? $2\frac{3}{4} : 2\frac{1}{5}$? *Ans.* $1\frac{1}{4}$.

9. What is the value of the compound ratio $\left\{\begin{matrix} 2:4 \\ 3:9 \end{matrix}\right\}$?

SOLUTION.—This compound ratio equals $(2:4) \times (3:9)$, which equals $\frac{2}{4} \times \frac{3}{9} = \frac{1}{6}$.

10. What is the value of the ratio $\left\{\begin{matrix} 4:3 \\ 6:8 \end{matrix}\right\}$? *Ans.* 1.

11. What is the value of the ratio $\left\{\begin{matrix} 5:2 \\ 8:6 \end{matrix}\right\}$? *Ans.* $3\frac{1}{3}$.

12. The antecedent is 24, the consequent 8; what is the ratio? *Ans.* 3.

13. The consequent is 8 and ratio 9; what is the antecelent? *Ans.* 72.

14. The antecedent is 36 and ratio 4; what is the consequent? *Ans.* 9.

15. The consequent is $\frac{15}{16}$ and ratio $\frac{18}{25}$; what is the antecedent? *Ans.* $\frac{27}{40}$.

16. The antecedent is $\frac{15}{16}$ and ratio $\frac{25}{32}$; what is the consequent? *Ans.* $1\frac{1}{5}$.

17. Can you express the ratio between \$24 and 6 lb.? Why not?

18. The antecedents of a ratio are 5 and 6, and the consequents 10 and 14; what is the ratio? *Ans.* $\frac{3}{14}$.

SIMPLE PROPORTION.

556. A **Proportion** is the expression of equality between equal ratios, the terms of the ratios being indicated.

557. The **Symbol** for proportion is the double colon, (: :), which expresses an equality of ratios; thus, 8 : 4 : : 6 : 3, means the same as 8 : 4 = 6 : 3.

558. A **Proportion** is *read* in two ways; thus, 8 : 4 : : 6 : 3 is read "the ratio of 8 to 4 equals the ratio of 6 to 3;" or "8 is to 4 as 6 is to 3."

559. The **Terms** of a proportion are the four numbers used in the comparison. The first and fourth terms are the *Extremes*; the second and third are the *Means*.

560. The **Couplets** are the two ratios compared. The *first couplet* consists of the first and second terms. The *second couplet* consists of the third and fourth terms.

561. **Proportion** may be *Simple* or *Compound*. In *Simple Proportion* both the ratios compared are simple; in *Compound Proportion* one or both of the ratios are compound.

562. A **Simple Proportion** is the expression of the equality of two simple ratios.

563. The **Principles** of proportion are the truths relating to a proportion. They enable us to find any one term when the other three are given.

PRINCIPLES.

1. *In every proportion the product of the means equals the product of the extremes.*

In any proportion, as 6 : 3 : : 8 : 4, we have $\frac{6}{3}=\frac{8}{4}$, and multiplying these equals by 4 and 3 we have $6\times 4=8\times 3$; that is, the product of the two means 8 and 3, equals the product of the two extremes 6 and 4.

2. *Either extreme equals the product of the means divided by the other extreme.*

For, from the proportion 6 : 3 : : 8 : 4, we have $6\times 4=3\times 8$; hence, $6=3\times 8\div 4$, or $4=3\times 8\div 6$. Therefore, etc.

3. *Either mean equals the product of the extremes divided by the other mean.*

For, from the proportion 6 : 3 : : 8 : 4, we have $6\times 4=3\times 8$; hence, $3=6\times 4\div 8$, or $8=6\times 4\div 3$. Therefore, etc.

4. *The first term of a proportion equals the second term multiplied by the ratio of the third to the fourth.*

For, from the proportion 8 : 6 : : 12 : 9, we have $\frac{8}{6} = \frac{12}{9}$; hence, $8 = \frac{12}{9} \times 6$, or 12 : 9 multiplied by 6. Therefore, etc.

5. *The fourth term of a proportion equals the third term divided by the ratio of the first to the second.*

For, from the proportion 8 : 6 : · 12 : 9, we have $8 \times 9 = 6 \times 12$, or $9 = 6 \times 12 \div 8$, which equals $12 \times \frac{6}{8}$, which equals $12 \div \frac{8}{6}$, or 12 ÷ (8 : 6). Therefore, etc.

NOTES.—1. Let the pupils be required to demonstrate these principles by using symbols of any numbers; that is, by letters. French authors usually represent the unknown term by *x*; the same is done in this work.

2. Principle 1 may be demonstrated by showing that in a proportion we have 2d term × ratio : 2d term : : 4th term × ratio : 4th term; in which we see the factors in the means are the same as the factors in the extremes.

MENTAL EXERCISES.

1. Write a proportion and point out the different terms and couplets. Write a proportion and show that the ratios are equal.

2. If we multiply the antecedent of one couplet, what must we do to the other couplet to make the ratios equal?

3. If we divide the antecedent of one couplet, what must we do to the other couplet to make the ratios equal?

4. Write a proportion and illustrate Prin. 1; Prin. 2; Prin. 3; Prin. 4; Prin. 5.

5. Show that if we change the two means one for the other, or the two extremes, the four numbers will still form a proportion.

6. Take some proportion and show that we may invert the terms of the couplets, and the four terms will still be in proportion.

WRITTEN EXERCISES.

Find the terms denoted by *x* in each of the following proportions:

1. $x : 4 :: 6 : 12$. SUG.—$x = \frac{4 \times 6}{12} = 2$. *Ans.* 2.
2. $x : 8 :: 18 : 9$. *Ans.* 16.
3. $18 : 72 :: 21 : x$. *Ans.* 84.
4. $16 : 48 :: 15 : x$. *Ans.* 45.
5. $21 : x :: 32 : 96$. *Ans.* 63.
6. \$6 : \$15 : : x : 95 yd. *Ans.* 38 yd.
7. \$17 : \$68 : : 35 lb. : x. *Ans.* 140 lb.
8. $x : \frac{6}{7} :: \frac{28}{35} : \frac{4}{5}$. *Ans.* $\frac{6}{7}$.
9. $.5 : 1.3 :: 6.8 : x$. *Ans.* 17.68

APPLICATION OF SIMPLE PROPORTION.

564. Simple Proportion is employed for the solution of problems in which three of four quantities are given, so related that the fourth may be determined from them, by equality of the ratios.

565. The required quantity must bear the same relation to a given quantity of the same kind that one of the remaining quantities does to the other. We can then form a proportion containing one unknown quantity, and find the unknown term by the principles of proportion.

NOTE.—Proportion was formerly called the "Rule of Three." Some of the old arithmeticians thought so highly of it that they called it "The Golden Rule of Three."

1. What will 20 yards of cloth cost, if 5 yards cost $15?

SOLUTION.—It is evident that the cost of 20 yd. bears the same relation to the cost of 5 yd. as 20 yd. bears to 5 yd., hence we have the proportion, *cost of* 20 yd. *is to* $15 *as* 20 yd. *is to* 5 yd., from which, by Prin. 2, we have the cost of 20 yd. $= \frac{20 \times 15}{5} =$ $60. Hence the

OPERATION.

$$\text{Cost of 20 yd.} : \overset{\$}{15} :: \overset{\text{yd.}}{20} : \overset{\text{yd.}}{5}$$

$$\text{Cost of 20 yd.} = \frac{20 \times 15}{5} = 60$$

Rule.—I. *Write the required quantity for the first term and the similar known quantity for the second term, and place the other two quantities for the third and fourth terms, so that the two ratios will be equal.*

II. *Find the first term by dividing the product of the second and third terms by the fourth.*

SOLUTION 2d.—It is evident that the relation of 5 yd. to 20 yd. is the same as the relation of the cost of 5 yd. to the cost of 20 yd.; hence, we have the proportion, 5 yd. is to 20 yd. as $15 is to the *cost of* 20 yd., from which, by Prin. 2, we have the *cost of* 20 yd. equals $60.

OPERATION.

$$\overset{\text{yd.}}{5} : \overset{\text{yd.}}{20} :: \overset{\$}{15} : \text{cost of 20 yd.}$$

$$\text{Cost of 20 yd.} = \frac{20 \times 15}{5} = 60$$

Rule 2d.—I. *Write the number which is of the same kind as the required quantity for the third term.*

II. *Place the other two numbers in the first and second terms, the greater in the second term when the result is to be greater than the third term, and the less in the second term when the result is less than the third term.*

III. *Find the fourth term by dividing the product of the second and third terms by the first.*

NOTES.—1. The author believes that the simplest method of using proportion is to put the *unknown* quantity in the *first term.* He gives the old method also, for teachers who prefer it. See *Brooks's Philosophy of Arithmetic.*

2. Pupils should be required to put the unknown quantity, which they may represent by x, in different terms, that they may thoroughly understand the subject.

WRITTEN EXERCISES.

2. What cost 78 hhd. of molasses, if 13 hhd. are worth $250? *Ans.* $1500.

3. How many yards of cloth will $144 buy, if 28 yd. cost $112? *Ans.* 36.

4. What cost 132 acres of land, if 110 acres are worth $8250? *Ans.* $9900.

5. If $100 gains $6 in a year, how much will $250 gain in a year? *Ans.* $15.

6. If 16 horses eat 26 bundles of hay in a week, how many will 36 horses eat in the same time? *Ans.* 58.50.

7. If 75 horses cost $9000, how many horses can be bought for $16200? *Ans.* 135.

8. If there are 84 privates in each company, how many companies in a brigade of 3360 men? *Ans.* 40.

9. If 25 oxen eat 36 acres of grass in a month, how many oxen would 468 acres keep the same time? *Ans.* 325.

10. If 79 men earn $395 in a week, how many men can earn $675 in the same time? *Ans.* 135 men.

11. How much will 34 lb. of tea cost, if 8 lb. 8 oz. of the same kind of tea cost $4½? *Ans.* $18.

12. If 19 bu. of rye make 4 bar. of flour, how many bushels will it require to make 19 barrels? *Ans.* 90¼ bu.

13. How much will 28 cwt. 75 lb. of sugar cost at the rate of 7 cwt. 50 lb. for $40.50? *Ans.* $155.25.

14. In what time will the cars go from Lancaster to Philadelphia, 68 miles, at the rate of 5 miles in 10 min. 45 sec.? *Ans.* 2 h. 26⅕ min.

15. If a man spends \$230 in the three spring months, at the same rate per day how much will he spend in a year? *Ans.* \$912.50.

16. If 12 men build a wall in 24 days, how long will it take 60 men to build it at the same rate? *Ans.* $4\frac{4}{5}$ da.

NOTE.—Here it is evident that the time in which 60 men do it, is to 24 days, the time in which 12 men do it, as 12 men is to 60 men.

17. If 28 men mow a field of grass in 12 days, how many men will be required to mow it in 8 days? *Ans.* 42 men.

18. What is the height of a staff which casts 41 ft. of shadow, if a staff 6 ft. 9 in. high cast a shadow 13 ft. 8 in.? *Ans.* $20\frac{1}{4}$ ft.

19. If $13\frac{1}{2}$ bu. of corn cost \$6.25, what will $16\frac{1}{5}$ bu. cost at the same price per bushel? *Ans.* \$7.50.

20. If $\frac{5}{6}$ of a barrel of flour cost $\frac{12}{25}$ of an eagle, how many dollars will $15\frac{5}{6}$ barrels cost? *Ans.* \$91.20.

21. If a person do a piece of work in 142 days, working 9 hours per day, in what time will he do it, working $6\frac{3}{4}$ hours a day? *Ans.* $189\frac{1}{3}$ days.

22. If 96 bushels of oats keep 42 horses 8 days, how long will 168 bushels keep them? *Ans.* 14 da.

23. If 4 A. 120 P. of land cost \$437, what will 16 A. 80 P cost at a rate of \$25 less an acre? *Ans.* \$1105.50.

24. If 17 men can mow a field in 9 days, how long would it take to reap half of it if 5 men refuse to labor? *Ans.* $6\frac{3}{8}$ days.

25. A failed and could pay only 75 cts. on each dollar he owed; how much did C receive, whom he owed \$1968? *Ans.* \$1476.

26. If a three-cent loaf weigh 9 ounces when flour is \$6 a barrel, how much should it weigh when flour is \$8 a barrel? *Ans.* $6\frac{3}{4}$ oz.

27. A lent me \$560 for 10 months; how long should I lend him \$800 to reciprocate the favor? *Ans.* 7 mo.

28. A has grain worth $\$1.12\frac{1}{2}$ a bushel, and B has flour worth \$6.25 a barrel; now if in an exchange A puts his grain at \$1.25 a bushel, what should B charge for his flour? *Ans.* $\$6.94\frac{4}{9}$.

29. Two cog-wheels, one having 28 and the other 20 cogs, run together; in how many revolutions of the larger wheel will the smaller gain 12 revolutions? *Ans.* 30.

30. A garrison of 2400 men has provisions sufficient to last them 20 days, at the rate of $1\frac{1}{2}$ lb. a day; how large a reinforcement could be received for the time if the allowance be reduced to 15 oz. a day? *Ans.* 1440 men.

COMPOUND PROPORTION.

566. **A Compound Proportion** is a proportion in which one or both ratios are compound.

567. Thus, $\left\{\begin{matrix}2 : 4\\5 : 15\end{matrix}\right\} :: 6 : 36$ and $\left\{\begin{matrix}4 : 12\\7 : 14\end{matrix}\right\} :: \left\{\begin{matrix}5 : 10\\6 : 18\end{matrix}\right\}$ are examples of compound proportion.

PRINCIPLES.

1. *The product of the simple ratios of the first couplet equals the product of the simple ratios of the second couplet.*

For, the value of a compound ratio is the product of the simple ratios, and these are equal, since a proportion expresses the equality of ratios. Thus, from the second of the above proportions we have, $\frac{4}{12}\times\frac{7}{14}=\frac{5}{10}\times\frac{6}{18}$.

2. *The product of all the terms in the extremes equals the product of all the terms in the means.*

For, from the nature of proportion, we have from the proportion above, $\frac{4}{12}\times\frac{7}{14}=\frac{5}{10}\times\frac{6}{18}$; and clearing of fractions, we have $4\times7\times10\times18=5\times6\times12\times14$, which by examination, we see is the product of the extremes equal to the product of the means.

3. *Any term in either extreme equals the product of the means divided by the product of the other terms in the extremes.*

For, since from the proportion above we have $4\times7\times10\times18=5\times6\times12\times14$, we will have $4=\frac{5\times6\times12\times14}{7\times10\times18}$, and similarly for any other term in either extreme.

4. *Any term in either mean equals the product of the extremes divided by the product of the other terms in the means.*

For, from the above proportion, we have $4\times7\times10\times18=5\times6\times12\times14$, hence $5=(4\times7\times10\times18)\div(6\times12\times14)$; and similarly for any other term in the means.

Find the term denoted by x in each of the following:

1. $x : 12 :: \left\{\begin{matrix} 7 : 18 \\ 9 : 3 \end{matrix}\right\}$. *Ans.* 14.

2. $\left\{\begin{matrix} 15 : 16 \\ 8 : 12 \end{matrix}\right\} :: x : 32$. *Ans.* 20.

3. $\left\{\begin{matrix} 12 : 7 \\ 5 : 18 \end{matrix}\right\} :: \left\{\begin{matrix} 20 : 9 \\ x : 14 \end{matrix}\right\}$. *Ans.* 3.

4. $\left\{\begin{matrix} x : 26 \\ 11 : 12 \end{matrix}\right\} :: \left\{\begin{matrix} 22 : 21 \\ 7 : 13 \end{matrix}\right\}$. *Ans.* 16.

5. $\left\{\begin{matrix} \frac{3}{4} : \frac{4}{5} \\ \frac{5}{9} : \frac{3}{7} \end{matrix}\right\} :: \left\{\begin{matrix} \frac{7}{16} : \frac{3}{11} \\ \frac{25}{36} : x \end{matrix}\right\}$. *Ans.* $\frac{11}{12}$.

APPLICATION OF COMPOUND PROPORTION.

568. Compound Proportion is used in the solution of problems in which the required term depends on a compound ratio.

569. In simple proportion the unknown quantity depends upon the relation of *one pair* of similar quantities; in compound proportion the unknown quantity depends upon *two or more pairs* of similar quantities.

NOTE.—Problems in compound proportion may be solved by two or more simple proportions, or by analysis.

1. If 4 men can earn \$24 in 7 days, how much can 14 men earn in 12 days?

OPERATION.

$$\text{The sum} : \$24 :: \left\{\begin{matrix} 14 : 4 \\ 12 : 7 \end{matrix}\right\}$$

$$\text{The sum} = \frac{\$24 \times 14 \times 12}{4 \times 7} = \$144, \textit{Ans.}$$

SOLUTION.—It is evident that the sum 14 men can earn in a given time is to the sum that 4 men can earn in that time as 14 to 4, and also the sum that they can earn in 12 days is to the sum that they can earn in 7 days as 12 is to 7; hence the sum that 14 men can earn in 12 days is to \$24 (the sum that 4 men will earn in 7 days), as 14 : 4 and 12 : 7; hence we have the proportion, *the sum* : \$24 :: $\left\{\begin{matrix} 14 : 4 \\ 12 : 7 \end{matrix}\right\}$, from which, by Prin. 3, we have *the sum* $= \frac{\$24 \times 14 \times 12}{7 \times 4}$, or \$144.

OPERATION.

$$\frac{\overset{3}{12}}{7} \times \frac{\overset{2}{14}}{4} \times \$24 = \$144$$

ANALYSIS.—If 4 men earn \$24 in 7 da., 1 man will earn $\frac{1}{4}$ of \$24, and 14 men will earn $\frac{14}{4}$ of \$24. If 14 men earn $\frac{14}{4}$ of \$24 in 7 da., in 1 day they will earn $\frac{1}{7}$ of $\frac{14}{4}$ of \$24, and in 12 da. they will earn $\frac{12}{7}$ of $\frac{14}{4}$ of \$24, which, by cancelling and multiplying, equals \$144.

NOTE.—The analysis may be abbreviated thus: If 4 men earn \$24, 14 men will earn $\frac{14}{4}$ of \$24. If they earn it in 7 da., in 12 da. they will earn $\frac{12}{7}$ of $\frac{14}{4}$ of \$24 = \$144.

Rule.—I. *Put the required quantity for the first term and the similar known quantity for the second term, and form ratios with each pair of similar quantities for the second couplet, as if the result depended upon each pair and the second term.*

II. *Find the required term by dividing the product of the means by the product of the fourth terms.*

NOTES.—1. Teachers may put the unknown quantity in the fourth term instead of the first, if they prefer it. The method of solution will be the same in principle, and the rule can be readily changed to correspond with it.

2. Pupils should be required to solve both ways, and to give the rule for both methods.

WRITTEN EXERCISES.

2. If 36 men earn \$324 in 18 days, how much will 42 men earn in 27 days? *Ans.* \$567.

3. If 58 cows eat 29 bundles of hay in 25 days, how many cows will eat 35 bundles in 14 days? *Ans.* 125 cows.

4. If \$600 in 4 yr. 6 mo. at 6% gain \$162 interest, how much will \$800 gain in 6 yr. 4 mo. at 8%? *Ans.* \$405$\frac{1}{3}$.

5. If 12 men in 35 days build a wall 140 rd. long, 6 ft. high, how many men can in 40 days build a wall of the same thickness 144 rods long, 5 ft. high? *Ans.* 9 men.

6. If 18 carpenters build a house in 45 days, working 12h. a day, in how many days would 36 carpenters have built it, working 10 h. a day? *Ans.* 27 days.

7. If 28 men dig a trench 120 rods long, 15 ft. wide, and 12 feet deep, how many men will dig a trench 360 rods long, 9 feet wide, and 10 feet deep? *Ans.* 42 men.

8. If 8 yd. of muslin, 1$\frac{1}{4}$ yd. wide, cost \$1.25, what cost 10 yd. of the same quality, 1$\frac{1}{8}$ yd. wide? *Ans.* \$1.40$\frac{5}{8}$.

9. If 35 horses can eat a lot of grain in 36 days, in what time will 3 times as much grain be consumed, if 5 horses are added when the grain is $\frac{3}{4}$ eaten? *Ans.* 104$\frac{5}{8}$ da.

10. If 32 men can build 60 rods of wall in 15 days, in what time can they build 75 rods, if 8 men leave when 40 rods have been built? *Ans.* 21$\frac{2}{3}$ da.

11. If 147 loaves of bread, weighing 6 oz. each, cost $8.10 when flour is $7 a barrel, what cost 96 loaves, of 7 oz. each, when flour is worth $9 a barrel? *Ans.* $7.93+.

12. If 17 plank, 35 ft. long, 28 in. wide, and 6 in. thick, cost $68, what cost 40 plank, 32 ft. long, 25 in. wide, and 7 in thick, lumber worth $\frac{1}{5}$ more per foot? *Ans.* 182.85\frac{5}{7}$.

13. If from a dairy of 24 cows, each giving 18 qt. of milk daily, 10 cheeses of 60 lb. each are made in 12 weeks, how many cows will produce 40 cheeses of 75 lb. in 9 weeks, if they give 12 qt. each? *Ans.* 240 cows.

14. If 24 men, in 15 da. of 12 h. each, dig a trench 300 rd. long, 5 yd. wide, 6 ft. deep, in what time can 45 men, working 10 h. a day, dig a trench 125 rd. long, 15 ft. wide, 8 ft. deep? *Ans.* 5$\frac{1}{3}$ da.

15. If 50 men can build 50 rods of wall in 75 days, how many men will be required to build 80 rods of wall $\frac{3}{2}$ as thick and $\frac{3}{5}$ as high in 40 days? *Ans.* 180 men.

PARTITIVE PROPORTION.

570. Partitive Proportion is the process of separating a number into parts which bear certain relations to each other.

571. There are several cases arising from the various relations which may exist between the parts into which a number is divided.

NOTE.—The method of solution is analytical, and no rule is given.

CASE I.

572. ***When one part is a number more or less than another.***

MENTAL EXERCISES.

1. The sum of two numbers is 42, and their difference 6; what are the numbers?

SOLUTION.—Since their difference is 6, the less plus 6 equals the greater, which added to the less, equals twice the less, plus 6, which equals 42. If twice the less, + 6 = 42, twice the less = 42 — 6, or 36, once the less = $\frac{1}{2}$ of 36, or 18, and the greater = 18 + 6, or 24.

2. Divide the number 45 into two such parts that one shall be 7 less than the other.

3. James and Joseph each earned a certain sum of money, and James received $5 as a Christmas present, when together they had $35 ; how much had each?

4. Mary and Susan had each the same number of peaches, but Mary having eaten 5 of hers, they together had 15 left; how many had each at first?

WRITTEN EXERCISES.

1. William and Henry have 693 oxen; how many has each, if William has 57 more than Henry?

SOLUTION.—By the conditions of the problem, Henry's number plus 57 equals William's number, which, added to Henry's number, equals two times Henry's number, plus 57, which equals what they both have, or 693; if twice Henry's number, plus 57, equals 693, twice Henry's number equals 693 minus 57, which equals 636, and once Henry's number equals ½ of 636, which is 318, etc.

OPERATION.

$$\begin{aligned} H\text{'s} + 57 &= W\text{'s} \\ 2\ H\text{'s} + 57 &= 693 \\ 2\ H\text{'s} &= 636 \\ H\text{'s} &= 318 \\ W\text{'s} &= 375 \end{aligned}$$

2. The number of sheep and cows belonging to a drover equals 427, and he has 125 more sheep than cows; how many has he of each? *Ans.* Cows, 151; sheep, 276.

3. The sum of two fractions is $\frac{112}{117}$, and their difference is $\frac{14}{39}$; required the two fractions. *Ans.* $\frac{35}{117}$; $\frac{77}{117}$.

4. Charles and William bought two houses, for which they paid $8976; but William's house cost $758 more than Charles's; what was the price of each? *Ans.* $4109; $4867.

5. Mr. Jones, by his will, divided $10000 among his three sons; to the eldest he gave $500 more than to the second, and to the second $1000 more than to the third; what did each receive? *Ans.* $4000; $3500; $2500.

CASE II.

573. ***When one part is a number of times another, or a fractional part of another.***

MENTAL EXERCISES.

1. Divide the number 24 into two such parts that 3 times one part shall equal the other.

SOLUTION.—By a condition of the problem, 3 times the first part equals the second, which added to the first, equals 4 times the first, which equals the number, or 24. If 4 times the first part equals 24, once the first part equals ¼ of 24, or 6, and 3 times the first part, or the second part, equals 3 times 6, or 18.

2. John has 5 times as many apples as George, and together they have 36; how many has each?

3. A tree was broken off by a high wind so that the part standing was $\frac{1}{5}$ of the fallen part; if the whole height was 60 feet, what was the length of the parts?

4. The sum of three numbers is 63; the second is twice the first, and the third three times the second; what are the numbers?

WRITTEN EXERCISES.

1. A and B together have 2538 acres of land, and B has 5 times as much as A; how many acres has each?

SOLUTION.—By the conditions, 5 times A's number equals B's number, which added to A's number equals 6 times A's number, which equals 2538 acres; once A's number equals $\frac{1}{6}$ of 2538 acres, or 423 acres, and 5 times A's number, or B's number, equals 5 times 423 acres, or 2115 acres.

OPERATION.

A's + 5 A's = 2538
6 A's = 2538
A's = 423
B's = 5 A's = 2115

2. The cost of a horse and harness was $328; and the horse cost 7 times as much as the harness; what was the cost of each? *Ans.* Horse, $287; Harness, $41.

3. A vessel and cargo, insured for $75,800, were lost at sea; what was paid on each if the cargo was valued at $\frac{1}{4}$ as much as the vessel? *Ans.* $15160; $60640.

4. A, B, and C gained by a speculation $11480, of which A's share was twice as much as C's, and B's 5 times as much as C's; how much did each gain?

Ans. A, $2870; B, $7175; C, $1435.

5. A man bought two Southern plantations for $30784, paying $\frac{5}{8}$ as much for one as for the other; what did each cost him? *Ans.* $11840; $18944.

CASE III.

574. ***When a number of times one part equals a number of times another part.***

MENTAL EXERCISES.

1. The sum of two numbers is 32, and 5 times the first equals 3 times the second; what are the numbers?

SOLUTION.—Since 5 times the first equals 3 times the second, once the first equals $\frac{3}{5}$ of the second, which, added to the second, equals $\frac{8}{5}$ of the second, which equals 32, hence $\frac{1}{5}$ of the second equals $\frac{1}{8}$ of 32, which is 4, and the numbers are 20 and 12

2. Jane has 10 plums more than Mary, and 5 times Jane's number equals 6 times Mary's; how many had each?

3. Two-thirds of John's and William's money is $60; and 4 times John's equals 5 times William's; what amount had each?

4. Four-fifths of the height of a tree is 80 feet, and 7 times the part below the first fork equals the part above it; what is the length of the two parts?

WRITTEN EXERCISES.

1. A and B together have $630, and 5 times A's share equals 4 times B's share; how much has each?

SOLUTION.—Since 5 times A's share equals 4 times B's, once A's share equals $\frac{4}{5}$ of B's, and adding to B's, we have $\frac{9}{5}$ of B's, which equals what both have, or $630; hence $\frac{1}{5}$ of B's is $70 B's is $350, and A's $280.

OPERATION.

$$\begin{aligned} \text{B's} + \tfrac{4}{5} \text{ of B's} &= \$630 \\ \tfrac{9}{5} \text{ of B's} &= 630 \\ \tfrac{1}{5} \text{ of B's} &= 70 \\ \text{B's} &= 350 \\ \text{A's} &= 280 \end{aligned}$$

2. Two partners gain $7560, and 4 times the share of the first equals 5 times the share of the second; required the share of each. *Ans.* $4200; $3360.

3. Martin and Nelson bought 8450 acres of Michigan woodland; how many did each own, if 6 times Martin's equal 7 times Nelson's? *Ans.*, M's, 4550; N's, 3900.

4. Mr. Judson shipped to market 234 bushels of corn and oats; how many bushels of each did he ship if 5 times the quantity of corn equaled 8 times the quantity of oats?

Ans. 144 bu. corn; 90 bu. oats.

5. An Illinois farmer raises 2500 bushels of corn and wheat, and 4 times the quantity of corn, + 500 bushels, equals 2 times the quantity of wheat; what was the quantity of each? *Ans.* Corn, 750 bu.; wheat, 1750 bu.

CASE IV.

575. ***When a fractional part of one part equals a fractional part of another part.***

MENTAL EXERCISES.

1. The sum of two numbers is 16, and $\frac{1}{3}$ of the smaller equals $\frac{1}{5}$ of the greater; required the numbers.

SOLUTION.—If $\frac{1}{3}$ of the smaller equals $\frac{1}{5}$ of the greater, $\frac{3}{3}$ of the smaller equals $\frac{3}{5}$ of the greater, which, added to $\frac{5}{5}$ of the greater, equals $\frac{8}{5}$ of the greater, which equals 16; hence $\frac{1}{5}$ of the greater equals $\frac{1}{8}$ of 16, or 2. the greater equals 10 and the smaller equals 6.

2. The difference between two numbers is 5, and $\frac{1}{3}$ of the first equals $\frac{1}{8}$ of the second; required the numbers.

3. James bought a coat and vest for $36, and $\frac{3}{13}$ of the price of the coat equals $\frac{3}{5}$ of the price of the vest; what was the price of each?

4. I have a cistern containing 165 gallons, which is filled by two pipes, and $\frac{3}{8}$ of what one pours in equals $\frac{3}{7}$ of what the other pours in; how much flows in through each?

WRITTEN EXERCISES.

1. A and B have 210 acres of land, and $\frac{3}{4}$ of A's share equals $\frac{6}{7}$ of B's; how many acres has each?

SOLUTION.—If $\frac{3}{4}$ of A's share equals $\frac{6}{7}$ of B's, $\frac{1}{4}$ of A's equals $\frac{1}{3}$ of $\frac{6}{7}$, or $\frac{2}{7}$ of B's, and $\frac{4}{4}$ of A's equals 4 times $\frac{2}{7}$, or $\frac{8}{7}$ of B's share. Then $\frac{7}{7}$ of B's share, plus $\frac{8}{7}$ of B's share, which is A's, equals $\frac{15}{7}$ of B's, which equals 210 acres; B's equals 98 acres, and A's equals 112 acres.

OPERATION.

$$\frac{3}{4}\text{ A's} = \frac{6}{7}\text{ B's}$$
$$\frac{1}{4}\text{ A's} = \frac{2}{7}\text{ B's}$$
$$\text{A's} = \frac{8}{7}\text{ B's}$$
$$\frac{7}{7}\text{ B's} + \frac{8}{7}\text{ B's} = \frac{15}{7}\text{ B's} = 210$$
$$\text{B's} = 98$$
$$\text{A's} = 112$$

2. Two neighbors raised 3800 bushels of wheat, and $\frac{2}{5}$ of what one raised equals $\frac{4}{9}$ of what the other raised; how much did each raise? *Ans.* 2000 bu.; 1800 bu.

3. At an election the number of votes cast was 510, and $\frac{2}{3}$ of the votes for one candidate equaled $\frac{3}{4}$ of the votes for another; how many votes were cast for each? *Ans.* 270; 240.

4. In a ton of mineral rock $\frac{1}{25}$ is lead and silver mixed in such proportions that $\frac{8}{9}$ of the quantity of silver equals $\frac{11}{12}$ of the quantity of lead; how many pounds of each in a ton?

Ans. $40\frac{8}{13}$ lb. silver; $39\frac{5}{13}$ lb. lead.

5. The combined salaries of a man and his wife, both teaching, amount to $2100, and $\frac{2}{3}$ of the man's salary equals $\frac{5}{6}$ of his wife's, plus $200; what is the salary of each?

Ans. $1300; $800.

CASE V.

576. ***When the parts are to each other as two or more numbers.***

MENTAL EXERCISES.

1. The sum of two numbers is 25, and the larger is to the smaller as 3 to 2; required the numbers.

SOLUTION.—Since the numbers are to each other as 3 to 2, if we divide 25 into 3 + 2, or 5 equal parts, 3 of those parts, or $\frac{3}{5}$ of 25, will be the first number, 2 of those parts, or $\frac{2}{5}$ of 25, will be the second number: $\frac{1}{5}$ of 25 is 5, and $\frac{3}{5}$ are 3 times 5, or 15, the first number, and $\frac{2}{5}$ are 10, the second number.

2. In a school of 60 pupils there are 7 girls to every 5 boys; how many of each sex in the school?

3. Divide 24 peaches among three girls so that their shares may be in the proportion of 1, 2, and 3.

4. Three drovers bought 1200 head of cattle, the first paying $3 as often as the second paid $4 and the third $5; how many head of cattle should each receive?

5. Divide the number 68 into two parts which are to each other as $\frac{2}{3}$ to $\frac{3}{4}$.

SUG.—Since the numbers are to each other as $\frac{2}{3}$ to $\frac{3}{4}$, or as $\frac{8}{12}$ to $\frac{9}{12}$, or as 8 to 9, if we divide 68 into 8 + 9, or 17 equal parts, etc.

6. Divide the fraction $\frac{7}{8}$ into two parts which are to each other as $\frac{4}{5}$ to $\frac{5}{6}$.

7. Divide the reciprocal of 4 into two parts which shall be to each other as the reciprocals of 3 and 5.

WRITTEN EXERCISES.

1. A father divided $8704 between his two sons so that their shares were as 7 to 9; what was the share of each?

SOLUTION.—Since their shares were as 7 to 9, if we divide $8704 into 7 plus 9, or 16 equal parts, 7 of these parts, or $\frac{7}{16}$ of $8704, equal the share of the first, and 9 of these parts, or $\frac{9}{16}$ of $8704, equal the share of the second; $\frac{1}{16}$ of $8704 equals $544, $\frac{7}{16} = \$3808$, and $\frac{9}{16} = \$4896$.

OPERATION.

$7 + 9 = 16$

$\frac{1}{16}$ of $8704 = 544$

$\frac{7}{16} = 3808$

$\frac{9}{16} = 4896$

2. Two dealers bought 32149 feet of hemlock boards and divided them in the proportion of 6 to 7; what quantity did each receive? *Ans.* 14838 ft.; 17311 ft.

3. The capital of a firm amounts to $64800, and the shares of the two partners are to each other as 12 to 15; what is the share of each? *Ans.* $28800; $36000.

4. A brother and sister inherited $26106, and the brother's share of the property was to the sister's as $\frac{3}{4}$ to $\frac{5}{6}$; how much had each? *Ans.* $12366; $13740.

5. Three partners divide $75280, the profits of their business, into three parts which are to each other as 4, 7, and 9; what is the profit of each?

Ans. $15056; $26348; $33876.

6. In a composition of zinc and copper consisting of 5764 lb., the two metals are to each other as the reciprocals of 5 and 6; what is the quantity of each?

Ans. 3144 lb.; 2620 lb.

7. Three Kansas farmers raised 15810 bushels of wheat; how many bushels did each raise if their amounts are to each other as 9, 10, and 12? *Ans.* 4590; 5100; 6120.

8. A firm lost one year $3800; what is the loss of each of the two partners, if their shares were to each other as $\frac{3}{4}$ to $\frac{5}{6}$? *Ans.* $1800; $2000.

9. A speculator bought three properties for $263069; and the money paid for each was as the fractions $\frac{4}{5}$, $\frac{5}{6}$, and $\frac{6}{7}$; what was the cost of each?

Ans. $84504; $88025; $90540.

10. An eccentric old schoolmaster made a will which read as follows: "I bequeath $4059 to my two sons in the proportion of $\frac{5}{6}$ to $\frac{7}{8}$; which amounts are respectively equal to $\frac{9}{10}$ and $\frac{11}{12}$ of the amounts I bequeath to my two daughters;" required the share of each. *Ans.* $1980; $2079; $2200; $2268.

CONJOINED PROPORTION.

577. Conjoined Proportion is the process of comparing numbers so related that each consequent is of the same kind as the next antecedent.

578. The method of treatment is analytical and presents one of the best illustrations of the beautiful process of *Arithmetical Analysis.*

NOTE.—Arbitration of Exchange, which has already been treated, is an application of Conjoined Proportion.

MENTAL EXERCISES.

1. If 4 oranges are worth 3 lemons, and 4 lemons cost 16 cents, what is the cost of 20 oranges?

SOLUTION.—If 4 lemons cost 16 cents, 1 lemon cost $\frac{1}{4}$ of 16 cents, or 4 cents, and 3 lemons cost 3 times 4 cents or 12 cents; if 4 oranges cost as much as 3 lemons, which is 12 cents, one orange costs $\frac{1}{4}$ of 12 cents, or 3 cents, and 20 oranges cost 20 times 3 cents, or 60 cents.

2. How many pigs can be obtained for 3 cows, if 10 pigs are worth 2 sheep, and 12 sheep are worth 2 cows?

3. How many oranges can you buy for 25 cents, if 4 oranges are worth 6 apples, and 3 apples are worth 5 cents?

4. If 12 sheep are worth 3 cows, and 12 cows are worth 6 horses, how many sheep could you exchange for 24 horses?

5. What will be the cost of 20 turkeys, if 2 turkeys are worth 4 hens, and 3 hens cost $6?

6. A can do twice as much in a day as B, and B can do 5 times as much as C; how long will it take A to do as much as C does in 20 days.

WRITTEN EXERCISES.

1. How many cents will 12 oranges cost, if 5 oranges are worth 4 lemons, and 6 lemons are worth 2 melons, and 1 melon is worth 10 cents?

OPERATION.

$$\frac{12}{5}\times\frac{4}{6}\times2\times10=32$$

SOLUTION.—If 1 melon is worth 10 cents, 2 melons are worth 2 times 10 cents, or 20 cents; if 6 lemons are worth 20 cents, 1 lemon is worth $\frac{1}{6}$ of 20 cents, and 4 lemons are worth $\frac{4}{6}$ of 20 cents; if 5 oranges are worth $\frac{4}{6}$ of 20 cents, 1 orange is worth $\frac{1}{5}$ of $\frac{4}{6}$ of 20 cents, and 12 oranges are worth $\frac{12}{5}$ of $\frac{4}{6}$ of 20 cents, or 32 cents.

OPERATION.

10 cents - 1 melon
2 melons - 6 lemons
4 lemons - 5 oranges
12 oranges - x

$$x=\frac{10\times2\times4\times12}{1\times6\times5}=32$$

SOLUTION 2D.—We will represent the term we wish to find by x. Now, if we arrange the quantities so that each stands opposite its equivalent, as in the margin, the product of the terms in the first column will equal the product of the terms in the second column; hence the product of the terms in the first column, divided by the product of all the terms in the second column, except x, will give the value of x. Hence the following

Rule.—I. *Place the antecedents in one column and the consequents in another, with a hyphen between them.*

II. *Divide the product of the terms in the column containing the odd term by the product of the terms in the other column.*

2. In a book bindery it was found 5 men do as much as 7 boys, and 7 boys do as much as 10 girls; how many girls will it require to do as much work as 28 men?

Ans. 56 girls.

3. What is the cost of 20 lb. of Singapore pepper, if 5 lb. are worth 2 lb. cloves, and 9 lb. cloves are worth 14 lb. white pepper, and 3 lb. white pepper are worth 10 lb. Calcutta ginger, and 4 lb. ginger cost 27¢? *Ans.* $2.80.

4. If 14 lb. Rio coffee costs as much as 11 lb. Java, and 6 lb. Java as much as 7 lb. Laguayra, and 17 lb. Laguayra as 18 lb. Jamaica, and 16 lb. Jamaica as 17 lb. Manila, what cost 20 lb. Rio if 15 lb. Manila cost $2.40? *Ans.* $3.30.

5. If 11 shares of United Co's of N. J. are worth 16 shares of Norristown, and 4 shares of Norristown are worth 10 of Pennsylvania, and 3 of Pennsylvania are worth 9 of Reading; how many shares of Reading are worth 22 of United Co's? *Ans.* 240.

6. What will be the cost of 3 kegs of 2 d. sheathing nails if 2 kegs of 2 d. nails are worth 3 kegs of 4 d. nails, and 5 kegs of 4 d. nails are worth 6 kegs of 8 d. nails, and 9 kegs of 8 d. nails are worth 10 kegs of 10 d. nails, and 2 kegs of 10 d. nails are worth $6.20? *Ans.* $18.60.

7. What will be the cost of 6 boxes of old layer raisins if 7 boxes are worth 120 lb. new Valencia raisins, and 4 lb. new Valencia raisins are worth 7 lb. currants, and 11 lb. currants are worth 2 lb. shelled Languedoc almonds, and 6 lb. shelled Languedoc almonds are worth $1.98? *Ans.* $10.80.

8. If 12 shillings in Boston equaled 15 shillings in Philadelphia, and 45 shillings in Philadelphia equaled 28 shillings in Charleston, S. C., and 14 shillings in Charleston equaled 24 shillings in New York, how many shillings in New York were equal to 72 shillings in Boston? *Ans.* 96.

9. If James earns as much in 6 months as John does in 8 months, and John earns as much in 4 months as Jonathan in 7 months, and Jonathan earns as much in 5 months as Josiah in 6 months, how long will it take Josiah to earn as much as James in 18 months? *Ans.* $50\frac{2}{5}$ months.

10. What is the cost of 200 lb. of buckwheat flour if 900 lb. cost as much as 11 barrels of rye flour, and 75 barrels of rye flour as 44 barrels Minnesota Extra, and 26 barrels Minnesota Extra as $37\frac{1}{2}$ barrels Extra Round Hoop Ohio; and 10 barrels Extra Round Hoop Ohio cost $52?

Ans. 10.75\frac{5}{9}$.

MEDIAL PROPORTION.

579. Medial Proportion is the process of combining two or more quantities of different values.

580. The **Mean Value** is the average value of the combination.

NOTE.—The subject has been called *Alligation*, from *alligo*, I bind, a name suggested by the method of linking the figures with a line in solving the problems.

CASE I.

581. ***Given, the quantity and value of each, to find the mean value.***

NOTE.—This case was formerly called *Alligation Medial.*

1. A merchant mixed 24 lb. of sugar at 10 cents a pound, 30 lb. at 14 cents, and 26 lb. at 20 cents; what is the average price of the mixture?

SOLUTION.—24 lb. at 10 cents a pound cost 240 cents, 30 lb. at 14 cents a pound cost 420 cents, 26 lb. at 20 cents cost 520 cents; taking the sum we find 80 lb. cost 1180 cents; hence 1 lb. cost $\frac{1}{80}$ of 1180 cents, which is $14\frac{3}{4}$ cents; hence the mean value of the mixture is $14\frac{3}{4}$ cents. From this solution we derive the following

OPERATION.

lb.		¢		¢
24	@	10	=	240
30	@	14	=	420
26	@	20	=	520
80		cost		1180
1		"		80)1180 = $14\frac{3}{4}$, *Ans.*

Rule.—*Find the sum of the values of the ingredients and divide it by the sum of the ingredients.*

WRITTEN EXERCISES.

2. A person mixed 25 lb. of tea at 50 cents a pound, 34 lb. at 80 cents, and 41 lb. at $1.10; what is the mean price or quality of the mixture? *Ans.* $$.84\frac{4}{5}$.

3. A person mixed 18 gal. of wine at $.50, 26 gal. at $.80, 20 gal. at $1.20, with 6 gal. of water; what was the value of a gallon of the mixture? *Ans.* $$.76\frac{6}{7}$.

4. A goldsmith combined 8 oz. of gold 21 carats fine, 12 oz. 22 carats fine, 18 oz. 20 carats fine, with 28 oz. of alloy; required the fineness of the composition. *Ans.* 12 carats.

5. A person mixed 12 gal. of alcohol 90% strong, 7 gal

80% strong, 10 gal. 75% strong, and 11 gal. 70% strong; what per cent. of alcohol in the mixture? *Ans.* 79%.

6. A drover bought 30 cows at \$20 a head, 40 at \$25 a head, 30 at \$28 a head; he sells them at a gain of 25%; what is the average price per head received? *Ans.* \$30.50.

CASE II.

582. ***Given, the mean value and the value of each ingredient, to find the proportional quantity of each.***

NOTE.—This and the following cases were formerly called *Alligation Alternate.*

1. A grocer wishes to mix sugars worth 5, 7, 12, and 14 cents a pound, forming a mixture worth 9 cents a pound; in what proportion must the sugars be mixed?

OPERATION.

Mean	Prices					Ans.
9	5	$\frac{1}{4}$		5		5
	7		$\frac{1}{2}$		3	3
	12		$\frac{1}{3}$		2	2
	14	$\frac{1}{5}$		4		4

SOLUTION.—If we take 1 lb. at 5 cents for the mixture worth 9¢, we gain on it 4¢, and to gain 1 cent we would take $\frac{1}{4}$ of a pound. If we take 1 lb. at 14¢, we will lose 5¢, and to lose 1 cent, what we have just gained, we would take $\frac{1}{5}$ lb.; hence we take $\frac{1}{4}$ lb. at 5¢ as often as $\frac{1}{5}$ lb. at 14¢, or in whole numbers, 20 times $\frac{1}{4}$, which is 5 of the first, as often as 20 times $\frac{1}{5}$, which is 4 of the fourth. In a similar manner we find that we must take 3 lb. at 7¢, as often as 2 lb. at 12¢; hence the quantities may be mixed in the proportion of 5, 3, 2, and 4.

Rule.—I. *Write the several prices or qualities in a column, and the mean price or quality of the mixture at the left.*

II. *Select two quantities, the one less and the other greater than the average, write the reciprocal of the difference between each quantity and the average opposite the quantity, and reduce these to integers by multiplying by the least common denominator, and proceed in the same manner until all the prices have been used.*

III. *Add two or more proportional numbers if they stand opposite a given quantity; the results will be the proportional numbers required.*

NOTES.—1. When there are three quantities, compare *the one* which is greater or less than the average with *both* the others, and take the sum of the two numbers opposite *this one.*

2. A common factor may be inserted in any couplet or omitted from it without changing the proportional parts; *it is thus seen that there may be any number of answers in the same proportion.*

WRITTEN EXERCISES.

2. A grocer has teas worth 7, 10, 16, and 18 dimes a pound; what relative quantities of each must be taken to form a mixture worth 12 dimes a pound? *Ans.* 6; 4; 2; 5.

3. A merchant has 4 pieces of muslin, worth 10, 14, 20, and 22¢ a yard, respectively; how many yards must he sell of each that the price may average 18¢? *Ans.* 1; 1; 2; 2.

4. How shall I combine gold 16 carats, 18 carats, and 22 carats, to make a mixture of 20 carats fine, if I wish to mix equal quantities of 1st and 2d? *Ans.* 1st, 1; 2d, 1; 3d, 3.

5. What relative quantities of rice worth $12\frac{1}{2}$, $18\frac{3}{4}$, and $20\frac{1}{2}$ cents a pound, must be taken to form a mixture worth $16\frac{1}{4}$ cents a pound? *Ans.* 27; 15; 15.

6. A farmer bought pigs at $\$4\frac{1}{2}$ each, sheep at $\$5\frac{1}{4}$ each, and calves at $\$6\frac{1}{3}$ each; how many must he sell of each so that the average price may be \$5 each? *Ans.* 9; 2; 3.

7. A man has a quantity of 3, 5, 25, and 50 cent pieces, which he wishes to exchange for 10 cent pieces; what is the relative number of pieces exchanged? *Ans.* 40; 15; 5; 7.

CASE III.

583. ***Given, the mean value, the value of each ingredient, and the quantity of one or more, to find the other quantities.***

1. A farmer bought 20 hens at 10 dimes each; how many must he buy at 4 and 5 dimes each, so that the average price may be 8 dimes each?

SOLUTION.—We find by Case II. that the number at 4 and 10 dimes are as 1 to 2, and at 5 and 10 as 2 to 3; hence, as often as we take 1 at 4 and 2 at 5, we take 2 + 3 = 5 at 10. But he bought 20, or 4 times 5, at 10 dimes, hence he must buy 4 times 1 or 4 at 4 dimes, and 4 times 2 or 8 at 5 dimes.

OPERATION.

Mean	Values						× 4 =
8	4	$\frac{1}{4}$		1		1	4
	5		$\frac{1}{3}$		2	2	8
	10	$\frac{1}{2}$	$\frac{1}{2}$	2	3	5	20

Rule.—I. *Find the proportional quantities by Case II.*

II. *Divide the given quantity by the proportional quantity limited, and multiply each of the other proportional quantities by the quotient.*

WRITTEN EXERCISES.

2. A merchant bought some hats for $5 each, some vests for $7 each, and 48 coats for $16 each; the average price was $12; how many vests and hats did he buy? *Ans.* 16 of each.

3. A publisher sold Mentals @ 15¢, Primaries @ 10¢, Grammars @ 30¢, and 360 Spellers @ 40¢; required the number of each, if the average price was 35¢. *Ans.* 36.

4. A merchant wishes to mix 40 lb. of sugar at 6¢ and 40 at 8¢, with some at 14¢ and 15¢, so that the mixture may be worth 10¢; how much of the latter kinds must he take?

Ans. 20 lb. at 14¢; 32 lb. at 15¢.

5. A grocer wished to mix 15 lb. of tea at $1½ a pound, 21 lb. at $1, with that worth 70 cents and 50 cents, so that the mixture may be worth 80 cents; how much must he take of the 3d and 4th? *Ans.* 42 lb. at 70¢; 35 lb. at 50¢.

6. A man has some 3 ct. pieces, some 5 ct. pieces, some 10 ct. pieces, and 290 fifty-cent pieces, which he exchanges for 25 ct. pieces; how many must he exchange of each kind?

Ans. 250 three-cent; 50 five-cent; 50 ten-cent.

CASE IV.

584. ***Given, the mean value, value of each ingredient, and entire quantity, to find the quantity of each ingredient.***

1. A person has a sum of money in ten-cent pieces, which he wishes to exchange for 3, 5, 25, and 50 cent pieces, having 255 pieces in all; how many of each will he obtain?

OPERATION.

10	3	$\frac{1}{7}$		40		40	×5=	200
	5		$\frac{1}{5}$		3	3		15
	25		$\frac{1}{15}$		1	1		5
	50	$\frac{1}{40}$		7		7		35
						51		255

$255 \div 51 = 5.$

SOLUTION.—We find by Case II. that we must have 40 three-cent pieces as often as 7 fifty-cent pieces, and also 3 five-cent pieces as often as one 25-cent piece. Taking the sum of these we have 51 in all; but we wished 255, which is 5 times 51, hence we must take 5 times as many of each, which gives respectively 200, 15, 5, and 35.

Rule.—I. *Find the proportional quantities by Case II.*

II. *Divide the required quantity by the sum of the pro-*

portional quantities, and multiply each proportional quantity by the quotient.

NOTE.—When the sum of the proportional parts is not an exact divisor of the quantity, each couplet must be multiplied by such numbers as will make the sum of the proportional parts a divisor of the entire quantity.

WRITTEN EXERCISES.

2. A grocer wished to mix teas worth 25, 30, 44, and 55¢ a pound, making a mixture of 68 pounds, worth 35¢ a pound; required the quantity of each kind.

Ans. 8; 36; 20; 4.

3. A merchant mixed sugars worth 5, 7, 11, and 12¢ a pound, in order to make a mixture of 66 lb. worth 8¢ a pound; how many pounds of each did it require?

Ans. 24; 18; 6; 18.

4. A man bought 100 apples for $1, some worth $\frac{1}{4}$ of a cent, some $\frac{1}{2}$, some $1\frac{1}{4}$, and some 2¢ apiece; how many did he purchase of each kind?

Ans. 40, 1st; 10, 2d; 20, 3d; 30, 4th.

5. A lady bought 102 yards of muslin at an average price of 15 ct., some at 8, some at 13, some at 18, and some at 20 ct. a yard; required the number of yards of each kind.

Ans. 30; 18; 12; 42.

6. A man has $134 in ten-cent pieces, which he wishes to exchange for pieces worth 3, 5, 25, and 50 cents respectively; how many of each kind will it require?

Ans. 800, 3 ct.; 300, 5 ct.; 100, 25 ct.; 140, 50 ct.

7. A person has some bank-notes whose denominations are respectively $1, $2, $5 and $20, which he wishes to exchange for 62 ten-dollar notes; how many must he exchange of each kind?

Ans. 20 ones; 10 twos; 4 fives; 28 twenties.

8. A person purchased 100 animals for $100; sheep at $\$3\frac{1}{2}$ apiece, calves at $\$1\frac{1}{3}$, and pigs at $\$\frac{1}{2}$; how many animals did he buy of each kind?

Ans.
Sheep, 5, 10, 15.
Calves, 42, 24, 6.
Pigs, 53, 66, 79.

NOTE.—This last problem is from Hackley's Algebra, 3d example, under Indeterminate Analysis. Its solution by Arithmetic is very simple.

PARTNERSHIP.

585. Partnership is the association of two or more persons for the transaction of business.

586. Partners are the persons associated in business, and are of three kinds, *General*, *Limited*, and *Special*.

587. The **Capital** of a firm is the money or property invested by the partners. The *Liabilities* are its debts.

588. The **Resources** or **Assets** of a firm are its property of any kind, together with the amounts due it. The excess of resources over liabilities is called the *Net Capital*.

589. Partnership is divided into *Simple* and *Compound Partnership* for convenience of treatment.

General Partners risk their whole property in the business; *Limited and Special Partners* risk only the amount of capital they agree to contribute. Partners whose names do not appear are sometimes called *Silent Partners*.

SIMPLE PARTNERSHIP.

590. In **Simple Partnership** the shares of the partners are employed for equal periods of time.

1. A, B, and C, went into partnership; A put in \$500, B put in \$700, and C, \$800; they gained \$600; what was each one's share of the gain?

SOLUTION.—The entire capital is \$2000. Since A put in \$500 he furnished $\frac{500}{2000}$, or $\frac{1}{4}$ of the capital, and hence should have $\frac{1}{4}$ of \$600, or \$150. B furnished $\frac{700}{2000}$, or $\frac{7}{20}$ of the capital, and should have $\frac{7}{20}$ of the gain, etc.

OPERATION.

$$\$500 \quad \frac{500}{2000} = \frac{1}{4} = \text{A's share.}$$
$$700 \quad \frac{700}{2000} = \frac{7}{20} = \text{B's share.}$$
$$800 \quad \frac{800}{2000} = \frac{2}{5} = \text{C's share.}$$
$$\text{Stock} = \$2000$$

$$\tfrac{1}{4} \text{ of } \$600 = \$150 = \text{A's share.}$$
$$\tfrac{7}{20} \text{ of } \$600 = \$210 = \text{B's share.}$$
$$\tfrac{2}{5} \text{ of } \$600 = \$240 = \text{C's share.}$$

Rule.—*Divide the gain or loss among the partners in proportion to their shares of the stock.*

WRITTEN EXERCISES.

2. A, B, and C form a partnership for shipping peaches; A puts in \$680, B \$720, C \$600; they gain \$600; what does each receive? *Ans.* A, \$204; B, \$216; C, \$180.

3. Three persons enter into partnership, with $6000, of which A contributes $\frac{1}{2}$, B $\frac{1}{3}$, and C the remainder; they gain $1800: what sum belongs to each?

Ans. A, $900; B, $600; C, $300.

4. Three men agree to share 60 gal. wine, A taking $\frac{1}{3}$, B $\frac{1}{4}$, and C $\frac{1}{5}$; but upon drawing off these parts they find there is a remainder; how should the wine be divided?

Ans. A, $25\frac{25}{47}$ gal.; B, $19\frac{7}{47}$ gal.; C, $15\frac{15}{47}$ gal.

5. A, B, and C were partners in the coal trade; A furnished $5000, B $7000, and C managed the business; they gained $2400; what was the share of each, if C received as much as both A and B?

Ans. A's, $500; B's, $700; C's, $1200.

6. Three persons engage in cotton speculation; A contributed $6400, B $7200, and C $5400; they lose $\frac{1}{5}$ of their stock by fire, and gained on the remainder $\frac{2}{5}$ of cost; what was the gain of each?

Ans. A's, $2048; B's, $2304; C's, $1728.

7. 'Squire Jones left by his will $5000 to his wife, $3500 to his son, and $4500 to his daughter; but upon settling his estate it was found to amount to only $10400; how much did each receive?

Ans. Wife, $4000; son, $2800; daughter, $3600.

8. A, B, and C go into the lumber trade with a joint capital of $9500; at the end of a year it is found that A's gain is $1650, B's $1500, and C's $1600; required each one's stock.

Ans. A's, $3300; B's, $3000; C's, $3200.

9. A shipping firm gained one year $4200; A's stock was $6500, B's stock $5300, and C's gain $1250; required C's stock and A's and B's gain.

Ans. A's, $1625; B's, $1325; C's stock, $5000.

10. A, B, and C form a partnership for carrying on a nursery; A contributes $800, B $600, and C 10 acres of land on which to establish the nursery; their first year's profits are $1500, of which C receives $660; what are A's and B's gain, and the value of C's land per acre?

Ans. A's, $480; B's, $360; $110 per acre.

COMPOUND PARTNERSHIP

591. In **Compound Partnership** the capitals of the partners are employed for different periods of time.

CASE I.

592. ***When the profits and losses are divided in proportion to capital and time.***

1. Two persons enter into partnership and gain $328; A put in $800 for 5 mo., and B $700 for 6 mo.; what was each man's share of the gain?

SOLUTION.—$800 for 5 months is equivalent to $4000 for 1 mo.; and $700 for 6 mo. equivalent to $4200 for 1 mo.; hence the entire capital is equivalent to $8200 for 1 mo. The rest of the solution may be given as in Simple Partnership.

OPERATION.

$\$800 \times 5 = \$4000 =$ A's for 1 mo.
$\$700 \times 6 = \$4200 =$ B's for 1 mo.
$\$8200 =$ whole for 1 mo.
$\frac{4000}{8200} = \frac{20}{41} =$ A's share of capital.
$\frac{4200}{8200} = \frac{21}{41} =$ B's share of capital.
$\$328 \times \frac{20}{41} = \$160$, A's gain.
$\$328 \times \frac{21}{41} = \$168$, B's gain.

Rule. *Multiply each partner's capital by the time it was employed, and divide the gain or loss in proportion to these products.*

WRITTEN EXERCISES.

2. A, B, and C engaged in partnership; A had $500 in trade for 9 mo., B $800 for 8 mo., and C $1200 for 7 mo.; they gain $488.70; what was each one's share of the gain?
Ans. A's, $113.94; B's, $162.06; C's, $212.70.

3. Four gentlemen rented a pasture-field for $62.40; the first put in 3 horses for 7 weeks, the second 4 horses for 8 weeks, the third 2 horses for 13 weeks, and the fourth 5 horses for 5 weeks; what should each pay?
Ans. $12.60; $19.20; $15.60; $15.

4. A, B, C, and D agree to clear a tract of woodland for $120; A worked 9 days of 10 hours each, B 15 days of 6 hours each, C 10 days of 9 hours each, and D 3 weeks 5 hours a day; what does each receive? *Ans.* $30.

5. Mr. Allen commenced business with $10000 capital; at

the end of 3 months he took in Mr. Green, with $7000 capital, and at the end of 6 months Mr. Handy with $3000 capital; at the end of the year they had gained $4020; required the share of each. *Ans.* $2400; $1260; $360.

6. Two persons, A and B, were in partnership 2 years; A at first put in $2500 and B $3000; at the end of 9 mo. A took out $800 and B put in $500; they lost in 2 yr. $3825; what was each one's share of the loss?

Ans. A's, $1440; B's, $2385.

7. Three drovers, A, B, and C, hire a pasture for $51.70 for 5 mo.; A put in 60 cows, B 80 cows, and C 90 cows; at the end of 3 mo. A sells $\frac{1}{3}$ of his, B $\frac{1}{4}$ of his, and C $\frac{1}{6}$ of his; how much rent ought each to pay?

Ans. A, $13.00; B, $18.00; C, $20.70.

8. A and B went into the hardware business, A's capital being to B's as 5 to 7; at the end of 6 months A withdraws $\frac{1}{5}$ of his capital, and B $\frac{1}{7}$ of his, and during the year they lose $1430; what was each man's share of the loss?

Ans. A's, $585; B's, $845.

9. Brown's capital was in trade 6 mo., Black's 8 mo., and White's 10 mo. Brown's gain was $750, Black's $1200, White's $800, and the whole capital, $19880; how much did each own? *Ans.* $7000; $8400; $4480.

REMARK.—Find each one's gain for 1 mo., and divide the whole capital in that proportion.

CASE II.

593. ***When the proportion of profits or losses is fixed, and interest is allowed for the difference between each partner's proportion of capital and the amount he actually contributes.***

1. A and B form a partnership; A contributes $1700, and is to have $\frac{2}{3}$ of the profits; B contributes $700, and is to have $\frac{1}{3}$ profits; each partner is to receive or pay interest at the rate of 6% per annum for any excess or deficit in his proportionate share of capital. At the end of a year the profits are $900. How much has each gained?

Solution.—Total capital is \$2400. A should contribute $\frac{2}{3}$ or \$1600, and is entitled to 1 year's interest, or \$6, on his excess. B should contribute $\frac{1}{3}$, or \$800, and must pay 1 year's interest, or \$6, on his deficit. A gained $\frac{2}{3}$ of \$900 = \$600, + \$6 interest = \$606. B gained $\frac{1}{3}$ of \$900 = \$300, — \$6 interest = \$294.

OPERATION.

\$1700	A contributed	\$1700
700	$\frac{2}{3}$ of 2400 =	\$1600
\$2400	Excess	\$100
		.06
	Int. for 1 year	\$6.00
	$\frac{1}{3}$ of \$2400 =	\$800
	B contributed	\$700
	Deficit	\$100
	Int. for 1 year	\$6.00

$\frac{2}{3}$ of 900 = 600, and + 6 = 606, A's gain.
$\frac{1}{3}$ of 900 = 300, and — 6 = 294, B's gain.

Rule.—I. *Find the interest on the excess or deficit of each partner's share of capital. If there are additions and withdrawals, subtract the interest on the former from the gross profits, and add the interest on the latter to the gross profits.*

II. *Divide the profits thus obtained in the required proportions, adding or subtracting the interest due to or by each partner respectively, and the result will be the net gain of each. For the present value of each share, add to each partner's original stock the net gain and the additions, and subtract the withdrawals.*

Note.—The interests on the excesses and deficits of capital exactly balance each other, and will not change the profits. Thus, if one partner puts in \$100 more than his share, the other partners must have \$100 less than theirs.

WRITTEN EXERCISES.

2. A and B form a partnership. A contributes \$4000 and is to have $\frac{2}{3}$ of the profits; B contributes \$2000 and is to have $\frac{1}{3}$ of the profits; each partner is to receive or pay interest at the rate of 6% per annum for any excess or deficit in his share of capital. At the end of the first year the profits are \$1500, and nothing is withdrawn. Required the worth of each share. *Ans.* A, \$5000; B, \$2500.

3. The following year A adds, during the year, \$1500 capital, averaging May 1; B adds \$900, averaging Sept 1. Profits \$2400, and nothing withdrawn. Required the worth of each share. *Ans.* A, \$8108; B, \$4192

4. The following year neither partner adds capital, but A

withdraws living expenses, $1500, averaging July 1, and B $1200, averaging Sept. 1. Profits $3000. Required the worth of each share. *Ans.* A, $8603.48; B, $3996.52.

5. Fourth year: on the first of the year A sells $\frac{1}{4}$ of his share to C, who thus becomes a partner: A's proportion of profits to be $\frac{1}{2}$; B's $\frac{1}{3}$ as before; and C's $\frac{1}{6}$. A adds capital Jan. 1, $1800, B, averaging March 1, $900, and C, averaging July 1, $300. A withdraws living expenses, averaging July 1, $1500, B, Sept. 1, $1200, and C, Nov. 1, $400. Profits $4200. Required the worth of each share.

Ans. A, $8880.27; B, $5075.64; C, $2744.09.

EQUATION OF PAYMENTS.

594. Equation of Payments is the process of finding the mean or equitable time for paying several sums, due at different times.

595. The **Term of Credit** is the time allowed for the payment of a debt.

596. The **Average Term of Credit** is the time to elapse before several debts due at different times may in equity be paid together.

597. The **Equated Time** is the date at which several debts due at different times may be paid in one sum.

598. The **Focal Date** is the date from which we begin the reckoning in averaging an account.

CASE I.

599. ***To find the average term of credit, when the terms of credit begin at the same time.***

1. A owes B $150 due in 5 months, and $250 due in 3 months; what is the average term of credit?

OPERATION.

$$\begin{array}{l} 150\times 5=750 \\ 250\times 3=750 \\ \overline{400} \quad)\overline{1500}(3\tfrac{3}{4}\text{ mo.} \end{array}$$

SOLUTION.—A credit on $150 for 5 months is regarded as equivalent to a credit on $1 for 750 months, and a credit on $250 for 3 months is equivalent to a credit on $1 for 750 months; and adding, we have the same as a credit on $1, for 1500 months; if $1 has a credit 1500 months, $400 would have a credit of $\frac{1}{400}$ of 1500 months, which is $3\frac{3}{4}$ months. Hence the

Rule.—*Multiply each payment by its term of credit, and divide the sum of the products by the sum of the payments; the quotient will be the average term of credit.*

NOTES.—1. If there are cents in any of the payments, they may be rejected when less than 50, and reckoned at $1 when more than 50. The fraction of a day in the answer is also rejected when less than $\frac{1}{2}$, and reckoned as 1 day if more than $\frac{1}{2}$.

2. It is objected to this rule that the interest on a certain sum not paid till after it is due, is more than the discount on the same sum paid an equal length of time before it is due. As practically, however, we generally reckon bank discount, which is the same as interest, the rule seems not really to lie open to this objection.

3. The time may also be found by dividing the sum of the interests on the payments, using any rate, by the interest on the sum of the payments for 1 month or 1 day, according to the unit of time used in the calculation. This method is preferred by some accountants.

WRITTEN EXERCISES.

2. Henry Smith owes Thomas Jones $6000, $\frac{1}{3}$ due in 3 mo., $\frac{1}{4}$ in 4 mo., and the remainder in 6 mo.; required the average term of credit. *Ans.* $4\frac{1}{2}$ mo.

3. I owe $1500, $\frac{1}{3}$ of which is due in 2 mo., $600 in 5 mo., and the remainder in $7\frac{1}{2}$ mo.; required the average term of credit. *Ans.* $4\frac{3}{5}$ mo.

4. A person owes $300 due in 4 mo., $400 due in 5 mo., $700 due in 6 mo., and $1000 due in 8 mo.; what is the average term of credit? *Ans.* $6\frac{5}{12}$ mo.

5. I bought merchandise April 1, 1876, as follows: $4200 for cash; $2800 on 4 mo., and $1400 on 6 mo.; what is the equated time of payment? *Ans.* June 11.

6. A gentleman bought a house, agreeing to pay $\frac{1}{2}$ in 4 mo., $\frac{1}{3}$ in 9 mo., and the remainder in 1 yr.; required the average term of credit. *Ans.* 7 mo.

REMARK.—Since the result will be the same whatever the sum owed, we may assume $1 as the capital, and proceed as before.

7. A merchant owes a certain sum, $\frac{1}{2}$ of which is due in 4 mo., $\frac{1}{4}$ in 6 mo., $\frac{1}{6}$ in 12 mo., and the remainder in 8 mo.; required the average term of credit. *Ans.* $6\frac{1}{6}$ mo.

8. A owes a sum, $\frac{1}{3}$ due on January 1, $\frac{1}{4}$ on May 1, $\frac{1}{6}$ on July 1, and the remainder on September 1; what is the equated time for the payment of the whole, estimating from the first of January? *Ans.* May 1.

CASE II.

600. ***To find the equated time when the credits begin at different dates.***

1. I purchased of Stewart & Co. the following bill of goods:—

Jan. 10,	a bill amounting to		$700 on 2 mo. credit.	
Jan. 20,	"	"	$500 on 3 mo.	"
Feb. 24	"	"	$800 on 3 mo.	"

Now, if I wish to make one payment of this bill, at what time in equity will it become due?

OPERATION.

Mar. 10, $\$700 \times 00 = 00000$
Apr. 20, $\$500 \times 41 = 20500$
May 24, $\$800 \times 75 = 60000$
$\overline{2000}$ $\quad \overline{)80500}(\ 40\frac{1}{4}$ da.

SOLUTION.—From the time the first is due to the time the second is due is 41 da., and to the time the third is due is 75 da.; hence, estimating from the time the first is due, the second has a credit of 41 da., and the third a credit of 75 da., and the first has a credit of no days. We then average it as in Case I., and find the term of credit to be $40\frac{1}{4}$ da. from March 10, the time at which the first debt is due; hence the equated time of payment is April 19th. From the above we derive the following

Rule.—I. *Select the date at which the first debt becomes due, and multiply each debt by its term of credit reckoned from the date selected.*

II. *Divide the sum of the products by the sum of the debts, and the quotient will be the average term of credit, estimated from the date selected.*

NOTE.—When the *earliest* date is not the *first* of the month, it is often more convenient to take the *first* of the month as the standard date.

WRITTEN EXERCISES.

2. Mr. Johnson sold goods to one of his customers at different dates, as stated in the following bill:

March 12,	to the amount of		$360 on 3 mo. credit.	
April 20,	"	"	$500 on 5 mo.	"
May 18,	"	"	$340 on 4 mo.	"
July 30,	"	"	$600 on 2 mo.	"

What is the average term of credit and also the equated time for the payment of this bill? *Ans.* Term of credit, 83 da from June 12; equated time, Sept. 3

3. Purchased of a merchant, at different times, the following bill of goods:

Feb. 14,	to the	amount of	$600 on 6 mo	credit.
March 16,	"	"	$600 on 6 mo.	"
May 10,	"	"	$600 on 6 mo.	"
June 18,	"	"	$600 on 6 mo.	"

Required the equated time for the payment of the bill.

Ans. Equated time, Oct. 15.

4. I sold goods to Mr. Bowman at different times and terms of credit, as follows:

July 16, 1862,	a bill	of	$800 on 3 mo.	credit.
Sept. 20,	"	"	$500 on 4 mo.	"
Oct. 12,	"	"	$350 on 6 mo.	"
Jan. 24, 1863,		"	$450 on 4 mo.	"
March 10,	"	"	$600 on 3 mo.	"

If he gives me his note for the amount, when, in equity, should it begin to bear interest? *Ans.* Feb. 23.

CASE III.

601. ***When a debt due at some future time has received partial payments, to find when the remainder should be paid.***

1. A borrows $3000 to be paid in 8 mo.; 5 mo. before it was due he paid $800, and 2 mo. before it was due he paid $600; how long after the expiration of the 8 mo. may the balance remain unpaid?

SOLUTION.—A credit on $800 for 5 mo. is equivalent to a credit on $1 for 4000 mo.; a credit on $600 for 2 mo. is equivalent to a credit on $1 for 1200 mo.; and adding, we have a credit on $1 for 5200 mo.; hence $1600, the sum which remains unpaid, should have a credit of $\frac{1}{1600}$ of 5200 mo., which is $3\frac{1}{4}$ mo. Hence

OPERATION.

$$800 \times 5 = 4000$$
$$600 \times 2 = 1200$$
$$1400 \qquad 5200$$
$$\frac{5200}{1600} = 3\tfrac{1}{4} \text{ mo.}$$

Rule.—*Multiply each payment by the time it was paid before it was due, and divide the sum of the products by the sum remaining unpaid.*

WRITTEN EXERCISES.

2. A borrowed $2400 to be paid in 6 mo.; 4 mo. before

being due he paid $600, and 3 mo. before due he paid $1200; at what time in equity should the remainder be paid?

Ans. In 10 mo.

3. I lent Mr. C. $1600 for 9 mo., ½ of which he paid in 5 mo., and ½ of the remainder in 6 mo.; how long, in equity, may the remainder remain unpaid?

Ans. 11 mo. after due.

4. I borrowed of Mr. W. $400 for 3 mo., $600 for 5 mo., and $800 for 6 mo.; at the end of 4 mo. I paid him $1200; at what time in equity should the remainder be paid?

Ans. 7 mo. after borrowing.

5. A milliner bought of Smith & Co., a bill of $240 for 20 days, and $560 for 30 days; at the end of 16 days she paid $300, and at the end of 24 days she paid $350; when, in equity, should the balance be paid? *Ans.* 56 days.

AVERAGING ACCOUNTS.

602. Averaging an Account is the process of finding the mean or equitable time for the payment of the balance of the account.

1. In the following account, required the balance and the time when due:

DR.	HENRY HARDY.						CR.
1876.				1876.			
May 10	To merchandise,	200	00	May 20	By Cash,	150	00
June 20	" "	400	00	July 16	" "	200	00
Aug. 28	" "	300	00	Sept. 20	" "	200	00

OPERATION.

Due.	Time.	Items.	Products.	Due.	Time.	Items.	Products.
May 10	00	$200	00000	May 20	10	$150	1500
June 20	41	$400	16400	July 16	67	$200	13400
Aug. 28	110	$300	33000	Sept. 20	133	$200	26600
		$900	49400			$550	41500
		- 550	41500				
		350	7900				

$7900 \div 350 = 22\frac{4}{7}$ da.

Hence the balance is $350, and is due 23 da. after May 10, that is, on June 2d.

SOLUTION.—Select the date of the item first due as the focal date, and find the time the others are due after it; then multiplying each item by the corresponding time, and taking the sums of the products, we find that if paid on the 10th of May the *Dr.* items must suffer a discount of $1 for 49400 days, and the *Cr.* items must suffer a discount of $1 for 41500 days. Subtracting the two sums we find that the *Dr.* side must suffer a discount of $1 for 7900 days more than the *Cr.* side, hence $350, the balance of the items, must suffer a discount of $\frac{1}{350}$ of 7900 days, which is $22\frac{4}{7}$ days. Hence the balance is due 23 days after May 10th, or June 1st. Hence we have the following

Rule.—I. *Find when each item is due, take the earliest date as the focal date, find the difference between the focal date and the remaining dates, and multiply each item by the corresponding difference.*

II. *Balance the columns of products and also the columns of items, and divide the former by the latter: the quotient added to the focal date will give the equated time.*

III. *If the two balances be on opposite sides of the account, the quotient obtained must be subtracted from the focal date.*

NOTES.—1. Other dates than the earliest might be selected as the focal date. If we reckon from the last date we have the interest instead of the discount.

2. Instead of *products* we may obtain the *interest* at any per cent. on each item, and divide the balance of interest by the interest on the balance of the account for one day; the quotient will be the number of days to be added to or subtracted from the focal date.

WRITTEN EXERCISES.

2. What is the balance of the following account and when is it due? *Ans.* Balance, $450; due Jan. 23d.

DR. CHARLES HARDING. CR.

1872.			1872.		
Jan. 8	To Mdse.,	$600	Jan. 27	By Cash,	$500
March 5	To Sundries,	$400	April 10	" 20 cows,	$400
April 20	To Mdse.,	$550	May 16	" Sundries,	$200

3. What is the balance of the following account, and when is it due? *Ans.* Balance, $205; due July 31st.

DR. HENRY T. OSBORN. CR.

1871.			1871.		
May 1	To Sund. on 3 mo.	$375	July 20	By Cash,	$300
May 18	" 3 mo.	$280	Aug. 10	" "	$250
June 20	" 3 mo.	$700	Aug. 31	" Merchandise,	$350
July 16	To Cash,	$350	Sept. 12	" Cash,	$600

4. The following account was settled by Mr. Kready giving his note for the balance; required the face of the note and the time when interest commenced.

Ans. Face, $475; Int. from March 8th.

Dr. BENJAMIN KREADY. **Cr.**

1872.			1872.		
April 14	To Mdse. on 2 mo.	$650	July 5	By Cash,	$500
May 20	" " 3 mo.	$550	Sept. 28	" "	$350
June 16	" " 2 mo.	$475	Oct. 12	" "	$450
July 12	" " 2 mo.	$350	Dec. 4	" "	$250

5. What is the balance of the following account, and if a note is given, when does interest begin?

Ans. Balance $900; Int. from Dec. 18, 1871.

Dr. SMITH, IN ACC'T WITH BRADFORD. **Cr.**

1872.			1872.		
March 19	To invoice,	$900	March 24	By Cash,	$300
" 29	" " 10 da.	$800	April 25	" Remittance.	$300
April 20	" "	$400	July 17	" Cash,	$200
May 4	" "	$200	Aug. 6	" "	$600

SETTLEMENT OF ACCOUNTS.

603. An **Account Current** is a written statement of the debit and credit items of business transactions between two parties.

604. The **Adjustment** of an account is the determining of the balance due at a specified date.

605. An account is **Settled** upon *payment* of the adjusted *balance*, or by *carrying* it to another account.

In finding the cash balance, interest should be allowed on each item for the time between the day it is due and the day of settlement.

Rule.—I. *Find the interest on each item from the time it becomes due to the date of settlement.*

II. *Add the interest to the item if due before the date of settlement, and subtract it when the item is due after the date of settlement. The difference of the sums of the results on both sides of the account will be the cash balance.*

NOTES.—1. An account may be adjusted by averaging it and finding the amount of the balance from the time it becomes due till the time of settlement.

2. In *averaging* an account, we find *at what date the balance* is due; in *adjusting* an account, we find *what balance* is due at a *specified date*.

WRITTEN EXERCISES.

1. Required the cash balance of the following account, July 16, 1875, interest at 6 per cent. *Ans.* $200.60.

CHARLES PANCOAST, IN ACCOUNT WITH MARCH & CO.

1875			1875		
Jan. 16,	To Mdse. on 3 mo.	450 00	March 12,	By Cash,	200 00
Feb. 24,	" " " 3 mo.	350 00	April 18,	" "	300 00
April 12,	" " " 3 mo.	300 00	July 3,	" "	400 00

2. Required the cash balance of the following account, Aug. 8, interest 6 per cent. *Ans.* $150.171.

WALTER ROSE, IN ACCOUNT WITH JAMES OSBORN.

1874			1874		
March 16,	To Mdse., 2 mo.	650 00	May 28,	By Cash,	500 00
April 20,	" " 2 mo.	750 00	July 12,	" "	800 00
May 24,	" " 3 mo.	375 00	July 20,	" "	200 00
June 12,	" " 3 mo.	575 00	Aug. 4,	" "	700 00

ACCOUNT SALES.

606. An **Account Sales** is a written statement, rendered by an agent or consignee to the consignor, of the sales of goods consigned, the charges, and the net proceeds.

607. Guaranty is a charge made for securing the owner against the risk of non-payment, when goods are sold on credit.

Expenses incurred in receiving the goods and all charges paid in cash are considered due the consignee when paid, but commission and after charges are due at the average maturity of the sales.

An account-sales is averaged to find when the net proceeds become due, in order that the consignor may draw a bill of exchange to fall due at the equated time. Except that the date of maturity of the commission and guaranty must be found by first averaging the sales, the account is averaged like an account current, the charges being the debits and the sales the credits.

1. Account sales of 400 barrels of pork received from Gibbs and Waterman of Chicago to be sold on their % and

1875					
Oct.	28	Sold 100 bbl. pork 20000 lb. @ 7¢ on 30 da.		1400	
Nov.	11	" 250 " " 50000 " " 6¢ cash		3000	
Nov.	30	" 50 " " 10000 " " 6¼¢ "		625	
		400		5025	—
		Charges			
Oct.	25	To Freight and Drayage	342.50		
Nov.	30	" Storage from Oct. 25	10.00		
Nov.	18	" Commission on $5025 @ 2¼%	113.06¼		
"	"	" Guaranty on $1400 @ 2½%	35.00	500	56

What are the net proceeds of the above account, and when is it due? *Ans.* $4524.44; Nov. 20.

2. A commission merchant in New York received a consignment from Milwaukee, Sept. 1, 1874, of 800 barrels of flour, paying for freight $250, and drayage $27.50. He sold Sept. 3, 200 barrels @ $6.50; Sept. 12, 150 barrels @ $7 at 30 days; Sept. 18, 250 barrels @ $6.75; Sept. 30, 200 barrels @ $7. The commission was 2½%, guaranty 2¼%, and storage $40; required the net proceeds and the equated time. *Ans.* $4960.44; Sept. 23.

SECTION X.

INVOLUTION AND EVOLUTION.

INVOLUTION.

608. **Involution** is the process of finding any power of a number.

609. A **Power** of a number is the product arising from using the number several times as a factor. The number itself is called the *first power.*

610. The **Second Power** of a number is the product obtained by using the number twice as a factor. Thus, 16 is the second power of 4, since $4 \times 4 = 16$.

611. The **Third Power** of a number is the product obtained by using the number three times as a factor. Thus 64 is the third power of 4, since $4 \times 4 \times 4 = 64$.

612. The **Fourth Power** of a number is the product obtained by using the number four times as a factor; the *Fifth Power,* five times as a factor, etc.

613. The **Degree** of a power is indicated by a small figure, called an *exponent,* placed at the right and a little above the number. Thus, 5^2 represents the 2d power of 5, 6^3, the third power of 6, etc.

614. The **Exponent** indicates how many times the number is used as a factor. Thus, 8^3 denotes that 8 is used as a factor three times; that is, $8 \times 8 \times 8$, which equals 512.

The second power of a number is called its *square,* because the area of a square equals the product of its two equal sides. The third power of a number is called its *cube,* because the product of the three equal sides of a cube gives its contents.

PRINCIPLES.

1. *A power of a number is obtained by using the number as a factor as many times as there are units in the degree.*

2. *The product of any two powers of a number equals a power of the number denoted by the sum of the exponents.*

For, if we multiply the cube of a number by the 4th power of the number, we will evidently have the number used seven times as a factor, or the 7th power of the number; thus, $5^3 \times 5^4 = (5 \times 5 \times 5) \times (5 \times 5 \times 5 \times 5) = 5^7$; and the same may be shown in any other case.

3. *A power of a number raised to any power equals a power of the number denoted by the product of the exponents.*

For, if we *square* the *cube* of a number, we will evidently use the number as a factor *two* times *three* times, or *six* times; thus, $(5^3)^2 = 5^3 \times 5^3$, which, by Prin. 2, equals 5^6, and the same may be shown in any other case.

NOTE.—By means of this principle we can abbreviate the operation of involution; thus we can raise a number to the sixth power by squaring its cube, or to the 12th power by squaring its sixth power, or cubing its 4th power, etc.

MENTAL EXERCISES.

1. The cube of 4 equals 4 used how often as a factor?

2. How often is 6 used as a factor in finding the 5th power of 6?

3. How often is 5 used as a factor in the cube of the square of 5?

4. What power of 6 is 6^2 multiplied by 6^3?

5. If we multiply 7^3 by 7^4, what power of 7 shall we have?

6. What power of 4 is equal to 4^3 multiplied by 4^5?

7. What power of 8 is equal to $8^2 \times 8^3 \times 8^4$?

8. What power of 2 is the square of the square of 2^3?

9. The square of a number equals 8 times that number; what is the number?

10. What number multiplied by 6 gives $\frac{1}{2}$ of the square of the number for a product?

11. What number multiplied by 16 gives $\frac{4}{5}$ of the square of the number for a product?

12. What fraction multiplied by $\frac{4}{5}$ equals $\frac{3}{4}$ of the square of the fraction?

13. What number multiplied by 12 and 9 gives $\frac{3}{4}$ of the cube of the number for a product?

WRITTEN EXERCISES.

1. Find the square of 16.

SOLUTION.—To find the square of 16 we multiply 16 by itself and we have 256. To find the cube of 16 we would multiply 256 by 16.

OPERATION.

$$\begin{array}{r} 16 \\ 16 \\ \hline 256 \end{array}$$

2. Square 15.	*Ans.* 225.	**6.** Cube 14.	*Ans.* 2744.
3. Square 32.	*Ans.* 1024.	**7.** Cube 35.	*Ans.* 42875.
4. Square 76.	*Ans.* 5776.	**8.** Cube 67.	*Ans.* 300763.
5. Square 205	*Ans.* 42025.	**9.** Cube 99.	*Ans.* 970299

Find the value of

10. 45^2.	*Ans.* 2025.	18. $(12\frac{1}{2})^3$.	*Ans.* $1953\frac{1}{8}$.
11. 24^3.	*Ans.* 13824.	19. $(3.8)^4$.	*Ans.* 208.5136
12. 38^4.	*Ans.* 2085136.	20. $(1.25)^3$.	*Ans.* 1.953125.
13. $(\frac{12}{13})^2$.	*Ans.* $\frac{144}{169}$.	21. $(15.5)^4$.	*Ans.* 57720.0625.
14. $8^2 \times 8^3$.	*Ans.* 8^5.	22. $4^2 \times 4^3 \times 4^4$.	*Ans.* 4^9.
15. $7^4 \times 7^5$.	*Ans.* 7^9.	23. $(\frac{3}{4})^3 \times (\frac{3}{4})^4$.	*Ans.* $(\frac{3}{4})^7$.
16. $12^3 \times 12^5$.	*Ans.* 12^8.	24. $(2.5)^4 \times (2.5)^6$.	*Ans.* $(2.5)^{10}$
17. $(\frac{1}{2})^2 \times (\frac{1}{2})^3$.	*Ans.* $(\frac{1}{2})^5$.	25. $(3.3)^2 \times (3.3)^3$.	*Ans.* $(3.3)^5$

SQUARING NUMBERS.

615. There are **Two Methods** of squaring numbers, called the *Analytic* or *Algebraic*, and the *Synthetic* or *Geometrical*.

616. The object of these methods is to find the law of forming the square, and thus prepare for corresponding methods of explaining Evolution.

NOTE.—Teachers who prefer the geometrical method of explaining evolution may allow pupils to omit explaining involution by the analytic method, and *vice versa*.

1. Find the square of 25 analytically and synthetically.

OPERATION.

$$\begin{array}{rl} 25 = & 20+5 \\ 25 = & 20+5 \\ \hline 125 = & 5\times 20+5^2 \\ 50 = & 20^2+5\times 20 \\ \hline 625 = & 20^2+2\times(5\times 20)+5^2 \end{array}$$

ANALYTICAL SOL.—Twenty-five equals 20 plus 5, or 2 tens plus 5 units. Writing this as $20+5$, and commencing at units to square, we have 5 times 5 equals 5^2, 5 times 20 equals 5×20, 20 times 5 equal 5×20, 20 times 20 equals 20^2, and adding, we have $20^2+2\times(5\times 20)+5^2$; hence the square of 25 equals the *square of the tens, plus twice the tens into the units, plus the square of the units*, which we find to be 625.

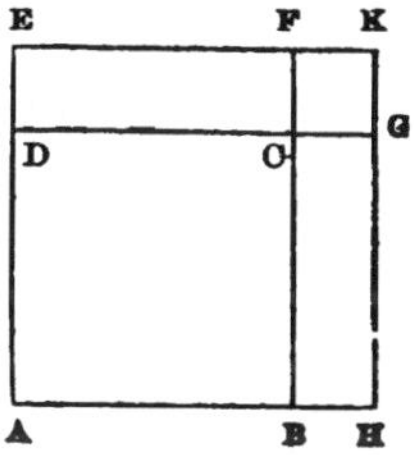

GEOMETRICAL SOL.—Let the line AB represent a length of 20 units, and BH, 5 units. Upon AB construct a square, the area will be $20^2=400$ square units. On the two sides DC and BC construct rectangles, each 20 units long and 5 broad, the area of each will be $5\times 20=100$ and the area of both will be $2\times 100=200$ square units. Now add the little square on CG, whose area is $5^2=25$ square units, and the sum of the different areas, $400+200+25=625$, is the area of a square whose side is 25.

NOTE.—When there are three figures, after completing the second square as above, we must make additions to it, as we did to the first square. When there are four figures, there are three additions, etc.

WRITTEN EXERCISES.

Square the following numbers:

2. 28.	*Ans.* 784.	8. 89.	*Ans.* 7921.
3. 34.	*Ans.* 1156.	9. 97.	*Ans.* 9409.
4. 39.	*Ans.* 1521.	10. 467.	*Ans.* 218089.
5. 46.	*Ans.* 2116.	11. 703.	*Ans.* 494209.
6. 57.	*Ans.* 3249.	12. 2005.	*Ans.* 4020025.
7. 78.	*Ans.* 6084.	13. 4628.	*Ans.* 21418384.

617. The following principles derived from the above solutions are important, and should be committed to memory:

PRINCIPLES.

1. *The square of a number of two figures, equals the* TENS^2 + 2 *times* TENS × UNITS + UNITS^2.

2. *The square of a number of three figures equals* HUNDREDS^2 + 2 *times* HUNDREDS × TENS + TENS^2 + 2(HUNDREDS + TENS) × UNITS + UNITS^2.

618. These principles may also be expressed in symbols. Let u represent units figure, t tens, h hundreds, and T thousands, and a period between two letters denote their multiplication; then we have

$$(t+u)^2 = t^2 + 2t.u + u^2.$$

$$(h+t+u)^2 = h^2 + 2h.t + t^2 + 2(h+t).u + u^2.$$

$$(T+h+t+u)^2 = T^2 + 2T.h + h^2 + 2(T+h).t + t^2 + 2(T+h+t).u + u^2.$$

CUBING NUMBERS.

619. There are **Two Methods** of cubing numbers, called the *Analytic* or *Algebraic*, and the *Synthetic* or *Geometrical* method.

620. The object of these methods is to find the law of forming the cube, and thus to prepare for corresponding methods of explaining Evolution.

1. Find the cube of 25 by the analytical method.

ANALYTICAL SOL.—Squaring 25 by the method already given, we have $20^2 + 2 \times (5 \times 20) + 5^2$. We then multiply this by 20+5. Five times 5^2 equals 5^3, 5 times $2 \times 5 \times 20$ equals $2 \times 5 \times 5 \times 20$, or $2 \times 5^2 \times 20$, five times 20^2 equals 5×20^2 We next multiply by 20. Twenty times 5^2 equals 20×5^2, twenty times $2 \times 5 \times 20$ equals $2 \times 5 \times 20^2$, twenty times 20^2 equals 20^3. Taking the sum of these products and we have first 5^3; next, *once* $5^2 \times 20$ plus *twice* $5^2 \times 20$ equals *three* times $5^2 \times 20$; next *twice* 5×20^2 plus *once* 5×20^2 equals *three* times 5×20^2, and next we have 20^3; hence $25^3 = 20^3 + 3 \times 5 \times 20^2 + 3 \times 5^2 \times 20 + 5^3$. Therefore the cube of 25 equals *the cube of the tens, plus three times the square of the tens into the units, plus three times the tens into the square of the units, plus the cube of the units.*

OPERATION.

$$
\begin{array}{rr}
25^2 = & 20^2 + 2 \times (5 \times 20) + 5^2 \\
25 = & 20 + 5 \\
\hline
 & 5 \times 20^2 + 2 \times 5^2 \times 20 + 5^3 \\
 & 20^3 + 2 \times 5 \times 20^2 + \quad 5^2 \times 20 \\
\hline
25^3 = & 20^3 + 3 \times 5 \times 20^2 + 3 \times 5^2 \times 20 + 5^3
\end{array}
$$

2. Find the cube of 45 by means of the cubical blocks.

Fig. 1.

Fig. 3.

Fig. 2.

Fig. 4.

GEOMETRICAL SOL.—Let A, Fig. 1, represent a cube whose sides are 40 units, its contents will be $40^3 = 64000$. To increase its dimensions by 5 units we must add, 1st, the three rectangular slabs, B, C, D, Fig. 2; 2d, the three corner pieces, E, F, G, Fig. 3; 3d, the little cube H, Fig. 4. The three slabs B, C, D, are 40 units long and wide and 5 units thick; hence their contents are $40^2 \times 5 \times 3 = 24000$; the contents of the corner pieces, E, F, G, Fig. 3, whose length is 40 and breadth and thickness 5, equal $40 \times 5^2 \times 3 = 3000$; and the contents of the little cube H, Fig. 4, equal $5^3 = 125$; hence the contents of the cube represented by Fig. 4 are $64000 + 24000 + 3000 + 125 = 91125$.

OPERATION.

$$
\begin{array}{rr}
40^3 = & 64000 \\
40^2 \times 5 \times 3 = & 24000 \\
40 \times 5^2 \times 3 = & 3000 \\
5^3 = & 125 \\
\hline
\text{Hence } 45^3 = & 91125
\end{array}
$$

NOTE.—When there are three figures in the number, complete the second cube as above, and then make additions and complete the third in the same manner; or let the first cube represent the cube already found, and then proceed as at first.

Cube the following numbers:

3. 12.	*Ans.* 1728.	**8.** 36.	*Ans.* 46656.
4. 16.	*Ans.* 4096.	**9.** 42.	*Ans.* 74088.
5. 18.	*Ans.* 5832.	**10.** 65.	*Ans.* 274625
6. 23.	*Ans.* 12167.	**11.** 84.	*Ans.* 592704
7. 34.	*Ans.* 39304.	**12.** 327.	*Ans.* 34965783.

621. The following principles are important, and should be committed to memory:

PRINCIPLES.

1. *The cube of a number consisting of two figures equals* TENS^3 + 3 *times* TENS^2 × UNITS + 3 *times* TENS × UNITS^2 + UNITS^3.

2. *The cube of a number consisting of three figures equals* HUNDREDS^3 + 3 *times* HUNDREDS^2 × TENS + 3 *times* HUNDREDS × TENS^2 + TENS^3 + 3 *times* (HUNDREDS + TENS)^2 × UNITS + 3 *times* (HUNDREDS + TENS) × UNITS^2 + UNITS^3

622. These principles may also be expressed in symbols as follows:

$$(t+u)^3 = t^3 + 3t^2.u + 3t.u^2 + u^3$$
$$(h+t+u)^3 = h^3 + 3h^2.t + 3h.t^2 + t^3 + 3(h+t)^2.u + 3(h+t).u^2 + u^3.$$

EVOLUTION.

623. Evolution is the process of finding a root of a number.

624. A Root of a number is *one* of its *equal* factors. Roots are of different degrees; as, *second, third,* etc.

625. The **Square Root,** or *second root,* of a number is *one* of its *two equal* factors. Thus, 8 is the square root of 64, since 8 × 8 = 64.

626. The **Cube Root,** or *third root,* of a number is *one* of its *three equal* factors. Thus, 4 is the cube root of 64, since 4 × 4 × 4 = 64.

627. The **Fourth Root**, is *one* of the *four equal* factors the *fifth root* is *one* of the *five equal* factors; etc.

628. The **Symbol of Evolution** is $\surd$; thus, $\sqrt[2]{64}$ or $\sqrt{64}$, denotes the square root of 64; $\sqrt[3]{64}$ denotes the cube root of 64.

629. The **Index** of the root is a small figure placed in the angle of the symbol. The *index* indicates the degree of the root.

Roots are also indicated by the denominator of a fractional exponent; thus $9^{\frac{1}{2}}$ denotes $\sqrt{9}$; $27^{\frac{1}{3}}$ denotes $\sqrt[3]{27}$, etc.

630. The following principles of involution are given to enable us to determine the number of figures in the root:

PRINCIPLES.

1. *The square of a number contains twice as many figures as the number itself, or twice as many, less one.*

DEM.—The square of 1 is 1, and the square of 9 is 81, hence the square of a number consisting of *one* figure is a number consisting of *one* or *two* figures. The square of 10, the smallest number of two figures, is 100, the square of 99, the largest number of two figures, is 9801, hence the square of a number consisting of *two* figures is a number consisting of *three* or *four* figures, that is, *twice two*, or *twice two less one*, etc. The same may be shown for the square of a number consisting of any number of figures.

$1^2 = 1$
$9^2 = 81$
$10^2 = 100$
$99^2 = 9801$

2. *The cube of a number contains three times as many figures as the number itself, or three times as many, less one or two.*

DEM.—The cube of 1 is 1, and the cube of 9 is 729, hence the cube of any number consisting of one figure is a number consisting of *one*, *two*, or *three* figures. The cube of 10 is 1000, a number of four figures, the cube of 99 is 970299, a number of six figures, hence the cube of a number consisting of two figures contains *four*, *five*, or *six* figures, that is, *three times* two, or *three times* two *less one* or *two*. The same may be shown for the cube of a number consisting of any number of figures.

$1^3 = 1$
$9^3 = 729$
$10^3 = 1000$
$99^3 = 970299$

EVOLUTION BY FACTORING.

631. When the number is a perfect power and the factors are easily found, the root of a number can be readily obtained by the following

Rule.—*Resolve the number into its prime factors, and for the square root form a product by taking* ONE *of every* TWO *equal factors; for the cube root* ONE *of every* THREE *equal factors*; etc.

WRITTEN EXERCISES.

1. Find the square root of 144.

SOLUTION.—We first resolve the number into its prime factors. Since the square root of a number is one of its two equal factors, we take *one* of every *two* equal factors and have $2 \times 2 \times 3 = 12$. Hence the square root of 144 is 12.

NOTE.—We have marked the factors taken with a little star, and it will be well for the student to do the same in his solutions.

OPERATION.

```
 2)144
*2)72
 2)36
*2)18
 3)9
  *3
```

Solve the following problems:

2.	$\sqrt{256}$.	*Ans.* 16.	**7.**	$\sqrt[3]{592704}$.	*Ans.* 84.
3.	$\sqrt{44100}$.	*Ans.* 210.	**8.**	$\sqrt[4]{20736}$.	*Ans.* 12.
4.	$\sqrt{32400}$.	*Ans.* 180.	**9.**	$\sqrt[4]{1679616}$.	*Ans.* 36.
5.	$\sqrt[3]{13824}$.	*Ans.* 24.	**10.**	$\sqrt[5]{248832}$.	*Ans.* 12.
6.	$\sqrt[3]{46656}$.	*Ans.* 36.	**11.**	$\sqrt[6]{5489031744}$.	*Ans.* 42.

MENTAL EXERCISES.

1. The square of a number is 64; what is the number?

2. $\frac{3}{4}$ of the square of a number equals 27; what is the number?

3. The square of twice a number equals 64; what is the number?

4. The square of $\frac{2}{5}$ of a number equals 100; what is the number?

5. The square of twice a number is 18 more than twice the square of the number; what is the number?

6. Twice the square of a number is 8 more than 6 times the square of half the number; what is the number?

7. Fifteen is 3 more than $\frac{3}{2}$ of the cube of a number; what is that number?

8. $\frac{3}{4}$ of the cube of a number is 10 more than the cube of $\frac{1}{2}$ of the number; what is the number?

9. The square of a number divided by the number equals 8; what is that number?

10. The square of a number divided by $\frac{1}{2}$ of the number equals 12 what is the number?

11. The cube of a number divided by the number equals 36; what is the number?

12. The 4th power of a number divided by the square of the number equals 49; what is the number?

13. The square of a number divided by $\frac{3}{8}$ of the number equals 27; what is the number?

14. A number divided by 6 gives double the square root of the number; what is the number?

15. The square of a number multiplied by one-half of the number equals 32; what is the number?

16. $\frac{2}{3}$ of $\frac{1}{6}$ of the square of a number, multiplied by $\frac{3}{4}$ of $\frac{2}{3}$ of the number, equals 4; what is the number?

SQUARE ROOT.

632. There are **Two Methods** of explaining the general process of extracting the square root, called the *Analytic* or *Algebraic Method*, and the *Synthetic* or *Geometrical Method.*

633. The **Analytic Method** of square root is so called because it analyzes the number into its elements and derives the process of evolution from the law of involution.

634. The **Geometrical Method** is so called because it makes use of a geometrical figure to explain the process of extracting the root.

NOTE.—With young pupils who have a difficulty in understanding evolution it will be well to drill them upon the method of doing the work, not requiring them to give the explanation until they are better prepared to understand it.

1. Extract the square root of 1225.

ANALYTICAL SOLUTION.—Since the square of a number *contains twice as many figures as the number itself, or twice as many less one*, the square root of 1225 will consist of two places, and hence will consist of tens and units, and 1225 consists of $tens^2 + 2 \times tens \times units + units^2$.

OPERATION.

$$t^2 + 2t.u + u^2 = 1225 \quad (30$$
$$t^2 = 30^2 \qquad 900 \quad 5$$
$$2t.u + u^2 = \qquad 325 \quad 35$$
$$2t = 30 \times 2 = 60$$
$$(2t + u).u = (60 + 5) \times 5 = 325$$

The greatest number of tens whose square is contained in 1225 is 3 tens; squaring the tens and subtracting, we have 325, which equals $2 \times tens \times units + units^2$. Now, since $2 \times tens \times units$ must be greater than $units^2$, 325 must consist principally of twice the tens into the units, hence if we divide by twice the *tens* we can ascertain the units. Twice the *tens* equals $30 \times 2 = 60$; dividing, we find the units to be 5; now finding $2 \times tens \times units + units^2$, or, what is the same, $2 \times tens + units$, both multiplied by *units*, which equals $(60 + 5) \times 5 = 325$, and subtracting, nothing remains. Hence the square root of 1225 is 3 tens and 5 units, or 35.

GEOMETRICAL SOL.—Let Fig. 1 represent a square which contains 1225 square units, then our object is to find the number of linear units in the edge. Since the square of a number consists of *twice as many places as the number itself, or twice as many less one,* the square root of 1225 will consist of two places, and hence will consist of tens and units.

OPERATION.

```
                 1225(30
        30² = 900     5
 30 × 2 = 60     325 35
(60 + 5) × 5 = 325
```

Fig. 1.

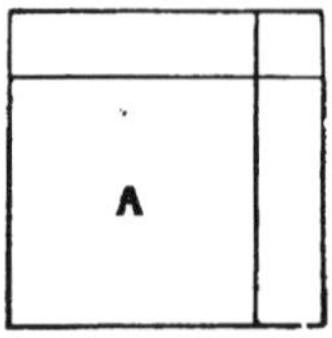

Fig. 2.

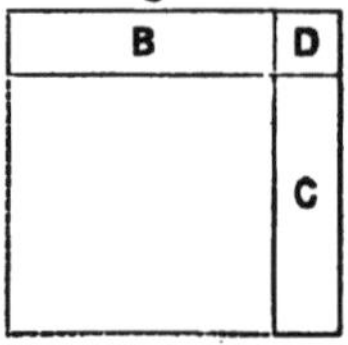

Fig. 3.

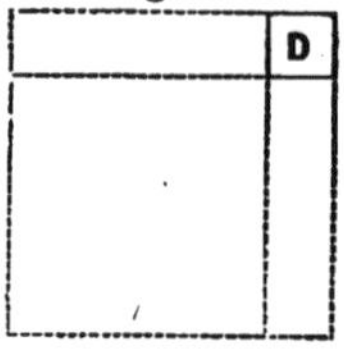

The greatest number of tens whose square is contained in 1225 is 3 tens. Let A, Fig. 1, represent a square whose sides are 30 units, its area will be 30^2, or 900 square units. Subtracting 900 from 1225, we find remaining a surface containing 325 square units. By inspection we find this surface to consist principally of the two rectangles B and C, Fig. 2, each of which is 30 units long, and since they nearly complete the square, their area is nearly 325 units; hence if we divide 325 by their length we will find their width. The length of both is $30 \times 2 = 60$; dividing 325 by 60 we find their width to be 5 units. Adding the length of the little corner square D, Fig. 3, whose sides are 5 units, we find the entire length of the surface remaining after the removal of the square A, is $60 + 5 = 65$ units, and multiplying this by the width, we find the whole area of the remainder to be $65 \times 5 = 325$ square units. Subtracting 325 square units from the square units left after subtracting 900 square units, nothing remains, therefore the side of the square whose area is 1225 square units is 35 units; hence the square root of 1225 is 35.

NOTES.—1. When there are three figures in the root, by the analytic method we use the formula for three terms; by the geometrical method, after removing the first rectangles and small square, we have two rectangles and a small square remaining, which we remove as before.

2. In practice, we determine the number of figures in the root by pointing off the number into periods of two figures each, beginning at the right; we also abbreviate the work by omitting ciphers and condensing the other parts, preserving only the *trial* and *true* divisors. For illustration see solution in the margin.

OPERATION

```
     10·49·76(324
3     9
62   140
644  124
      2576
      2576
```

3. This can also be explained by building up the square instead of separating it into its parts, for which see *Manual.*

Rule.—I. *Begin at units, and separate the number into periods of two figures each.*

II. *Find the greatest number whose square is contained in the left hand period, place it at the right as a quotient, subtract its square from the left hand period, and annex the next period to the remainder for a dividend.*

III. *Double the root found and place it at the left for a* TRIAL DIVISOR; *divide the dividend, excluding the right hand term, by this divisor; the quotient will be the second term of the root.*

IV. *Annex the second term of the root to the trial divisor for the* TRUE DIVISOR, *multiply the result by the second term of the root, subtract the product from the dividend, and bring down the next period for the next dividend.*

V. *Double the root now found for a second* TRIAL DIVISOR, *find the third term of the root as before, and thus proceed until all the periods have been used.*

NOTES.—1. If the product of a true divisor by a term of the root exceeds the dividend, the term must be diminished by a unit.

2. When a cipher occurs in the root, annex a cipher to the trial divisor, bring down the next period, and proceed as before.

3. The square root of a common fraction is evidently the square root of each term. When these terms are not perfect squares, reduce the fraction to a decimal, and extract the root. When a number is not a perfect square, annex periods of ciphers and carry the root on to decimals.

4. By squaring 1, .1, .01, etc., we see that the *square of a decimal contains twice as many decimal places as the decimal,* hence to extract the square root of a decimal, we *point off the decimals into periods of two figures each, counting from the decimal point,* and proceed as in whole numbers.

$1^2 = 1$
$.1^2 = .01$
$.01^2 = .0001$

WRITTEN EXERCISES.

Extract the square root of

1. 256.	*Ans.* 16.	7. 59049.	*Ans.* 243.
2. 625.	*Ans.* 25.	8. 46656.	*Ans.* 216.
3. 729.	*Ans.* 27.	9. 117649.	*Ans.* 343.
4. 1296.	*Ans.* 36.	10. 262144.	*Ans.* 512.
5. 2401.	*Ans.* 49.	11. 390625.	*Ans.* 625.
6. 4096.	*Ans.* 64.	12. 5764801.	*Ans.* 2401.

Find the square root of

13. $\frac{121}{144}$.	*Ans.* $\frac{11}{12}$.	21. .065536.	*Ans.* .256.
14. $\frac{196}{225}$.	*Ans.* $\frac{14}{15}$.	22. 53.1441.	*Ans.* 7.29.
15. $\frac{729}{1024}$.	*Ans.* $\frac{27}{32}$.	23. 167.9616.	*Ans.* 12.96.
16. $\frac{1369}{2209}$.	*Ans.* $\frac{37}{47}$.	24. 4304.6721.	*Ans.* 65.61.
17. .2209.	*Ans.* .47.	25. 5.	*Ans.* 2.236+.
18. .3136.	*Ans.* .56.	26. $\frac{2}{3}$.	*Ans.* .81649+.
19. .0729.	*Ans.* .27.	27. $\frac{16}{111}$.	*Ans.* .37966+
20. .015625.	*Ans.* .125.		

APPLICATIONS OF SQUARE ROOT.

635. The **Applications** of **Square Root** to problems involving geometrical figures are extensive.

636. The **Side** of a square is equal to the square root of its area.

WRITTEN EXERCISES.

1. A man owns a farm in the form of a square which contains 10 acres; how many rods in length or breadth is it?

SOLUTION.—The 10 acres equal 10 × 160, or 1600 sq. rd.; extracting the square root, we have 40 rods.

2. A man has a square lot of land containing 1440 acres; how many rods in length or breadth? *Ans.* 480 rods.

3. A man has a rectangular board 128in. long and 32 in. wide, from which he makes a square table as large as possible; required its length. *Ans.* 64 in.

4. A general trying to mass his army into a solid square of 80 on each side, found he lacked 500 men to complete the square; how many men in his army? *Ans.* 5900.

5. A general attempting to draw his army of 9480 men into a square found he had 71 men over; required the number of men in rank and file. *Ans.* 97.

6. What would it cost to fence a square lot, containing 160 acres, at the rate of $4 per rod? *Ans.* $2560.

7. A general drew up his army of 27175 men in three equal squares, and found he had 168 over in the first, 132 in the second, and lacked 200 in the third; what was the number of men in the side of each square? *Ans.* 95.

RIGHT-ANGLED TRIANGLES.

637. A **Right-Angled Triangle** is a triangle which has one right angle.

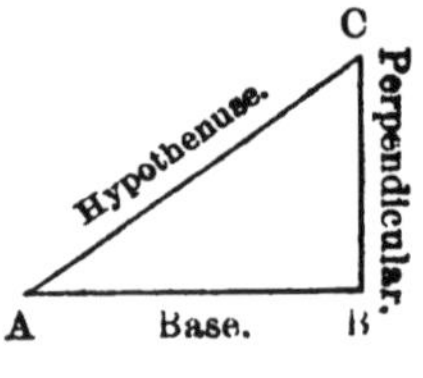

638. The **Base** of a triangle is the side on which it stands; as AB.

639. The **Perpendicular** is the side which forms the right angle with the base; as BC.

640. The **Hypothenuse** is the side opposite the right angle; as AC.

641. The **Principles** of right-angled triangles are as follows:

PRINCIPLES.

1. *The square of the hypothenuse equals the sum of the squares of the other two sides.*

2. *Hence, the square of either side equals the square of the hypothenuse, diminished by the square of the other side.*

NOTE.—The smallest integers which can express the relation of the three sides of a right-angled triangle are 3, 4, and 5. We may have an infinite number of right-angled triangles with their sides in this relation. Other relations are 5, 12, and 13; 8, 15, and 17, etc.

WRITTEN EXERCISES.

1. The two sides of a right-angled triangle are 51 and 68 inches respectively; required the hypothenuse.

SOLUTION.—Hypothenuse $=\sqrt{51^2+68^2}=\sqrt{7225}=85$, *Ans.*

2. The hypothenuse of a right-angled triangle is 115, the base 92; what is the perpendicular? *Ans.* 69.

3. A ladder 65 ft. long is placed against a house so that its foot is 25 ft. from the house; how high does it reach? *Ans.* 60 ft.

4. A rectangular lot of land is 1080 rods long and 810 rods broad; what is the distance between two opposite corners? *Ans.* 1350 rods.

5. Two vessels sail from the same port, one sails north 3 miles an hour, the other west 4 miles an hour; how far are they apart in 2 days? *Ans.* 240 miles.

6. A ladder 82 ft. long stands close against a building; how far must it be drawn out at the bottom that the top may be lowered 2 feet? *Ans.* 18 ft.

7. A pole was broken 52 ft. from the bottom, and fell so that the end struck 39 ft. from the foot; required the length of the pole. *Ans.* 117 ft.

8. A ladder 130 ft. long, with its foot in the street, will reach on one side to a window 78 ft. high, and on the other to a window 50 ft. high; what is the width of the street? Ans. 224 ft.

SIMILAR FIGURES.

642. Similar Figures are those which have the same form. Thus, circles are similar figures; also squares, etc.

643. The **Principles** of similar figures derived from geometry are as follows:

PRINCIPLES.

1. *The areas of all similar figures are to each other as the squares of their like dimensions.*

2. *Hence, the like dimensions of similar figures are to each other as the square roots of their areas.*

WRITTEN EXERCISES.

1. The area of a rectangle is 270 and one side is 18; what is the area of a similar rectangle, the longer side being 24?

SOLUTION.—Since the rectangles are similar, their areas are as the squares of their corresponding sides; hence we have the proportion in the margin. Cancelling and multiplying, we have 480.

OPERATION.

Area of 2d : 270 : : 24^2 : 18^2

$$\text{Area of 2d} = \frac{270 \times 24^2}{18^2} = 480, \textit{ Ans.}$$

2. There are two circular gardens, one 5 rods in diameter and the other 30 rods; the second is how many times as large as the first? *Ans.* 36 times.

3. I have a lot 20 rd. long and 16 rd. broad; what are the dimensions of a similar lot 4 times as large? *Ans.* 40; 32.

4. The area of a circle, whose diameter is 20 feet, is 314.16 square feet; what is the diameter of a circle whose area is 78.54 square feet? *Ans.* 10 ft.

5. A farmer has a rectangular field 80 rods long and 60 wide; what are the dimensions of another similar field containing $13\frac{1}{3}$ acres? *Ans.* $53\frac{1}{3}$ rd.; 40 rd.

6. If a horse tied to a post in the centre of a field by a rope 1 ch. 78 li., can graze upon an acre, what length of rope would allow it to graze upon $5\frac{4}{9}$ acres? *Ans.* 4 ch. $15\frac{1}{3}$ li.

7. If a pipe $1\frac{1}{2}$ in. in diameter pour in a cistern 45 gal. in a given time, how much will a pipe 2 in. in diameter pour in, in the same time? *Ans.* 80 gal.

8. If a pipe whose diameter is 1.5 in. fill a cistern in 5 hours, in what time will a pipe whose diameter is 3 in. fill the same cistern? *Ans.* $1\frac{1}{4}$ hours.

SUG.—It pours in 4 times as much, and fills it in $\frac{1}{4}$ of 5 hours.

9. Required the distance between a lower corner and the opposite upper corner of a room 48 feet long, 20 feet wide, and 39 feet high. *Ans.* 65 feet.

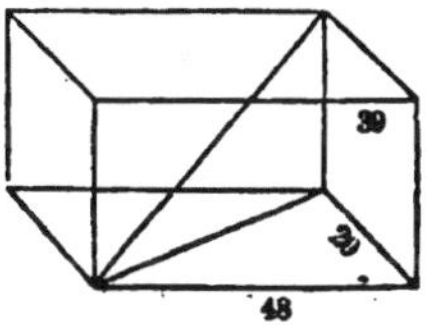

CUBE ROOT.

644. There are **Two Methods** of explaining the general process of extracting the Cube Root, called the *Analytic* or *Algebraic Method*, and the *Synthetic* or *Geometrical Method.*

645. The **Analytic Method** of cube root is so called because it analyzes the number into its elements, and derives the process from the law of involution.

646. The **Geometrical Method** of cube root is so called because it makes use of a cube to explain the process.

1. Extract the cube root of 91125.

ANALYTIC SOLUTION.—Since the cube of a number consists of *three times as many places as the number* itself, or of *three times as many less one or two*, the cube root of 91125 will consist of two places, or of tens and units, and the number itself will consist of *tens*^{3} + 3 × *tens*^{2} × *units* + 3 × *tens* × *units*^{2} + *units*^{3}.

The greatest number of tens whose cube is contained in 91125 is 4 tens. Cubing the tens and subtracting, we have 27125, which equals 3 × *tens*^{2} × *units* + 3 × *tens* × *units*^{2} + *units*^{3}. Now, since 3 × *tens*^{2} × *units* is much greater than 3 × *tens* × *units*^{2} + *units*^{3}, 27125 consists principally of 3 times *tens*^{2} × *units*; hence, if we divide by 3 times *tens*^{2}, we can ascertain the *units*.

OPERATION.

$$\begin{array}{lrl} & 91125 & (40 \\ 40^3 = & 64000 & 5 \\ & 27125 & 45 \\ \textit{trial divisor, } 3 \times 40^2 = 4800 & & \\ 3 \times 40 \times 5 = 600 & & \\ 5^2 = 25 & & \\ \textit{true divisor, } 5425 & 27125 & \end{array}$$

SHOWN BY LETTERS.

$$t^3 + 3t^2 \times u + 3t \times u^2 + u^3 = 91125(45$$
$$t^3 = 40^3 \qquad = 64000$$
$$3t^2 \times u + 3t \times u^2 + u^3 = 27125$$
$$3t^2 \times 3 \times 40^2 \qquad = 4800$$
$$3t \times u = 3 \times 40 \times 5 = 600$$
$$u^2 = 5^2 \qquad = 25$$
$$(3t^2 + 3t \times u + u^2) \times u = 5425 \times 5 = 27125$$

3 times *tens*² equals $3 \times 40^2 = 4800$; dividing by 4800, we find the *units* to be 5. We then find 3 times *tens* × *units* equal to $3 \times 40 \times 5 = 600$, and *units*² $= 5^2 = 25$, and adding these and multiplying by *units* we have (3 *tens*² + 3 *tens* × *units* + *units*²) × *units*, which equals $5425 \times 5 = 27125$; subtracting, nothing remains, hence the cube root of 91125 is 45

Fig. 1.

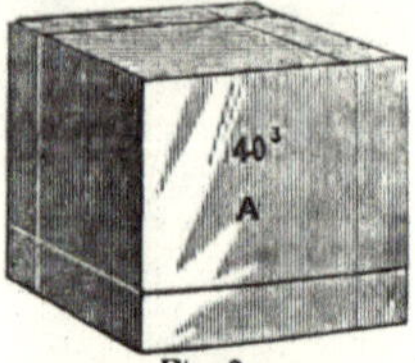

Fig. 3.

Fig. 2.

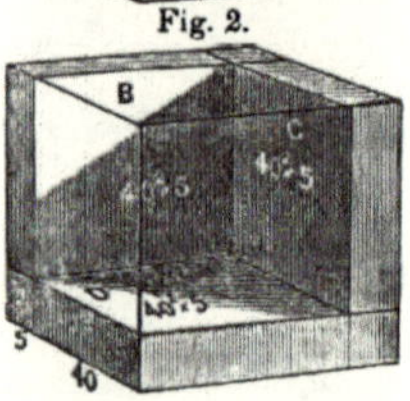

Fig. 4.

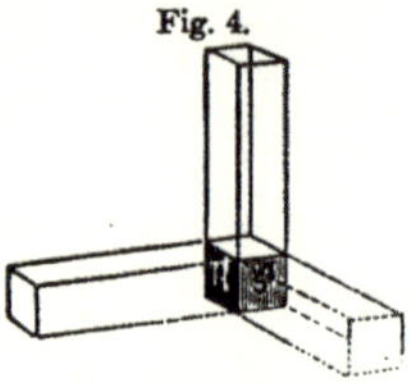

GEOMETRICAL SOLUTION.—Let Fig. 1 represent the cube which contains 91125 cubic units, then our object is to find the number of linear units in its edge. The number of terms in the root, found as before, is two. The greatest number of tens whose cube is contained in the given number is 4 tens. Let A, Fig. 1, represent a cube whose sides are 40, its contents will be $40^3 = 64000$. Subtracting 64000 from 91125 we find a remainder of 27125 cubic units, which by removing the cube A from Fig. 1, leaves a solid represented by Fig. 2.

OPERATION.

$$
\begin{array}{rr|l}
 & 91125(& 40 \\
40^3 = & 64000 & 5 \\
 & 27125 & 45 \\
3 \times 40^2 = 4800 & & \\
3 \times 40 \times 5 = 600 & & \\
5^2 = 25 & & \\
5425 & 27125 &
\end{array}
$$

Inspecting this solid, we perceive that the greater part of it consists of the three rectangular slabs, B, C, and D, each of which is 40 units in length and breadth, hence if we divide 27125 by the sum of the areas of one face of each regarded as a base, we can ascertain their thickness. The area of a face of one slab is $40^2 = 1600$, and of the three, 3×1600, $= 4800$, and dividing 27125 by 4800 we have a quotient of 5, hence the thickness of the slab is 5 units.

Removing the rectangular slabs, there remain three other rectangular solids, E, F, G, as shown in Fig. 3, each of which is 40 units long and 5 units thick, hence the surface of a face of each is $40 \times 5 = 200$ square units, and of the three it is $3 \times 40 \times 5 = 600$ square units.

Finally removing E, F, and G, there remains only the little corner cube H, Fig. 4, whose sides are 5 units, and the surface of one of its faces $5^2=25$ square units. We now take the sum of the surfaces of the solids remaining after the removal of the cube A, and multiply this by the common thickness, which is 5, and we have their solid contents equal to $(4800+600+25)\times5=27125$ cubic units, which, subtracted from the number of cubic units remaining after the removal of A, leaves no remainder. Hence the cube which contains 91125 cubic units is $40+5$, or 45 units on a side.

NOTE.—This can also be explained by building up the cube instead of separating it into its parts, for which see *Manual*.

647. We will now solve a problem with three figures in the root, indicating the solution by means of letters, and abbreviating the operation as in practice. A point like a period indicates the multiplication of the letters.

SHOWN BY LETTERS.

	htu	
	14348907(243	
$h^3=200^3=$	8000000	
		6348907
$3h^2=3\times200^2=$	120000	
$3h.t=3\times200\times40=$	24000	
$t^2=40^2=$	1600	
	145600	5824000
		524907
$3(h+t)^2=3\times240^2=$	172800	
$3(h+t).u=3\times240\times3=$	2160	
$u^2=3^2=$	9	
	174969	524907

OPERATION AS IN PRACTICE.

	14·348·907(248	
$2^3=$	8	
		6348
$2^2\times300=$	1200	
$2\times4\times30=$	240	
$4^2=$	16	
	1456	5824
		524907
$24^2\times300=$	172800	
$24\times3\times30=$	2160	
$3^2=$	9	
	174969	524907

NOTES.—1. By the geometric method, when there are more than two figures we remove the first cube, rectangular slabs and solids, and small cube, and we have remaining three slabs, three solids, and a small cube, as before.

2. The method employed in actual practice is derived from the other by omitting ciphers, using parts of the number instead of the whole number each time we obtain a figure of the root, etc. It will also be seen that by separating the number into *periods of 3 figures each*, we have the *number of places in the root*, the *part of the number* used in obtaining each figure of the root, etc.

Rule.—I. *Begin at units and separate the number into periods of three figures each.*

II. *Find the greatest number whose cube is contained in the left hand period, write it for the first term of the root, subtract its cube from the left hand period, and annex the next period to this remainder for a dividend.*

III. *Multiply the square of the first term of the root by* 300 *for a* TRIAL DIVISOR; *divide the dividend by it, and the result will be the second term of the root.*

IV. *To the trial* divisor add 30 times the *product of the second term of the root by the first term, and also the square of the second term; their sum will be the* TRUE DIVISOR.

V. *Multiply the true divisor by the second term of the root; subtract the product from the dividend, and annex the next period for another dividend. Square the root now found, multiply by 300, and find the third figure as before, and thus continue until all the periods have been used.*

NOTES.—1. If the product of the true divisor by the term of the root exceeds the dividend, the root must be diminished by a unit.

2. When a dividend will not contain a trial divisor, place a cipher in the root and two ciphers at the right of the trial divisor, bring down the next period, and proceed as before.

3. To find the cube root of a common fraction, extract the cube root of both terms. When these are not perfect cubes, reduce to a decimal and then extract the root.

4. By cubing 1, .1, .01, etc., we see that the *cube of a decimal contains three times as many decimal places as the decimal;* hence, to extract the cube root of a decimal, we *point off the decimal in periods of three figures each, counting from the decimal point.*

$1^3 = 1$
$.1^3 = .001$
$.01^3 = .000001$

WRITTEN EXERCISES.

Find the cube root of

1.	15625.	*Ans.* 25.	**12.**	2571353.	*Ans.* 137.
2.	19683.	*Ans.* 27.	**13.**	1124864.	*Ans.* 104.
3.	42875.	*Ans.* 35.	**14.**	41063625.	*Ans.* 345.
4.	54872.	*Ans.* 38.	**15.**	130323843.	*Ans* 507.
5.	74088.	*Ans.* 42.	**16.**	$\frac{1728}{4096}$.	*Ans.* $\frac{12}{16}$.
6.	175616.	*Ans.* 56.	**17.**	389.017.	*Ans.* 7.3.
7.	300763.	*Ans.* 67.	**18.**	259.694072.	*Ans.* 6.38.
8.	405224.	*Ans.* 74.	**19.**	4.	*Ans.* 1.5873+.
9.	571787.	*Ans.* 83.	**20.**	5.	*Ans.* 1.7099+.
10.	857375.	*Ans.* 95.	**21.**	6.	*Ans.* 1.8171+.
11.	1860867.	*Ans.* 123.	**22.**	$\frac{2}{3}$.	*Ans.* .873+.

NEW METHOD OF CUBE ROOT.

648. The **New Method** of extracting cube root is shorter and more convenient than the ordinary method. The abbreviation consists in obtaining the true and trial divisors by a law which enables us to use our previous work.

NOTE.—This method seems to have been approximated by several writers, although I have not found any who present it in the form in which it is here given.

1. Extract the cube root of 14706125.

Solution.—We find the number of figures in the root, and the first term as before. We write 2, the first term of the root, at the left, at the head of Col. 1st; 3 times its square, with two dots annexed, at the head of Col. 2d; its cube under the first period; then subtract and annex the next period for a dividend and divide it by the number in Col. 2d, as a *trial divisor*, for the second term of the root.

OPERATION.

1st col.	2d col.	
2	12 . . t. d.	14·706·125(245
4	256	8
64	1456 c. d.	6706
8	16	5824
725	1728 . . t. d.	882125
	3625	
	176425 c. d.	882125

We then take 2 times 2, the first term, and write the product 4 in Col. 1st, under the 2, and add; then annex the second term of the root to the 6 in Col. 1st, making 64, and multiply 64 by 4 for a correction, which we write under the trial divisor; and adding the *correction* to the *trial divisor*, we have the *complete divisor*, 1456. We then multiply the complete divisor by 4, subtract the product from the dividend, and annex the next period for a new dividend.

We then square 4, the second figure of the root, write the *square* under the *complete divisor*, and add the *correction*, the *complete divisor* and the *square* for the next *trial divisor*, which we find to be 1728. Dividing by the trial divisor, we find the next term of the root to be 5.

We then take 2 times 4, the second term, write the product 8 under the 64, add it to 64, and annex the third term of the root to the sum, 72, making 725, and then multiply 725 by 5, giving us 3625 for the next *correction*. We then find the *complete divisor*, by adding the *correction* to the *trial divisor*; multiply the complete divisor by 5, and subtract, and we have no remainder.

Note.—The correctness of this method may readily be seen by using letters, and following the changes indicated in the solution.

Rule.—I. *Separate the number into periods of three figures each; find the greatest number whose cube is contained in the first period, and write it in the root.*

II. *Write the first term of the root at the head of the* 1st *Col.;* 3 *times its square, with two dots annexed, at the head of* 2d *Col., and its cube under the first period; subtract, and annex the next period to the remainder for a dividend; divide by the number in* 2d *Column as a* TRIAL DIVISOR, *and place the quotient as the second term of the root.*

III. *Add twice the first term of the root to the number in the first column; annex the second term of the root, multiply the result by the second term, and write the product under the trial divisor for a* CORRECTION; *add the* CORREC-

TION *to the* TRIAL DIVISOR, *and the result will be the* COMPLETE DIVISOR; *multiply the* COMPLETE DIVISOR *by the last term of the root, subtract the product from the dividend, and annex the next period to the result for a new dividend.*

IV. *Square the last term of the root, and take the sum of this* SQUARE, *the last* COMPLETE DIVISOR, *and the last* CORRECTION, *annexing two dots for a new* TRIAL DIVISOR; *divide the dividend by this for the next term of the root.*

V. *Add twice the second term of the root to the last number in the first column; annex the last term of the root to the sum, multiply the result by the last term, and write the product under the last trial divisor for a* CORRECTION; *add the* CORRECTION *to the* TRIAL DIVISOR, *and the result will be the* COMPLETE DIVISOR; *use this as before, and thus continue until all the periods have been used.*

NOTE.—This rule is indicated in the following formula:

1. TRUE DIVISOR = TRIAL DIVISOR + PRODUCT.
2. TRIAL DIVISOR = PRODUCT + TRUE DIVISOR + SQUARE.

Extract the cube root of

2. 12326391.	*Ans.* 231.	5. 277167808.	*Ans.* 652.
3. 34965783.	*Ans.* 327.	6. 633839.779.	*Ans.* 85.9
4. 41063625.	*Ans.* 345.	7. 4.080659192.	*Ans.* 1.598.

NOTE.—For other methods of extracting the cube root see *Brooks's Higher Arithmetic.*

APPLICATIONS OF CUBE ROOT.

649. The **Applications** of cube root to problems involving geometrical volumes, such as cubes, parallelopipedons, spheres, etc., are extensive.

650. The **Edge** of a cube is equal to the cube root of its contents.

WRITTEN EXERCISES.

1. What are the dimensions of a cubical chest which shall contain 64000 cubic feet? *Ans.* 40 ft.

2. Required the number of square feet in one face of a cubical block whose contents are 405224 cu. ft. *Ans.* 5476.

3. What is the entire surface of a cube whose cubical contents are 91125 cubic feet? *Ans.* 12150 sq. ft.

4. What is the edge of a cube which shall contain as much as a solid 20 ft. 6 in. long, 10 ft. 8 in. wide, and 6 ft. 9 in. high? *Ans.* 11.4 ft.—.

5. What is the depth of a cubical cistern which shall contain 200 gal. (231 cu. in.) of water? *Ans.* 35.9 in.—

6. A farmer had a cubical bin which contained 50 bushels of grain; what was its depth? *Ans.* 3.962 ft.+.

7. What would it cost to plaster the bottom and sides of a cubical reservoir which contains 100 barrels of water, at 6 cents a square foot? *Ans.* $16.85.

SIMILAR VOLUMES.

651. Similar Volumes are such as have the same shape, but differ in size; as, cubes, spheres, etc.

652. A **Dimension** of a volume is a length, breadth, height, diameter, radius, circumference, etc.

653. The **Principles** of similar volumes are derived from geometry.

PRINCIPLES.

1. *Similar volumes are to each other as the cubes of their like dimensions.*

2. *Like dimensions of similar volumes are to each other as the cube roots of those volumes.*

WRITTEN EXERCISES.

1. A man has two balls, one 6 in. in diameter, the other 2 in.; the first is how many times as large as the second?

OPERATION.

1st : 2d : : $6^3 : 2^3$
1st $= 2d \times (\frac{6}{2})^3 = 2d \times 3^3$
1st $= 2d \times 27$

SOLUTION.—By the principle above we have the proportion 1st : 2d : : $6^3 : 2^3$; and since the first term equals the 2d term multiplied by the ratio of the 3d to the 4th, we have 1st term = 2d term multiplied by the ratio of 6^3 to 2^3, which is 2d term $\times (\frac{6}{2})^3$, or 2d $\times 3^3$, or 2d$\times$27. Hence the first is 27 times as large as the second.

2. Required the relation of two cubes whose dimensions are 3 in. and 15 in. respectively. *Ans.* 2d is 125 times 1st.

3. If a ball 3 in. in diameter weigh 9 pounds, how much will a ball 4 in. in diameter weigh? *Ans.* 21⅓ lb.

4. If a cubical box 4 ft. long, hold 51.43 bu. of grain, how much will a cubical box 6 ft. long hold? *Ans.* 173.58 bu.

5. If a haystack 12 feet in diameter contain 15 tons, what is the diameter of a similar stack of 120 tons? *Ans.* 24 ft.

6. If a man 5 ft. high weigh 150 lb., what is the weight of a man of similar build whose height is 6 ft? *Ans.* 259⅕ lb.

7. The sun is 885680 miles in diameter, and the earth 7912 miles; the sun is how many times as large as the earth? *Ans.* About 112^3.

8. There are two balls whose diameters are respectively 3 in. and 4 in.; what is the diameter of a ball whose contents are equal to them both? *Ans.* 4.5 in. nearly.

SUG.—Cube 3 and 4, take their sum, and then compare this with either of the given balls.

HIGHER ROOTS.

654. Any root whose index contains only the factors 2 or 3 can be extracted by means of the square and cube root according to the following principle:

PRINCIPLE.

A root of a number equals a root of a root of the number, in which the product of the indices of the two latter roots equals the index of the former.

Since the square of the cube of a number equals the sixth power, the sixth root of a number equals the square root of the cube root of the number, and the same is true in any other case.

WRITTEN EXERCISES.

1. Extract the sixth root of 4096.

SOLUTION.—To find the sixth root of 4096 we first extract the square root, which we find to be 64, and then find the cube root of 64, which is 4. Hence the sixth root of 4096 is 4.

Required the value of

2. $\sqrt[4]{625}$.	*Ans.* 5.	5. $\sqrt[8]{390625}$.	*Ans.* 5
3. $\sqrt[6]{729}$.	*Ans.* 3.	6. $\sqrt[9]{262144}$.	*Ans.* 4.
4. $\sqrt[4]{20736}$.	*Ans.* 12.	7. $\sqrt[12]{16777216}$.	*Ans.* 4.

NOTE.—For a general rule of Evolution see *Brooks's Higher Arithmetic.*

SECTION XI.

ARITHMETICAL AND GEOMETRICAL SERIES.

655. A **Series** is a succession of numbers, each derived from the preceding by some fixed law.

656. The **Terms** of a series are the numbers which compose it. The *Extremes* are the first and last terms; the *Means* are the terms between the extremes.

657. An **Ascending Series** is one in which the terms increase from left to right; a *Descending Series* is one in which the terms decrease from left to right.

NOTE.—There are many different kinds of series; the only two suitable for arithmetic are Arithmetical and Geometrical Series. These series are usually called *Progressions*.

ARITHMETICAL PROGRESSION.

658. An **Arithmetical Progression** is a series of numbers which vary by a common difference; as 3, 5, 7, 9, etc.

659. The **Common Difference** is the difference between any two consecutive terms; thus, in the above series the common difference is 2.

660. The **Quantities** considered are five, any three of which being given, the others may be found.

QUANTITIES CONSIDERED.

1. The first term.
2. The last term.
3. The common difference.
4. The number of terms.
5. The sum of all the terms.

CASE I.

661. ***Given, the first term, the common difference, and the number of terms, to find the last term.***

1. The first term is 3, the common difference 2, and number of terms 10; required the last term.

Solution.—The first term is 3, the second term equals 3 plus *once* the common difference, the third term equals 3 plus *twice* the common difference, etc.; hence the tenth term equals the first term plus *nine* times the common difference, which equals $3+2\times9=21$.

OPERATION.

$$\begin{aligned} 2d &= 3+2 = 5 \\ 3d &= 3+2\times2=7 \\ 4th &= 3+2\times3=9 \\ \text{hence } 10th &= 3+2\times9=21. \end{aligned}$$

Rule.—*The last term equals the first term increased by the common difference multiplied by the number of terms less one.*

Note.—In a descending series we must subtract instead of adding.

2. Given the first term 4, the common difference 3, to find the 12th term. *Ans.* 37.

3. The first term is 3, the common difference 4; what is the 22d term? *Ans.* 87.

4. Required the 76th term of a descending series, the 1st term being 80 and common difference $\frac{2}{3}$. *Ans.* 30.

5. A man bought 50 yards of muslin at $\frac{1}{2}$ cent for the first yard, 1 cent for the second, $1\frac{1}{2}$ for the third, and so on; what did the last yard cost? *Ans.* 25 cents.

6. The amount of $100 at 5 per cent. for 1, 2, 3, etc., years, is respectively $105, $110, $115, etc.; what is the amount for 25 years? *Ans.* $225.

CASE II.

662. ***Given, the last term, the common difference, and the number of terms, to find the first term.***

1. Required the first term, the last term being 41, the number of terms 20, and the common difference 2.

Solution.—From the rule in Case I., we have $41 = 1\text{st term} + 19$ times 2, hence we find first term $= 41 - 19\times2 = 3$.

OPERATION.

$$\begin{aligned} 41 &= 1st + 2\times19 \\ 1st &= 41 - 2\times19 = 3. \end{aligned}$$

Rule.—*The first term equals the last term, diminished by the common difference multiplied by the number of terms less one.*

2. Required the first term, the last term being 95, common difference 5, and number of terms 18. *Ans.* 10.

3. A woman bought 25 yards of muslin at the rate of 25 cents for the last yard, $24\frac{1}{2}$ for the next to the last, and so on, what did the first yard cost? *Ans.* 13 cents.

4. A man traveled for 10 days, traveling $2\frac{1}{2}$ miles further each day, and on the last day he went $32\frac{1}{2}$ miles; how far did he travel the first day? *Ans.* 10 miles.

CASE III.

663. ***Given, the first term, the last term, and the number of terms, to find the common difference.***

1. What is the common difference, if the first term is 7, the last 799, and the number of terms 100?

SOLUTION.—By Case I. we have 799 = 7 + (99 times the *common difference*); hence the *common difference* equals (799 − 7) ÷ 99, which equals 8.

OPERATION.

$$799 = 7 + 99 \times \text{diff.}$$

$$\text{diff.} = \frac{799 - 7}{99} = 8$$

Rule.—*The common difference equals the difference of the extremes divided by the number of terms less one.*

2. What is the common difference, if the first term is 4, the last 76, and the number of terms 25? *Ans.* 3.

3. The amount of $100 at 5% for 25 years is $225; what is the annual interest? *Ans.* $5.

4. The youngest of 11 children is 26 and the oldest 46 years old, their ages being in arithmetical progression; what is the common difference of their ages? *Ans.* 2 years.

CASE IV.

664. ***Given, the first term, the last term, and the common difference, to find the number of terms.***

1. What is the number of terms, if the first term is 100, the last term 8, and the common difference 4?

SOLUTION.—By Case II. we have 100 = 8 + (No. of terms − 1) × 4; from which we have (No. of terms − 1) × 4 = 100 − 8; and No. of terms − 1 = (100 − 8) ÷ 4, or No. of terms = (100 − 8) ÷ 4 + 1 = 24.

OPERATION.

$$100 = 8 + (n - 1) \times 4$$

$$(n - 1) \times 4 = 100 - 8$$

$$n = \frac{100 - 8}{4} + 1 = 24$$

Rule.—*The number of terms equals the difference between the extremes divided by the common difference, plus one.*

2. What is the number of terms, if the first term is 5, the last 397, and the common difference 7? *Ans.* 57.

3. In what time will $200 at 6 per cent. simple interest amount to $344? Ans. 12 years

4. A laborer received 75 cents the first day, 85 cents the second, and so on till he received $3.75 a day; how many days did he work? *Ans.* 31 days.

CASE V.

665. ***Given, the first term, the last term, and the number of terms, to find the sum of the series.***

1. Given the first term 2, the last term 18, and the number of terms 5, to find the sum of the terms.

OPERATION.

$$\text{Sum} = 2 + 6 + 10 + 14 + 18$$
$$\text{Sum} = 18 + 14 + 10 + 6 + 2$$
$$2 \times \text{Sum} = 20 + 20 + 20 + 20 + 20$$
$$2 \times \text{Sum} = 20 \times 5 = (2 + 18) \times 5$$
$$\text{Sum} = \frac{2+18}{2} \times 5 = 50$$

SOLUTION.—To derive the rule, we find by Case III. the common difference to be 4 Writing the series in its natural, and then in an inverted order, we take the sum of the two series, and we have *twice the sum*, equal to 20 taken 5 times, that is, $(2 + 18) \times 5$; hence, the sum equals $\frac{1}{2}$ of $(2 + 18) \times 5$, which equals 50. Now $(2 + 18)$ is the *sum of the extremes*, and 5 is the *number of terms*; hence we have the following

Rule.—*The sum of an arithmetical series equals half the sum of the extremes multiplied by the number of terms.*

2. The first term equals 3, the last term 65, and number of terms 20; required the sum of the terms. *Ans.* 680.

3. How far can I walk in 8 days, going 25 miles the first day, and increasing the rate 4 miles a day? *Ans.* 312 mi.

4. A merchant paid 2 cents for the first yard of cloth, 5 for the second, 8 for the third, etc.; how much did he pay for 75 yards? *Ans.* $84.75.

5. How many strokes, beginning at 1 o'clock, does the hammer of a clock strike in 6 hours? How many in 12 hours? *Ans.* 21; 78.

6. 100 apples are placed in a row 2 yards apart, the first being 2 yards from a basket; how far will a boy travel, starting from the basket, to gather them singly into the basket? *Ans.* 11 mi. 152 rd. 4 yd.

7. A body will fall $16\frac{1}{12}$ ft. in 1 second, 3 times as far the next, 5 times as far the third, etc.; how far will it fall in a minute? *Ans.* 10 mi. $309\frac{1}{11}$ rd.

GEOMETRICAL PROGRESSION.

666. A **Geometrical Progression** is a series of numbers which vary by a common multiplier; as, 2, 6, 18, 54, etc.

667. The **Rate** or *Ratio* is the common multiplier; thus, in the above series, the rate is 3.

668. In an **Ascending** series, the rate is greater than a unit; in a **Descending** series, the rate is less than a unit.

669. The **Quantities** considered are five, any three of which being given, the others may be found.

QUANTITIES CONSIDERED.

1. The first term,
2. The last term,
3. The number of terms,
4. The rate,
5. The sum of the terms.

CASE I.

670. ***Given, the first term, the rate, and the number of terms, to find the last term.***

1. The first term equals 2, the rate 3, and the number of terms 8; required the last term.

OPERATION.

$2d = 2 \times 3$
$3d = 2 \times 3^2$
$4th = 2 \times 3^3$
hence $8th = 2 \times 3^7 = 4374$

SOLUTION.—The 2d term equals 2×3; the 3d term equals 2×3 multiplied by 3; or 2×3^2, which is the 1st term into the second power of the rate; the 4th term equals 2×3^2 multiplied by 3, or 2×3^3, which is the 1st term into the 3d power of the rate; hence the 8th term equals the first term into the 7th power of the rate, or 2×3^7, which equals 4374. Hence the

Rule.—*The last term equals the first term multiplied by the rate raised to a power one less than the number of terms.*

NOTE.—The table of compound interest, Art. 511, can be derived from this case, 1 plus the rate per cent. being the *rate*.

2. The first term of a geometrical series is 3 and the rate 2; what is the 10th term? *Ans.* 1536.

3. The first term of a descending series is 64, and the rate is $\frac{1}{2}$; what is the 14th term? *Ans.* $\frac{1}{128}$.

4. The first term of a series is 2187, the rate is $\frac{1}{3}$; required the 12th term. *Ans.* $\frac{1}{81}$.

5. The first term of a geometrical series is 1 and the rate 2; what is the 30th term? *Ans.* 536870912.

6. The first term of a geometrical series is 1 and the rate 3; what is the 21st term? *Ans.* 3486784401.

7. A merchant doubles his capital every 4 yr.; if he begins with $5000, how much has he at the end of 16 yr.?

Ans. $80000.

8. A man bought 20 horses, agreeing to pay for them all as much as the last horse would cost, at the rate of 2 cents for the first, 4 cents for the second, 8 cents for the third, etc.; what did they cost? *Ans.* $10485.76.

CASE II.

671. ***Given, the first term, the rate, and the last term or number of terms, to find the sum of the terms.***

1. The first term is 2, the rate is 3, and number of terms 5; required the sum of the terms.

OPERATION.

$$\text{Sum} = 2 + 6 + 18 + 54 + 162$$
$$\text{Sum} \times 3 = \quad 6 + 18 + 54 + 162 + 486$$
$$\text{twice the sum} = \quad 486 - 2$$
$$\text{Sum} = \frac{486 - 2}{2} = 242, \textit{ Ans.}$$

SOLUTION.—Writing the series expressing the sum, and then multiplying by the rate and taking the difference of the two series, we have *twice* the *sum* equals 486 — 2, and the sum equals ½ of 486 — 2, which is 242. In this solution we observe that 486 is the last term, multiplied by the rate, and that this is diminished by the first term, and the difference divided by the rate minus one; hence we have the following

Rule.—*To find the sum, multiply the last term by the rate, subtract the first term, and divide the remainder by the rate diminished by unity.*

NOTE.—In a decreasing series we subtract the last term multiplied by the rate from the first term, and divide by 1 minus the rate.

2. The first term is 4, the rate 5, and the last term 500; required the sum of the terms. *Ans.* 624.

3. The first term of a decreasing series is 64, the rate ½, the last term 2; what is the sum? *Ans.* 126.

4. The first term of a decreasing series is 243, the rate ⅓; what is the sum of 5 terms? *Ans.* 363

5. A man bought 12 yards of cloth, giving 1 cent for the

first yard, 3 cents for the second, 9 cents for the third, etc.; how much did it cost? *Ans.* \$2657.20.

6. A mother gave her daughter 1 cent at birth, doubling it on each succeeding birthday; how much was the daughter worth when she became 21 yr. of age? *Ans.* \$20971.51.

7. A lady thinking \$1½ a yard too much for a silk dress containing 15 yards, agreed to pay 1 cent for the first yard, 3 cents for the second, etc.; which price was the greater and how much? *Ans.* 2d, \$71722.03 more.

8. A man wishing to buy a horse, refused to give \$250, but agreed to pay 1 ¢ for the 1st nail in his shoes, 2 ¢ for the 2d, 4 ¢ for the third, etc.; what did the horse cost, there being 32 nails in his shoes? *Ans.* \$42949672.95

INFINITE SERIES.

672. An **Infinite Series** is a series in which the number of terms is infinite.

673. In a descending series of an infinite number of terms, the last term becomes so small that it may be considered zero; hence the above rule becomes

Rule.—*The sum of an infinite series equals the first term, divided by a unit minus the rate.*

1. What is the sum of the infinite series $1, \frac{1}{2}, \frac{1}{4}, \frac{1}{8}$, etc.

SOLUTION.—In this series the first term is 1, and the rate ½, and the last term is regarded as zero; hence we have the sum of the series equal to 1 divided by 1 — ½ or 1 ÷ ½, which is 2.

OPERATION.

$$\text{Sum} = \frac{1}{1-\frac{1}{2}} = \frac{1}{\frac{1}{2}} = 2, \textit{ Ans.}$$

2. Sum of the infinite series $\frac{1}{2}, \frac{1}{4}, \frac{1}{8}, \frac{1}{16}$, etc.? *Ans.* 1.

3. Sum of the infinite series $\frac{1}{3}, \frac{1}{9}, \frac{1}{27}, \frac{1}{81}$, etc.? *Ans.* ½.

4. Sum of the infinite series $\frac{1}{4}, \frac{1}{8}, \frac{1}{16}, \frac{1}{32}$, etc.? *Ans.* ½.

5. A ball falls 8 ft. to the floor and bounds back 4 ft., then falling bounds back 2 ft., and so on; how far will it move before coming to rest? *Ans.* 24 feet.

6. A hound and fox, 10 rods apart, run so that when the hound runs 10 rods the fox runs 1 rod, etc.; how far will the hound run to catch the fox? *Ans.* 11⅑ rods.

SECTION XII.

HIGHER PERCENTAGE

ANNUITIES.

674. An **Annuity** is a sum of money to be paid annually, or at some other regular interval of time.

675. An **Annuity Certain** begins and ends at fixed times. A *Perpetuity* is an annuity which continues for ever

676. A **Contingent Annuity** begins or ends with some uncertain event, such as the birth or death of one or more persons.

677. An **Immediate Annuity** begins immediately. A *Deferred Annuity*, or Annuity in Reversion, begins at some future time.

678. The **Forborne**, or **Final Value**, is the sum of the amounts of all the payments on interest from the time each is due to the end of the annuity.

679. The **Present Value** is such a sum as put at interest for the given time and rate, will amount to the final value.

680. Annuities are estimated at both simple and compound interest. We will give two cases under each.

NOTES.—1. An Annuity is a periodical income. Such incomes may be secured by the payment of a certain sum of money. They may be obtained of *Trust Companies*. It is a popular form of investment in the National Debt of England.

2. The advantage of such an investment is that a larger rate of interest is received, since the capital invested is not to be returned. An old person may receive a very high rate on such an investment.

ANNUITIES AT SIMPLE INTEREST.

CASE I.

681. ***To find the amount, or final value of an annuity at simple interest.***

1. What will be the amount, or final value of an annuity of $200 in 5 years, at 6 per cent.?

SOLUTION.—If left unpaid until the end of 5 years, the last payment will be $200 without interest, the 4th payment will have become $200 plus the interest for one year, which is $212; the 3d payment will have become $200 plus the interest for 2 yr., which is $224; the 2d payment $236, the 1st payment, $248, and the sum of these payments will be the final value or amount, which we find is $1120. These sums form an arithmetical series, of which the first term is the annuity, $200, the common difference is the interest for 1 year, the number of terms is the time in years, and the sum of the terms is the final value. Hence we have the following

OPERATION.

$$\text{Last term} = \$200 + 12 \times 4 = \$248$$

$$\text{Sum} = \frac{\$200 + \$248}{2} \times 5 = \$1120$$

Rule.—*Take the annuity for the first term, the interest for 1 year for the common difference, and the time for the number of terms; find the last term and then the sum of the terms; this sum will be the final value.*

NOTE.—When the payments are made semi-annually, quarterly, etc., the number of such periods will be the number of terms, and the interest for such time will be the common difference.

WRITTEN EXERCISES.

2. An annuity of $400 was unpaid for 7 yr.; what was then due, interest at 6%? *Ans.* $3304.

3. An annuity of $600 was unpaid for 5 yr.; what was then due, interest at 4 per cent.? *Ans.* $3240.

4. What is the final value of an annuity of $580 for 8 yr., at 5 per cent.? *Ans.* $5452.

5. A bought a house for $4000 and agreed to pay $500 annually; now if he neglects to pay what is due yearly, what will be the sum due at the end of 8 yr., interest 6%? *Ans.* $4840.

CASE II

682. ***To find the present value of an annuity at simple interest.***

1. What is the present value of an annuity of $200 for 4 yr., at 6 per cent.?

SOLUTION.—Since the present worth of an annuity is the present worth of the final value, we first find the final value by Case I., and then find the present worth of this by Art. 496. The final value we find is $872, and the present value of $872 is $703.22. Hence we have the following

OPERATION.

Final value = $872

$872 ÷ 1.24 = $703.22 +

Rule.—*Find the final value of the annuity by Case I., and then find the present worth of that sum.*

WRITTEN EXERCISES.

2. What must I pay for an annuity of $400 for 6 yr., at 5 per cent.? *Ans.* $2076.92.

3. What is the present value of an annuity of $800 for 10 yr., at 6 per cent.? *Ans.* $6350.

4. I have an annuity of $750 to run 12 yr., at 8%; what is its present value? *Ans.* $6612.24.

5. A rented his house for $80 a quarter; what sum paid at the beginning of the year would pay it, interest at 8 per cent.? *Ans.* $305.19.

NOTE.—Since the amounts form an arithmetical series, the time, rate, and yearly payment may be found by the different cases under Arithmetical Progression, and it is therefore unnecessary to treat them separately here.

ANNUITIES AT COMPOUND INTEREST.

683. Annuities are usually reckoned at compound interest instead of simple interest.

CASE I.

684. ***To find the final value of an annuity certain at compound interest.***

1. What is the final value of an annuity of $300 for 4 yr., at 6%?

SOLUTION.—At 6% $1 gives an annual income of $.06, hence to give an income of $300 it will require as many times $1 as .06 is contained times in 300, which is $5000; and if the annuity remains unpaid for 4 yr. at 6%, the amount due will be the compound interest of $5000 for 4 yr. at 6%, which we find is $1312.39. Hence the following

OPERATION.

300 ÷ .06 = $5000

$0.262477, Com. Int. of $1

5000

$1312.385

Rule.—*Divide the annuity by the rate, and find the compound interest of the quotient for the given time and rate.*

NOTES.—1. Use the table, Art. 511, for finding the compound interest.

2. An annuity at compound interest is really a geometrical progression, the periodical payment being the *first term*, 1 plus the rate per cent., the *rate*, and the final value the *sum* of the *series*. The examples under this case could therefore be solved by Case II., in Geometrical Progression, but much work is saved by using the table of compound interest.

17

WRITTEN EXERCISES.

2. What is due me on an annuity of $250, unpaid for 6 yr. at 5 per cent.? *Ans.* $1700.48.

3. What is the final value of an annuity of $360 for 7 yr., at 6 per cent.? *Ans.* $3021.78.

4. Mr. B pays $40 a year for cigars; how much more would he be worth at the end of 20 years, by investing this sum at 8 per cent. compound interest? *Ans.* $1830.48.

5. Mr. A put $120 in a savings bank on the day his daughter was 10 years old, and the same sum on each subsequent birthday; what will the daughter be worth when 21 years old, compound interest 6 per cent.? *Ans.* $1904.39.

CASE II.

685. ***To find the present value of an annuity certain at compound interest.***

1. What is the present value of an annuity of $300 for 4 yr., at 6 per cent.?

OPERATION.

Final value = $1312.39
$1312.39 ÷ $1.262477 = $1039.53

SOLUTION.—The final value of this annuity, as found in Case I., is $1312.39, and the present worth of this sum is the present worth required. The compound amount of $1 for the given rate and time, as given in the table, Art. 511, is 1.262477, hence the present worth of the sum is $1, and the present worth of $1312.39 is as many times $1 as $1.262477 is contained times in $1312.38, which is $1039.53. Hence the following

Rule.—*Find the final value as in the preceding case, and divide this sum by the amount of* $1 *at compound interest for the given rate and time.*

WRITTEN EXERCISES.

2. What must I pay for an annuity of $600, running 8 yr., at 6% comp. int.? *Ans.* $3725.88.

3. Find the present value of an annuity of $480 running 10 yr., at 8% comp. int. *Ans.* $3220.84.

4. A can rent his house for $450, payable at the beginning of the year, or $120 payable quarterly; which is the more profitable, money worth 8 per cent.?

Ans. 2d, $6.93.

5. B bought a house for $6000 down, or equal install ments of $1200 a year, for 6 yr.; which is the better for B, money being worth 6 per cent.? *Ans.* 2d, $99.21.

NOTE.—For a fuller discussion of Annuities see *Brooks's Higher Arith metic.*

INSURANCE.

686. Insurance is a contract of indemnity for loss or damage within a given time. It is of two kinds: *Property Insurance* and *Personal Insurance.*

687. Property Insurance is security against loss by fire or transportion. Insuring anything is called "taking a risk."

688. Property Insurance is of two kinds: *Fire Insurance* and *Marine Insurance.*

689. Fire Insurance is security against loss by fire; *Marine Insurance* is security against loss by navigation.

690. The **Insurer** or **Underwriter** is the party or company taking the risk. The *Insured* or *Assured* is the party protected.

691. The **Policy** is the written agreement or contract between the insurers and the insured.

692. The **Premium** is the sum charged for insurance; it is a certain rate per cent. of the amount insured.

693. The **Sum Covered** by insurance is the amount in sured on a property.

694. The **Base** is the amount insured on a property The *Rate* varies with the risk.

The *Rate* of insurance is quoted as so many cents on the $100, or, as so much per cent. Policies are renewed annually or at stated periods, and the premium is paid in advance. Risks are usually *rated* per annum. The rate for more than 1 yr. is determined by the following table:

The	rate	for	2	yr.	is	$1\frac{1}{2}$	times	the	annual	rate.
"	"	"	3	"	"	2	"	"	"	"
"	"	"	4	"	"	$2\frac{1}{2}$	"	"	"	"
"	"	"	5	"	"	3	"	"	"	"
"	"	"	7	"	"	$4\frac{1}{2}$	"	"	"	"

Insurance is generally done by *stock companies*. When an *individual* takes a risk, it is called an "out-door" business. A *Mutual Insurance* Company is one in which the profits and losses are shared by those who are insured.

To prevent fraud, companies will seldom insure the full value of property. In cases of loss, the *underwriters* may either replace the property insured, or pay its value. Only the amount of actual loss can be recovered; and often claims are *adjusted* for a part of the amount insured.

695. Short Rate Tables are tables prepared for reckoning the insurance when the time is less than one year.

The rate in short periods is quoted *per annum*, and the actual rate for a short period is given in the table. Such a table is given in the appendix, and is used in solving some of the problems in Cases I. and IV.

696. Perpetual Policies are sometimes issued, the rate being usually equal to that of *ten* annual premiums.

In *Perpetual Policies* the premium is considered merely a deposit with the Insurance Company; for at any time, at the instance of either party, the policy may be cancelled, and 90% of the premium or deposit must be returned to the policy holder.

697. The **Quantities** considered are: 1. The *Amount Insured;* 2. The *Rate of Insurance;* 3. The *Premium;* 4. The *Valuation of Property.*

CASE I.

698. ***Given, the amount insured and the rate, to find the premium.***

1. I insured my house for $5680 at $1\frac{1}{2}\%$; required the amount of the premium.

OPERATION.

$$\begin{array}{r} \$5680 \\ .01\frac{1}{2} \\ \hline \$85.20 \end{array}$$

SOLUTION.—The premium on $5680 at $1\frac{1}{2}\%$, is $.01\frac{1}{2}$ times $5680, which we find to be $85.20.

Rule.—*Multiply the amount insured by the rate, to find the premium.*

WRITTEN EXERCISES.

2. A insured his store valued at $8500, for $6500, at $1\frac{1}{2}\%$; required the amount of the premium. *Ans.* $97.50.

3. What is the premium for an insurance of $7500 on a house and furniture for 5 yr., at $3\frac{1}{4}\%$? *Ans.* $243.75.

4. Insured my house for $3000, furniture for $1500, and

library for $750, at $1\frac{1}{4}\%$, the policy costing $1.25; what is the cost of insuring? *Ans.* $66.87½.

5. What is the premium on a $975 policy, dated April 18th, 1877, and expiring Sept. 30th, 1877, annual rate on the risk being $\frac{1}{2}\%$? *Ans.* $3.41.

6. On a vessel there was a fire insurance of $75000 at $\frac{3}{4}\%$, and a marine insurance of $\frac{1}{3}$ as much on the cargo, at $1\frac{1}{8}\%$; in a storm $\frac{1}{4}$ of the cargo was thrown over, and the vessel was afterward destroyed by fire; what was the actual loss to the underwriters? *Ans.* $80406.25.

7. A store in Boston worth $10000 and a stock of goods worth $15000 were insured for 75% of their value, at $\frac{3}{4}\%$; what was the loss to the owners, and what the loss of the company, if they were entirely consumed in the fire of '73?
Ans. Owner, $6390.62½; Co., $18609.37½.

8. Mr. Levan orders insurance as follows: $2500 on wool storage for 1 mo., $2500 on do. for 2 mo., $2500 on do for 3 mo., and $2500 on do. for 4 mo., all in same warehouse, the annual rate being $.85 on the hundred dollars; also at the same time orders a policy for $2500 on his frame dwelling for 3 yr., annual rate $\frac{1}{2}\%$. For what must he draw his check to the insurance agent? *Ans.* $54.50.

9. Mr. Smythe takes out an insurance of $18000 for 1 mo., on cotton stored in a warehouse, rated at 1% per annum; at the expiration of this time, not having sold, he has the policies renewed for 1 mo. longer; how much would he have saved by taking out the insurance for 2 mo., at first? *Ans.* $18.

NOTE.—The rates in problems 5, 8, and 9 are found by the table in Appendix.

CASE II.

699. ***Given, the rate and the premium or the value of the property, to find the amount insured.***

1. A man paid $122.50 to insure a house, at $1\frac{3}{4}\%$; what was the value of the house?

SOLUTION.—At a premium of $1\frac{3}{4}\%$, $.01\frac{3}{4}$ times the amount insured equals the premium, which is $122.50; hence the amount insured equals $122.50 ÷ $.01\frac{3}{4}$, or $7000.

OPERATION.

$$\frac{\$122.50}{.01\frac{3}{4}} = \$7000.$$

Rule.—*Divide the premium by the rate, to find the amount insured.*

NOTE.—To find what amount must be insured to cover the premium in case of loss, we *divide the valuation of the property by* 1 *minus the rate.*

WRITTEN EXERCISES.

2. I paid $58.12½ to insure the transportation of goods at 2½%; what sum was covered on the goods? *Ans.* $2325.

3. The premium for insuring ⅘ of the value of a house, for 3 years, at 1¾%, was $86.36¼; what was the value of the house? *Ans.* $6168.75.

4. B's house, worth $15880, is insured for ⅘ of the value at 2% for 5 yr., so as to include the premium if burned; required the sum stated in the policy. *Ans.* $12963.27.

5. A merchant insured his store for ¾ of the value, at 1¼%, but soon after the store was burned down, and his loss over the insurance was $4150; what was the value of the store? *Ans.* $16000.

CASE III.

700. ***Given, the premium and the amount insured, to find the rate.***

1. The premium for effecting an insurance of $6000 on a house, was $135; what was the rate?

SOLUTION.—Since the premium equals the amount insured multiplied by the rate, the *rate* equals the premium, $135, divided by $6000, the amount insured, which we find to be .02¼, or 2¼%.

OPERATION.

$$\frac{\$135}{\$6000} = .02\tfrac{1}{4}$$

Rule.—*Divide the premium by the amount insured, to find the rate.*

WRITTEN EXERCISES.

2. A merchant pays $55 for the insurance of $2500 on his store; what is the rate of insurance? *Ans.* 2⅕%.

3. The premium for insuring ⅚ of the cargo of a ship, valued at $89520, was $2424½; required the rate. *Ans.* 3¼%.

4. I effected an insurance of $5700 on my store, paying $79.25, including the cost of the policy, $3.25; what was the rate of insurance? *Ans.* 1⅓%.

5. I insured $2400 on my house, $1200 on my furniture and $350 on my library, for 3 yr., paying a premium of 49.37\frac{1}{2}$; how would it be rated annually? *Ans.* $\frac{5}{8}$%.

CASE IV.

701. ***To find the return premium on a cancelled policy.***

702. To Cancel a Policy is to annul the agreement between the party insured and the insurers.

When the policy is cancelled at the instance of the company, a *pro rata* proportion of the premium paid is returned; when done at the request of the policy holder, the company pay back a return premium governed by what are known as *Short Rate Tables.*

When a partial loss has been paid, the return premium is to the whole premium as the balance of the policy after deducting the partial losses paid is to the whole amount of the policy as first issued.

1. Mr. A effects an insurance on his stock of mdse. to the amount of $5000 for 6 mo., at short rates, his risk being rated at 85¢ on a hundred dollars; in consequence of a reduction of stock at the end of 4 mo. he wishes his policy cancelled; to how much return premium is he entitled?

SOLUTION.—The rate for 6 mo. as found in the table is $.0059, and for 4 mo., $.0042; hence the return premium is the difference between $.0059 and $.0042 multiplied by 5000, or $8.50.

OPERATION.

.0059 — .0042 = .0017
$.0017 × 5000 = $8.50

Rule.—*Multiply the amount insured by the difference of the rates for the two periods, to find the return premium.*

2. Mr. B takes out a perpetual insurance on his marble dwelling to the amount of $5500, his risk being rated at $\frac{1}{4}$% annually; what is the deposit premium, and if he afterward surrenders his policy for cancellation, how much return premium should he get? *Ans.* $137.50; $123.75.

3. Mr. C has an annual policy of insurance of $2500 on his house; at the end of 7 mo. a fire occurs which damages his property to the amount of $500, which the insurance company pays and indorses the payment on his policy; 2 mo. afterward Mr. C sells his house and surrenders the policy for cancellation in full; what is his return premium, the annual rate being $\frac{3}{10}$% on his risk? Ans. $0.80.

CASE V.

703. ***To adjust the loss on a risk between several different insurance companies.***

704. When several companies are interested in a risk, a loss is shared by the companies in proportion to the amounts of the several policies.

Companies usually attach on different items in the same proportion.

1. W. & Bro. hold a policy of insurance on their mill for \$5000 in the Delaware Mutual Fire Ins. Co. and also one for \$4500 in the Fire Association; a fire causes a loss on the property to the amount of \$1875; what amount does each of the companies pay?

SOLUTION.—The whole amount insured is \$9500; the amount of the loss to be paid by the Del. Mutual is to \$1875 as \$5000 is to \$9500, which we find by proportion gives \$986.84$\frac{4}{19}$: the amount to be paid by the Fire Asso. is to \$1875 as \$4500 is to \$9500, which gives \$888.15$\frac{15}{19}$.

Rule.—*Divide the loss between the several companies in proportion to the amounts of the several policies.*

WRITTEN EXERCISES.

2. The Ins. Co. of North America issued a policy to Green & Co., covering \$800 on their hotel building and \$1700 on the furniture therein; the Sun Fire Ins. Co. also issued a policy to same parties, covering \$2500 on the hotel, but nothing on the furniture. By a fire the building is damaged to the amount of \$2200 and the furniture is damaged to the amount of \$500; what proportion of the total damage does each company bear?

Ans. N. Am., \$1033.33$\frac{1}{3}$; Sun, \$1666.66$\frac{2}{3}$.

3. M has a policy in the Ætna Fire Ins. Co. for \$2000, covering \$1200 on his mill building and \$800 on machinery therein; he also has a policy in the Lancaster Fire Ins. Co. for \$1500, covering \$1000 on his mill building and \$500 on machinery therein. By a fire which occurs next door his property is damaged by water to the amount of \$230 as follows: \$80 loss on the building and \$150 on machinery; how much can he claim from each company?

Ans. Ætna, \$135.94$\frac{58}{143}$; Lancaster, \$94.05$\frac{85}{143}$.

LIFE INSURANCE.

705. Life Insurance is a contract by which a company in consideration of payments made by the insured, stipulates to pay a certain sum of money to his heirs at his death, or to himself if he attains a certain age.

706. The **Policies** of Life Insurance most frequently used are the following:

1. *Term Policies*, payable at the death of the insured, if it occur within a certain number of years, premium payable annually.
2. *Life Policies*, payable at the death of the insured, premium payable annually during life, or in one, five, or ten annual payments.
3. *Endowment Policies*, payable to the insured at the end of a certain number of years, or to his heirs if he dies sooner, premium payable either annually during the continuance of the policy, or in one, five, or ten annual payments.

707. The rates of premium, as fixed by different companies, are based on the *expectation of life*, determined by a table of mortality, the probable rates of interest, and the "loading," or margin for expenses.

NOTES.—1. Policies in many companies are forfeited on non-payment of premium. The laws of Massachusetts, however, provide that the companies chartered by that State shall allow the policy to run on a certain time, proportioned to the number of premiums that have been paid, and if the insured dies within this time, the company will pay the amount insured, deducting for the premiums omitted.

2. A table given by the New England Mutual Life Insurance Company of Boston will be found in the Appendix, on which most of our examples are reckoned.

708. The **Quantities** considered in Life Insurance, using the tables in our calculations, are, 1. The *Premium* on \$1000; 2. The *Gain* or *Loss*; 3. The *Amount of the Policy*; 4. The *Age*; 5. The *Period of Insurance*.

CASE I.

709. ***Given, the amount of policy, the age, and the period of insurance, to find the premium.***

1. What annual premium must a man aged 45 years pay for a life policy of \$2500?

SOLUTION.—The premium for life, in the table, at the age of 45, is \$38 for \$1000; hence, for \$2500 it will be 2.500 times \$38, which is \$95. Hence the following

OPERATION.

$\$38 \times 2.500 = \$95.$

Rule.—*Find in the table the premium corresponding to the given age and time, and multiply this sum by the amount of the policy, considering all terms of the policy below thousands decimally.*

WRITTEN EXERCISES.

2. Mr. Tappan takes out an endowment policy in the New England Mutual Insurance Company for $5000, payable to himself in 10 yr., or to his heirs at his death; what annual premium will he pay, his age being 52 yr.? *Ans.* $568.50.

3. Mr. Amory wished to insure his life at the age of 40 years for $10000; but not being able conveniently to spare the money, he deferred it till he was 45 years old, and then took an endowment policy for 15 years; how much more would the premiums amount to at the maturity of the policy than if he had taken the same kind of policy when he first intended? *Ans.* $360.

4. Nathan Foster took out an endowment policy for $17500, payable in 20 years, his age being 35 years; if he lives to receive the endowment, will he have paid more or less than if he had taken a policy of the same amount at 40 years of age for 15 years? *Ans.* $993.12½ less.

CASE II.

710. ***Given, the amount of policy, the age, and the period of insurance, to find the gain or loss by insuring.***

1. A man 43 years of age takes a life policy for $3500 premium payable during life; he dies after making 10 payments; how much will the amount of the policy exceed the payments?

SOLUTION.—Having found the premium by Case I. to be $123.20, 10 payments will amount to $1232, and the excess is the difference between $3500, the amount of the policy, and $1232, the amount of the payments, which is $2268.

OPERATION.

$35.20 × 3.500 = $123.20

$123.20 × 10 = $1232

$3500 − $1232 = $2268

Rule.—*Multiply the premium, as found by Case I., by the number of payments, and subtract this product from the amount of the policy.*

NOTE.—When interest is reckoned on the payments, as in problems 3, 4, and 5, we obtain the amount as we obtain the final value of an annuity, Art. 681.

WRITTEN EXERCISES.

2. John Gilbert, aged 35 years, takes out a life policy for $5000, premium payable during life; he dies at the age of 50; how much will his heirs receive above the amount of the premiums? *Ans.* $2880.

3. James Gibbons, aged 44 years, takes out an endowment policy for $8000, payable in 15 years; reckoning interest at 6% on his payments, will he gain or lose if he lives to receive the endowment? *Ans.* $4,485.28 loss.

4. Charles Marshall, at the age of 37 years, took out a life policy of $7000, premiums to cease in 10 years; he died aged 45 years 3 months; what was his gain by insuring, reckoning interest on premiums at 6%? *Ans.* $2472.74.

5. George Dwight, 32 years of age, took out an endowment policy for $11000, payable in 20 years. In two years and a half he died; what was the gain, reckoning interest on premium at 7%, and how much greater profit would it have been to take a life policy, premiums payable during life?

Ans. Gain, $9256.97; $864.22.

NOTE.—For a more extended discussion of this subject see *Brooks's Higher Arithmetic.*

BUILDING ASSOCIATIONS.

711. Building Associations are coöperative corporations instituted to receive small deposits at regular periods and to invest these in loans among the depositors or members, on mortgages given by the borrower.

These associations enable many persons of moderate earnings and incomes to erect or buy buildings, and to invest their savings securely and profitably. The regular installments form the capital of the association, which is loaned to members only. The business is managed directly by the depositors, and the profits are equitably divided among them.

712. The **Members** of an association are those who subscribe for shares. They are of two classes, *borrowers*, or those who borrow money of the association, and *non-borrowers*, who subscribe for shares as an investment.

713. The **Shares** are usually issued periodically in *series*, thus producing a constant succession of shares, each series successively reaching its value and being wound up, and a new series taking its place. Many associations have only one series.

When the installments and profits on any series have raised the value of its shares to par, it is wound up by returning to the non-borrowing members the value of their shares (though in some associations the paid-up shares are allowed to remain and draw cash dividends), and to the borrowing members their mortgages and cancelled obligations.

Thus, supposing $200 to be the value of a share and the payments $1 a month, if the capital is accumulated in one hundred months, the non-borrowing member will receive $200 on a share, and the borrowing member's debts will be cancelled, and his mortgage for $200 a share returned. The installments in each case have amounted to only $100, making a profit of $100, or 100% for the time. Many series are closed before their shares are fully equal to $200 in value.

714. The **Dues** are the fixed periodical installments, and are usually $1 a month. *Contingent Dues* for current expenses are assessed annually by some associations. In case of non-payment of dues, fines are levied. It is illegal in Pennsylvania to charge fines on unpaid fines.

At the regular monthly meetings of associations, the aggregate installments or dues, interest, fines, etc., paid in, are loaned to the highest bidder, or sometimes in the order of application, in which latter case there is a fixed or stated premium to be paid by the borrower.

715. The **Premium** is a percentage paid per share, in excess of interest, on money which is "bought" or borrowed of the association. It is *quoted* for the *beginning of the series*.

716. The **Stated Premium** is the minimum rate fixed by associations, at which money will be sold on shares, each year of a series.

The *Stated Premium* is fixed at $50, or 25% of a share, for the 1st year; $45, or 22½% for the 2d year; $40, or 20% for 3d year, etc.; decreasing 10% yearly to the 7th year, when it becomes $20, or 10%. Money is seldom loaned after the 7th year, or at a lower "stated premium." The *entire premium* on a loan equals the *stated premium* at that point of the series plus the *amount bid*.

Some associations have no *stated premium* to regulate the difference of premium between different series, but deduct, for each expired year of the series, 10% from the *premium bid*. This is avoided by the Installment plan, in which a number of cents a month is bid as premium, thus making no difference in what series the borrower holds shares.

717. There are **Three Modes** of loaning money and fixing the interest, adopted by different associations, called the *Installment Plan*, the *Net Plan*, and the *Gross Plan*.

By the first plan, the *par value* of a share is *loaned* on each share, and the premium is paid in monthly installments, together with the dues and interest. By the second plan, the premium is deducted from the par value, and interest is charged on the net amount of the loan. By the third plan, the premium is deducted from the par value, but interest is charged on the par value of the share.

Thus, by the Installment Plan, the net loan is $200, the par value of the share and the full amount of the mortgage; the payments are $1 a month dues, $1 interest, and — cents premium. By the Net Plan, if the premium is $50, the net loan is $150, and payments $1 a month dues and 75¢ a month interest. By the Gross Plan, the net loan is $150, but payments are $1 dues and $1 interest. The monthly premium in cents by the first plan corresponds nearly to the total premium in dollars on a new series by the other plans, on the basis of 100 months.

In Pennsylvania, where these associations are most numerous, the number of shares at any one time is limited to 5000, and the periodic payments of borrowers to $2. Thus, by the Installment and Gross Plans, the dues and interest at 6% on $200, par value of a share, are each $1 a month, which brings the payments up to the limit, $2.

If loans are paid before the termination of a series, an equitable part of the premium paid is refunded, by the *Gross* and *Net Plans*. No premium is returned by the *Installment Plan*, since none is paid in advance.

The Installment and Net Plans are more favorable to the borrower than the Gross Plan. Of the three, the Installment Plan is the simplest, and seems worthy of general adoption.

718. A **Withdrawal** is made by returning the stock certificates to the association, and making settlement.

In case of withdrawal, a non-borrower receives the dues paid in, and an equitable part of the accrued profits. By the Installment Plan, a borrower pays the difference between the withdrawal value of the shares and the gross amount of the loan. By the Net or Gross Plans, a borrower pays the difference between the sum of the withdrawal value of the shares, increased by the premium for the unexpired years of the series, and the gross amount of the loan.

The *profits* of an association accrue from *interest* and *premiums*. The *True Profit* at any date of a series is the *legal interest* on the payments, plus that part of the profit on premiums which the present value of a share is of the par value, $200. The *Withdrawal Profit* is the True Profit less a Withdrawal Discount fixed by the Association By-Laws.

NOTE.—Building Associations are not, as often supposed, builders of houses. They are corporations organized *to enable their members to build houses*, or buy them in their individual capacity, and might perhaps as appropriately be called Savings Fund and Loan Associations.

CASE I.

719. ***To find the actual cost of any amount of stock.***

1. What would be the annual aggregate dues on 20 shares of stock at $1 a month per share?

SOLUTION.—Since the dues on 1 share for 1 month are $1, on 20 shares they will be $20; and for 1 year, 12 times $20, or $240.

OPERATION.

$\$1 \times 20 \times 12 = \$240.$

Rule.—*Multiply the periodical dues by the number of periods, and to this product add the sum of the fines, if any have been levied.*

WRITTEN EXERCISES.

2. I buy 8 shares in the first series, 12 in 2d, and 27 in 3d of Franklin Building Association; if these series are $8\frac{1}{3}$, 9, and $10\frac{1}{2}$ years respectively in "running out," how much money in monthly dues will have been paid in on the three series when closed out? *Ans.* $5498.

3. I subscribed for 18 shares of Investment Building Association, new series, and was twice fined 10% for unpunctual payment of dues; at the end of the year I subscribed for 20 shares in the second series; how much had my subscriptions cost at the end of the second year? *Ans.* $675.60.

4. Mr. Allen pays dues on 20 shares for 2 years, and then discontinues his payments; if his fine is 10% of dues each month, what will be the amount to his credit at the end of the third year? *Ans.* $84.

SUG.—The installments and fines form an arithmetical progression for the third year.

CASE II.

720. ***To find the amount of a loan, and the monthly payments and entire payment of a borrower.***

1. I buy money on 10 shares of stock, Installment Plan, new issue, and bid 40¢ a month premium; what is the amount of my loan, and what are my monthly payments?

SOLUTION.—On the Installment Plan, the full value of a share is loaned; hence the loan on 1 share is *$200*, and on 10 shares it is *10 times* $200, or $2000.

OPERATION.

$\$200 \times 10 = \$2000$, Loan.

$\$200 \times .005 = \$1.00$, Int. per mo.

$\$1 + \$1 + \$0.40 = \$2.40$, Pay't on 1 share.

$\$2.40 \times 10 = \$24$; Amt. of Pay'ts.

The interest on $200 for 1 month at 6% is $1, and this added to $1, the dues, and 40¢, the premium, gives $2.40, the monthly payments on 1 share; and on 10 shares the payments will be $2.40×10, or $24.

Rule I.—*Multiply the loan on one share by the number of shares, to find the amount of the loan.*

Rule II.—*Multiply the sum of the dues, interest, and premium, on 1 share, by the number of shares, for the monthly payment. Multiply the monthly payment by the number of months the series has to run, for the entire payment.*

NOTES.—1. In the Installment Plan, the loan on 1 share is $200; in the Gross and Net Plans, *the loan on one share is $200 minus the premium.*

2. When a loan is bought after the beginning of a series, the *dues* must be reckoned from the *beginning of a series*, but the *interest* and *premium* only from the *beginning of the loan.*

WRITTEN EXERCISES.

2. Mr. Wilson bought a loan of a Building Association on 15 shares, at the beginning of the series, for 65 cents a month premium; what was his loan and what did he pay for it, if the series runs out in 9 years? *Ans.* $3000; $4293.

3. I bought a loan of a Building Association on 16 shares at the beginning of the series, at $50 premium, Gross Plan; what was the loan, and what did I pay for it, if the series runs out in $8\frac{2}{3}$ years? *Ans.* $2400; $3328.

4. I bought a loan of a Building Association on 23 shares, at the beginning of the 3d year, for 67 cents a month premium; what was the loan, and what did I pay for it if the series runs out in $8\frac{1}{2}$ years? *Ans.* $4600; $5341.98.

5. Mr. Brown built a house for $3500, and to pay for it borrowed money of the Quaker City Building Association on 20 shares of the 4th year of the series, at $15 and "stated premium," Net Plan; what balance remains due on the house, and if the series runs out in $8\frac{1}{3}$ yr., what will he pay for his loan? *Ans.* $500; $2960.

CASE III.

721. ***To find the actual cost of a loan to a borrower.***

1. I bought a loan on 15 shares in a new series of a Building Association, Gross Plan, at $12, and "stated premium;" if the series runs out in $8\frac{1}{3}$ years, what will be the actual cost of my loan?

*H**

SOLUTION.—The monthly payment on 1 share equals $1 dues and $1 interest, or $2, and on 15 shares the payment is $30. The first installment is on interest 100 months, the second installment 99 months, and so on; hence the interest of a payment of $1 for the different periods equals the interest of $1 for a number of months represented by an arithmetical series whose first term is 1, last term 100, and number of terms 100, or (Art. 665) ½ of $(100+1)\times 100$. The interest of $1 for 1 month is ½¢, and for the aggregate months, ½ of $100\times 101\times \frac{1}{2}¢ = \frac{101\times 100}{4}¢ = \$25.25$; and on $30 it is $\$25.25\times 30 = \$757.50$. The sum of the payments equals $\$30\times 100$, or $3000; and the cost of the loan equals $3000+$757.50, or $3757.50.

OPERATION.

$\$2.00\times 15 = \$30$, Monthly payment.

$30\times \frac{101\times 100}{4}¢ = \$757.50$, Int. 6%.

$\$30\times 100 = \$3000$, Sum of pay'ts.

$\$3000+\$757.50 = \$3757.50$, Cost.

Rule.—I. *Multiply the number of months by the number of months increased by* 1, *and divide by* 4, *to find the interest at* 6% *on the aggregate monthly payments of* $1.

II. *Multiply the interest on the aggregate payments of* $1, *by the monthly payment, to find the interest on the payments. Add this interest to the sum of the payments; the result will be the cost of the loan.*

NOTE.—We have assumed that the monthly payments are entitled to *simple interest* from the *time of their payment* until the *close of the series*, in determining the actual cost of a loan. It would be more correct to reckon *annual interest*, but this makes the calculation rather difficult. To be strictly accurate, we should reckon *compound interest*.

WRITTEN EXERCISES.

2. Mr. Thomas bought a loan on 12 shares, new series, of El Paso Building Association, at $85 premium, Net Plan; if the series runs out in 9 years, what is the actual cost of his loan? *Ans.* $2597.427.

3. Mr. Burton bought a loan of a Philadelphia building association on 10 shares of a new series at 65¢ a month premium; what is the actual cost of the loan if the series runs out in 9⅓ years? *Ans.* $3806.46.

4. A rents a house at $12 a month, and at the end of 10 years buys it for $1200; B buys a house for $1200, borrowing money of a building association on 8 shares of a new series at $50 premium, Gross Plan, which runs out in 10 yr., and paying an annual tax of $24 at the beginning of each year; which house cost the most? *Ans.* A's, $255.60.

CASE IV.

722. ***To find the rate of interest received by a non-borrower.***

1. What rate of interest do I receive on 5 shares, dues $1 per share, if the series runs out in $9\frac{1}{2}$ years?

OPERATION.

$$\$200-\$114=\$86$$

$$\frac{115\times114}{24}=\text{equated time.}$$

$$\$86\div\frac{115\times114}{24}=15.7+\%$$

SOLUTION.—The sum of the installments paid on 1 share for $9\frac{1}{2}$ years or 114 months, is $114; and the difference between $200, the final value, and $114, the amount paid, equals $86, which is the gain, or interest on the investment. $1, the first payment, is on interest for 114 months, the second payment is on interest for 113 months, etc.; hence the interest on the installments for the different periods is equivalent to the interest on $1 for a number of months represented by the sum of an arithmetical series whose first term is 1 and last term 114, or (Art. 665) $\frac{1}{2}$ of $(1+114)\times114$, months $=\frac{1}{24}$ of $(1+114)\times114$, years; hence the interest on $1 for 1 year, or the *rate*, is $\$86\div\frac{115\times114}{24}=\$.157+$, or 15.7%.

Rule.—I. *Subtract the sum of the installments paid on one share from the final value of the share, and the difference will be the interest on the investment.*

II. *Multiply the number of payments by the number of payments increased by* 1, *and divide by* 24, *to find the equated time, or the number of years in which* $1 *will produce the same interest as the installments.*

III. *Divide the interest on the investment by the equated time; the quotient will be the equated rate per cent.*

WRITTEN EXERCISES.

2. By the annual report of the Investment Building and Loan Association made at the end of the eighth year, the present value of the first series is $186.90; what is the equated rate of legal interest at that time? *Ans.* 23.43%.

3. It was estimated that the first series, including dues, would be worth $180.75 when $8\frac{1}{2}$ years old, but at the end of 8 years the association canceled the series by paying the estimated value, less the unpaid dues on each share; what rate % was realized by the stockholders? *Ans.* 20.29%.

NOTE.—In Prob. 2, $186.90 is regarded as the final value of the share

CASE V.

723. ***To find the rate of interest paid by a borrower.***

1. A buys a loan on 10 shares, Net Plan, at the beginning of a series, at $60 premium per share, and pays $10 dues and $7 interest on net sum received, for $8\frac{1}{3}$ years; what is the average or equated rate of interest?

SOLUTION.—The loan was 10×($200 − $60) = $1400; $10 + $7 int. = $17, the monthly payment, which in 100 mo. equals $1700. Now the interest on the monthly payments (Case III.) is equivalent to the interest on $17 for $\frac{101\times100}{24}$ years at 6%, or $429.25; hence the actual cost of the loan is $1700 + $429.25, or $2129.25; therefore $2129.25—$1400, or $729.25, is the interest on the loan for $8\frac{1}{3}$ years; and the interest for 1 year is $729.25 ÷ $8\frac{1}{3}$, or $87.51; hence the rate is $87.51 ÷ $1400 = .0625 + or $6\frac{1}{4}$%.

OPERATION.

$$10\times(\$200-\$60)=\$1400.$$
$$100\times(\$10+\$7)=\$1700.$$
$$\$17\times\frac{101\times100}{24}\times.06=\$429.25.$$
$$\$1700+\$429.25=\$2129.25.$$
$$\$2129.25-\$1400=\$729.25.$$
$$\$729.25\div8\tfrac{1}{3}=\$87.51.$$
$$\$87.51\div\$1400=.0625+.$$

Rule.—I. *Find the sum of the installments, and the interest on the installments for the equated time at 6%; their sum will be the entire cost of the loan.*

II. *Subtract the amount of the loan from its entire cost; the remainder will be the interest on the loan for the period, from which the rate is readily found by the method of simple interest.*

WRITTEN EXERCISES.

2. Mr. Jay borrows $4600, at 56 cents premium a month, on the Installment Plan; what sum do his monthly payments aggregate, and what equated rate % will he pay if the series runs out in $9\frac{1}{2}$ years? *Ans* $58.88; 9.25%.

3. I buy a loan of 10 shares, new series, in an association on the Installment Plan, at 60 cents a month premium, and in another, a loan of 10 shares on the Gross Plan at $60 premium; what rate % do I pay for each loan if each series runs out in $8\frac{1}{3}$ years? *Ans.* Inst., 7.54%; Gross, 9.47%.

NOTE.—A more complete discussion of this subject will be found in *Brooks's Higher Arithmetic.*

SECTION XIII.

MENSURATION.

724. Mensuration treats of the measurement of geometrical magnitudes.

725. Geometrical Magnitudes consist of the *Line, Surface, Volume,* and *Angle.*

726. A **Line** is that which has length without breadth or thickness. Lines are either *straight* or *curved.*

727. A **Straight Line** is one that has the same direction at every point.

728. A **Curved Line** is one that changes its direction at every point. The word *line* used alone means a *straight line.*

729. Parallel Lines are those which have the same direction. Parallel lines, it is thus seen, will never meet.

730. One line is said to be *perpendicular* to another when the adjacent angles formed by the two lines are equal.

731. An **Angle** is the opening between two lines which diverge from a common point.

732. A **Right Angle** is an angle formed by one line perpendicular to another; as, ABC.

733. An **Acute Angle** is an angle less than a right angle; as, DEF. An **Obtuse Angle** is one larger than a right angle; as, DEG.

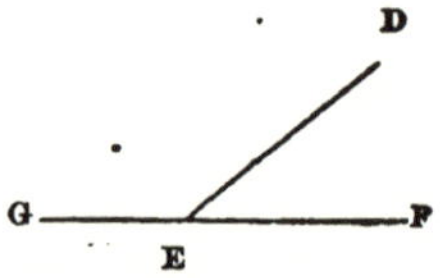

MENSURATION OF SURFACES.

734. A **Surface** is that which has length and breadth without thickness. Surfaces are *plane* or *curved.*

735. A **Plane Surface** is a surface such that if any two of its points be joined by a straight line, every part of that line will lie in the surface.

736. A **Plane Figure** is a plane surface bounded by lines, either straight or curved.

737. A **Polygon** is a figure bounded by straight lines; as, ABCDE. A Polygon of three sides is called a *Triangle,* of four sides, a *Quadrilateral,* etc.

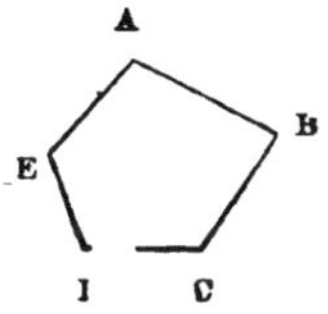

738. A **Diagonal** of a polygon is a line joining the vertices of two angles not consecutive.

739. The **Perimeter** of a polygon is the sum of its sides.

740. The **Area** of a plane figure is the number of square units in its surface.

NOTE.—The principles of mensuration are derived from geometry; their application to practical purposes is usually given in arithmetic.

THE TRIANGLE.

741. A **Triangle** is a polygon of three sides and three angles; as, ABC.

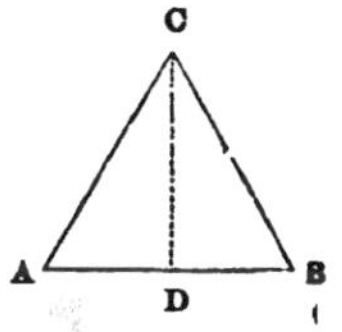

742. The **Base** is the side upon which it seems to stand; as, AB. The *Altitude* is a line perpendicular to the base, drawn from the angle opposite; as, CD.

743. An **Equilateral Triangle** is a triangle which has its three sides equal; when two sides are equal it is called *isosceles;* when its sides are unequal it is called *scalene.*

Rule.—*To find the area of a triangle, multiply the base by one-half of the altitude.*

NOTE.—If the three sides are given and not the altitude, *take half the sum of the sides, subtract from it each side separately, multiply the half sum and these remainders together, and take the square root of the product.*

1. What is the area of a triangle whose base is 25 rods and altitude 18 rods? *Ans.* 225 sq. rd., or 1 A. 65 P.

2. Required the area of a triangle whose base is 75 rods and altitude 57 rods. *Ans.* 13 A. $57\frac{1}{2}$ P.

3. Required the area of a triangular field whose base is 965 rods and altitude 576 rods. *Ans.* 1737 A.

4. What is the area of a field whose sides are respectively 20, 30, and 40 chains? *Ans.* 29 A. 8 P.—.

THE QUADRILATERAL.

741. A **Quadrilateral** is a polygon having four sides and therefore four angles There are three classes, the *parallelogram, trapezoid,* and *trapezium.*

745. A **Parallelogram** is a quadrilateral whose opposite sides are parallel. The *altitude* is the perpendicular distance between its opposite sides.

746. A parallelogram which is right-angled is called a *Rectangle.* When the four sides are equal it is called a *Square.*

747. An oblique-angled parallelogram is called a *Rhomboid.* An equilateral rhomboid is called a *Rhombus.*

Rule.—*To find the area of a parallelogram, multiply the base by the altitude.*

1. What is the area of a parallelogram 20 feet long and 18 feet wide? *Ans.* 40 sq. yd.

2. A has a rectangular lot 192 chains long and 65 chains wide; what is its area? *Ans.* 1248 acres.

3. What is the difference in the area of two lots, one being 245 rd. long, 42 rd. wide, and the other 85 chains long and 18 chains wide? *Ans.* 88 A. 110 P.

748. A **Trapezoid** is a quadrilateral which has two of its sides parallel. Its *altitude* is the perpendicular distance between its parallel sides.

Rule.—*To find the area of a trapezoid, multiply one-half the sum of the parallel sides by the altitude.*

1. Required the area of a trapezoid, one side being 120 in., the other 96 in., and the altitude 48 in. *Ans.* 36 sq. feet.

2. What is the area of a trapezoid, the sides being 365 and 124 in., and the altitude 86 in.? *Ans.* 146 sq. ft. 3 sq. in.

3. What is the area of a plank 12 feet long, 18 inches wide at one end, and 12 inches at the other end?

Ans. 15 sq. ft.

4. A farmer has a field in the form of a trapezoid, the two parallel sides being 95 and 75 rods respectively, and the perpendicular distance between them being 65 rods; how much land in the field? *Ans.* 34 A. 85 P.

749. A **Trapezium** is a quadrilateral which has none of its sides parallel. A diagonal, as AB, divides the trapezium into two triangles.

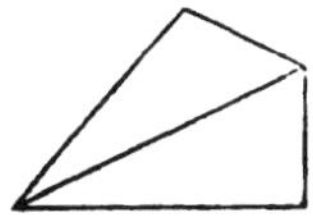

Rule.—*To find the area of a trapezium, divide the trapezium into two triangles by a diagonal, find the area of each triangle and take the sum.*

1. What is the area of a trapezium whose diagonal is 145 in., and the altitudes of the triangles, the diagonal being the base, are 30 and 40 inches respectively?

Ans. 35 sq. ft. 35 sq. in.

2. Required the area of a trapezium, the length of whose sides are respectively 20, 30, 25, and 35 chains, and the length of the diagonal 40 chains. *Ans.* 72 A. 56 P.—.

THE CIRCLE.

750. A **Circle** is a plane figure bounded by a curved line, every point of which is equally distant from a point within, called the *centre*.

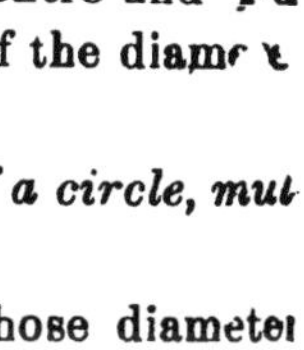

751. The curved line is called the *circumference*, and a line passing through the centre and ending in the circumference is the *diameter*. Half the diameter is called the *radius*.

752. Rule.—*To find the circumference of a circle, multiply the diameter by* 3.1416.

1. What is the circumference of a circle whose diameter is 25 inches? *Ans.* 78.54 in.

2. What is the distance around a circular fish-pond, the diameter of which is 16 rods? *Ans.* 50.2656 rd.

3. A man has a garden in the form of a circle, the diameter of which is 45 rods; what is the distance around it? *Ans.* 141.372 rd

753. Rule.—*To find the diameter of a circle, multiply the circumference by* .3183.

1. What is the diameter of a circle whose circumference is 40 feet? *Ans.* 12.732 feet.

2. What is the diameter of a water-wheel whose circumference is 78.54 feet? *Ans.* 25 feet.

754. Rule I.—*The area of a circle equals the circumference multiplied by one-fourth of the diameter, or the square of the circumference multiplied by* .07958.

Rule II.—*The area of a circle equals the square of the radius multiplied by* 3.1416, *or the square of the diameter multiplied by* .785398.

NOTE.—The area will vary slightly in the decimal figures as we use the different rules.

1. What is the area of a circle whose diameter is 25 and circumference 78.54? *Ans.* 490.875.

2. What is the area of a circle whose diameter is 36 inches? *Ans.* 1017.8784 sq. in.

3. What is the area of a circular garden whose circumference is 180 rods? *Ans.* 2578.23 sq. rd.

755. A square is inscribed in a circle when each of its angles is in the circumference.

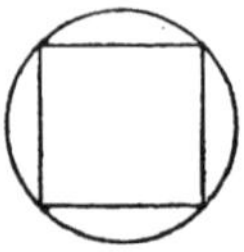

Rule.—*To find the side of an inscribed square, multiply the diameter by* .707106, *or multiply the circumference by* .225079.

1. What is the side of a square that can be cut out of a circular board whose diameter is 14 inches? *Ans.* 9.899 in.

2. How large a square can be cut out of a circular board whose circumference is 200 inches? *Ans.* 45.0158 in.

THE ELLIPSE.

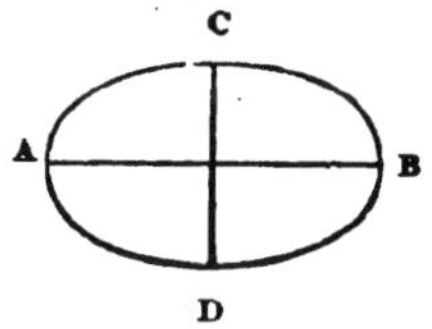

756. An **Ellipse** is a plane figure bounded by a curved line, the sum of the distances from every point of which to two fixed points is equal to the line drawn through those points and terminated by the curve. The two fixed points are called *foci:* the line through the foci is the *transverse* axis, and a line perpendicular to this passing through the centre and terminated by the curve, is the *conjugate* axis.

Rule.—*To find the area of an ellipse, we multiply half of the two axes together, and that product by* 3.1416.

1. What is the area of an ellipse whose transverse axis is 20 inches and conjugate axis is 16 inches?

Ans. 251.328 sq. in.

2. Required the area of an elliptical mirror whose length is 6 feet and breadth 5 feet. *Ans.* 23.562 sq. ft.

MENSURATION OF VOLUMES.

757. A **Volume** is that which has length, breadth, and thickness.

THE PRISM.

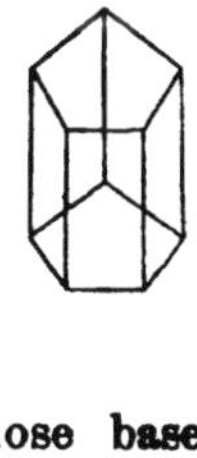

758. A **Prism** is a volume whose ends are equal polygons and whose sides are parallelograms.

759. The polygons are called *bases*, the parallelograms form the *convex surface*, and the prism takes its name from the form of its bases.

760. The **Parallelopipedon** is a prism whose bases are parallelograms. A *cube* is a parallelopipedon all of whose sides are squares.

761. Rule.—*To find the convex surface of a prism, multiply the perimeter of the base by the height.*

Note.—To find the entire surface we add the area of the bases.

1. What is the convex surface of a triangular prism, the three sides of whose base are respectively 6, 7, and 8 inches, and height 50 inches? *Ans.* 1050 sq. in.

2. What is the entire surface of the triangular prism given in the first problem? *Ans.* 1090.66 sq. in.

762. Rule.—*To find the contents of a prism, multiply the area of the base by the altitude of the prism.*

1. What are the contents of a square prism whose altitude is 30 feet, and the side of the base 3 feet? *Ans.* 270 cu. ft.

2. Required the contents of a triangular prism, the sides of whose base are each 16 inches, and whose altitude is 20 inches. *Ans.* 2217.02 cu. in.

THE PYRAMID.

763. The **Pyramid** is a volume bounded by a polygon and several triangles meeting in a common point. The polygon is called the *base*, and the triangles form the *convex surface.*

764. The point at the top is called the *vertex*, the distance from the vertex to the base is the *altitude*, and from the vertex to the middle of a side is the *slant height.*

765. Rule.—*To find the convex surface of a pyramid, multiply the perimeter of the base by one-half the slant height.*

1. What is the convex surface of a triangular pyramid whose sides are each 4 ft. and slant height 27 ft.?

Ans. 162 sq. ft.

2. Required the convex surface of a pentangular pyramid whose sides are each 5 ft. and slant height 60 ft.

Ans. 750 sq. ft.

766. Rule.—*To find the contents of a pyramid, multiply the area of the base by one-third of the altitude.*

1. Required the contents of a pyramid whose base is 8 ft. square, and whose altitude is 69 ft. *Ans.* 1472 cu. ft.

18

2. Required the contents of a pyramid whose base is a triangle, each side of which is 8 ft., and the altitude of the pyramid 69 ft. *Ans.* 637.376 cu. ft.

THE CYLINDER.

767. The **Cylinder** is a round body of uniform diameter with circles for its ends. The two circular ends are called *bases.*

768. The **Altitude** of a cylinder is the distance from the centre of one base to the centre of the other.

769. Rule.—*To find the convex surface of a cylinder, multiply the circumference of the base by the altitude.*

1. What is the convex surface of a cylinder, altitude 12 ft. and diameter of base 6 ft.? *Ans.* 226.1952 sq. ft.

2. What is the convex surface of a cylinder 40 feet long and 15 feet in diameter? *Ans.* 1884.96 sq. ft.

770. Rule.—*To find the contents of a cylinder, multiply the area of the base by the altitude.*

1. Required the contents of a cylinder 60 feet long and 8 feet in diameter. *Ans.* 3015.936 cu. ft.

2. Required the contents of a cylindrical log 12 feet long and 6⅔ feet in diameter. *Ans.* 418.88 cu. ft.

THE CONE.

771. A **Cone** is a volume whose base is a circle, and whose convex surface tapers uniformly to a point called the *vertex.*

772. The **Altitude** of a cone is the distance from the vertex to the centre of the base, and the *slant height* is the distance from the vertex to the circumference of the base.

773. Rule.—*To find the convex surface of a cone, multiply the circumference of the base by one-half the slant height.*

1. What is the convex surface of a cone, the circumference of whose base is 64 inches and slant height 40 inches? *Ans.* 1280 sq. in.

2. I have a conical haystack whose slant height is 8.25 ft., and the diameter of the base 6.5 ft.; how many square yards of canvas will cover it completely? *Ans.* 9.35935 sq. yd.

774. Rule.—*To find the contents of a cone, multiply the area of the base by one-third of the altitude.*

1. Required the contents of a sugar-loaf, diameter of the base being 8 in. and height 18 in. *Ans.* 301.5936 cu. in.

2. How many cubic feet in a conical haystack 6 ft. high and 20 ft. in circumference? *Ans.* 63.664 cu. ft.

THE FRUSTUM OF A PYRAMID AND CONE.

775. The **Frustum of a Pyramid** is the part of a pyramid which remains after cutting off the top by a plane parallel to the base.

776. The **Frustum of a Cone** is the part of a cone which remains after cutting off the top by a plane parallel to the base.

777. Rule.—*To find the convex surface of a frustum, take the sum of the perimeters or circumferences of the two bases, and multiply it by one-half the slant height.*

1. Required the convex surface of the frustum of a square pyramid whose slant height is 24 feet, the side of the lower base 12 feet, and upper base 8 feet. *Ans.* 960 sq. ft.

2. Required the surface of a frustum of a cone whose slant height is 20 feet, diameter of lower base 12 feet, and upper base 8 feet. *Ans.* 628.32 sq. ft.

778. Rule.—*To find the contents of a frustum, take the sum of the two bases and the square root of their product, and multiply this sum by one-third of the altitude of the frustum.*

1. What are the contents of the frustum of a square pyramid the sides of whose bases are 2 and 3 feet, and whose altitude is 15 feet? *Ans.* 95 cu. ft.

SUG.—$2^2+3^2+\sqrt{2^2\times 3^2}=4+9+6=19$, and this multiplied by 5 equals 95 cu. ft.

2. What is the amount of timber in a log which measures 8C feet in length, the radius of one base being 6 feet and of the other 3 feet? *Ans.* 5277.888 cu. ft.

THE SPHERE.

779. A **Sphere** is a volume bounded by a curved surface, every point of which is equally distant from a point within called the *centre*.

780. The **Diameter** of a sphere is a line passing through its centre and ending in the surface. The *radius* is half the diameter.

781. Rule.—*To find the surface of a sphere, we multiply the circumference by the diameter, or square the radius and multiply it by 4 times* 3.1416.

1. Required the surface of a sphere whose diameter is 24 inches. *Ans.* 1809.5616 sq. in.

2. Required the surface of a sphere whose diameter is 96 inches. *Ans.* 28952.9856 sq. in.

782. Rule.—*To find the contents of a sphere, we multiply the cube of the diameter by* $\frac{1}{6}$ *of* 3.1416.

1. Required the contents of a sphere whose diameter is 6 inches. *Ans.* 113.0976 cu. in.

2. If the diameter of the earth is 8000 miles, what are its surface and solid contents? *Ans.* Sur., 201062400 sq. mi.

783. Rule.—*To find the size of a cube which may be cut from a given sphere, we square the diameter, divide by* 3, *and extract the square root of the quotient.*

1. What is the side of a cube which may be cut from a sphere 21 inches in diameter? *Ans.* 12.124 in.

GAUGING.

784. Gauging is the process of finding the capacity of casks and other vessels.

Barrels and casks differ from cylinders in bulging out in the middle. By ascertaining the approximate mean diameter of the cask or barrel, the capacity can be obtained like that of a cylinder.

Rule I.—*To find the mean diameter of a barrel or cask, add to the head diameter* $\frac{2}{3}$, *or, if the staves are not much curved,* $\frac{3}{5}$ *of the difference between the head and bung diameters.*

Rule II.—*To find the capacity in gallons, multiply the square of the mean diameter by the length (both expressed in inches), and this product by* .0034.

1. How many gal. in a cask whose head diameter is 28 in., bung diameter 36 in., and length 40 in.? *Ans.* $151\frac{1}{9}$ gal.

2. How many gallons in a barrel of slight curvature, 3 ft. long, the head diameter being 26 in., and the bung diameter 29 in.? *Ans.* 94.595616 gal.

REVIEW OF MENSURATION.

1. Two towns, 42 mi. apart, are on a map located $10\frac{1}{2}$ in. apart; what is the scale on which the map is drawn? *Ans.* $\frac{1}{4}$ in. to the mi.

2. How many feet of boards will cover the gable end of a house 34 ft. wide, the ridge being 18 feet high? *Ans.* 306 sq. ft.

3. The rafters of a roof are 18 ft. long, and the distance between the eaves is 24 ft.; what is the height of the ridge? *Ans.* 13.41+ ft.

4. I have a triangular building lot whose sides measure 25, 35, and 40 feet respectively; if I sell it at $5 per square foot, what do I receive? *Ans.* $2165.05.

5. How many Belgian blocks, averaging 6 in. × 12 in. on the surface, will be required to lay a pavement on the roadway of a street 500 yd. long and 15 yd. wide? *Ans.* 135,000.

6. How many bricks, 8 in.×4 in., will be required to lay a pavement on a sidewalk 7 feet wide, extending along 4 lots, each having 18 ft. 6 in. front? *Ans.* 2331.

7. What is the expense of sodding a plot of ground 45 yd. long and 95 ft. wide, with sods 15 in.×24 in., the sods when laid costing $1.50 per hundred? *Ans.* $76.95.

8. How much will it cost to fence a rectangular garden 20 rods long and 15 rods wide, with pickets 4 inches wide and 3 inches apart, at $9 ₱ M.? *Ans.* $17.82.

9. Required the length of a hand rail for a flight of stairs of 18 steps, each step being 7 in. high and 9¼ in. wide? *Ans.* 17½ ft.

10. What will be the cost of flooring at $33.25 per M., of a three story house, the inside measure being 58 ft.×34 ft., deducting 15 ft. 6 in. by 8 ft. 3 in. for the stairs? *Ans.* $183.95.

11. What will be the cost of a thousand tiles in the shape of a rhombus 15 in. on a side, a line drawn from an obtuse angle perpendicular to the opposite side, meeting it 9 in. from the acute angle, at 75¢ a square foot? *Ans.* $937.50.

12. How much will it cost to roof a warehouse with slate 48 ft.× 60 ft., the height of the ridge being 10 ft. and the eaves projecting 6 inches, at $14.75 per square (100 sq. ft.)? *Ans.* $469.05.

13. A yard 36 feet square has in the centre a fountain, the basin of which is 12 feet in diameter; there is a flower-bed, 4 feet wide, around 3 sides of the yard; what will be the expense of paving the remainder at $2.25 per sq. yard? *Ans.* $195.73.

14. The pressure of the atmosphere is 15 lb. to the square inch what is the pressure on a globe 4 ft. in diameter? *Ans.* 10857.3696.

15. A horse is fastened by a rope 10 ft. long to the top of a post 6 ft. high; over how much space can he graze? *Ans.* 201.0624 sq ft.

16. The circular course of a riding-school is 110 feet in its outer diameter, and 10½ feet wide; what was the expense of its construction, at 10¢ per sq. foot.? *Ans.* $328.22.

17. A room 27 ft. 6 in. long, and 16 ft. 3 in. wide, has a semi-circular bow, 22 feet in diameter, thrown out on one side; find the area of flooring in the whole room. *Ans.* 636.9418 sq. ft.

18. I have a fish pond in the form of an ellipse, 20 ft. long, 15 ft. wide; how many hogsheads of water are required to fill it to the depth of 4 ft.? *Ans.* 111.90$\frac{6}{7}$ hhd.

19. In a circular grass-plot whose diameter is 50 yd., there is a gravel walk 1 yd. wide, running round it 1 yd. within the edge; what will be the cost of sodding the plot at 12¢ per sq. yd.? *Ans.* $217.90.

20. The steeple of a church in the form of a cone is 30 feet in diameter at the base, the slant height being 90 ft.; what will it cost to paint it at 25¢ per square yard? *Ans.* $117.81.

21. In a tin funnel, one part is conical, the slant height of the conical part is 4 in., the circumference at one end 10 in., and at the other end 1 in.; the other part is cylindrical, the length being 5 in.; required the number of sq. in. of tin in it. *Ans.* 27 sq. in.

22. A bin has the bottom 2 ft. 6 in. square, the top 3 ft. 6 in. square, the height 2 ft. 6 in.; what is the cost of lining it with zinc at 20¢ per sq. ft.? How many bu. will it hold? *Ans.* $7.37; 18.24 bu.

23. How many cu. ft. in a telegraph pole 40 ft. long, 18 in. in diameter at the base, and 8 in. in diameter at top? *Ans.* 38.688⅚ cu. ft.

24. How much map-surface on a school globe 12 inches in diameter; and how many cubic inches of material are there in it, if it is a hollow sphere 1 in. thick? *Ans.* 452.3904 sq. in.; 381.1808 cu. in.

25. A room 24 ft. 6 in. long, 13 ft. 4 in. wide, is flooded with water 8 in. deep; what is the weight of the water? *Ans.* 5104⅛ lb.

26. A circular room, of which the diameter is 25 ft., and height of wall 14 ft., is covered with a hemispherical dome, what is the cost of plastering the wall and dome at 25¢ per sq. ft.? *Ans.* $520.32¾.

27. How many dozen boxes of perfumery, 2 in. on each side, can be packed in a rectangular box whose dimensions are respectively 1 ft. 6 in., 1 ft., and 8 in.? *Ans.* 18 doz.

28. A cubic inch of gold is hammered out to cover a *square* (100 sq. ft.); what is its thickness? *Ans.* .000069 inch.

29. An ice house is 40 ft. long, 30 ft. wide, and 20 ft. deep; what area of ice 6 in. thick, will be required from a pond to fill it? What would it weigh at 56 lb. the cu. ft.? *Ans.* 48000 sq. ft.; 1344000 lb.

30. A company wish to excavate a canal 25 miles long, with an average width of 9 yards, and an average depth of 5 ft. 6 in.; how long will it take 500 men to do the work, if each man averages 12 cubic yards a day? *Ans.* 121 days.

31. If the men are paid $1.50 each per day, and the contractor estimates 5% profit, what proposal will he make for digging the above canal? *Ans.* $95287.50.

32. If 50 cubic feet of air are required per person in a well-ventilated room, how many persons can safely remain in a room 60 ft. long, 40 ft. wide, and 12 ft. high? *Ans.* 576 persons.

33. The cost of a cube of metal, at $12 per cubic inch, is $4116; find the cost of gilding it over at 2¢ per square inch. *Ans.* $5.88.

34. A cast-iron garden roller is 40 in. long, 22½ in. in diameter, and the iron is ¾ in. thick; required its weight, if a cubic inch of iron weighs 4¼ ounces. *Ans.* 576.532$\frac{1}{16}$ lb.

35. If a cannon ball, 6 in. in diameter, is melted and cast in a conical mould 6 in. in diameter at the base, what is the length of the cone? *Ans.* 12 in.

36. A railroad embankment has an average depth, for 5 miles, of 12 feet, width at top 8 yards, and at bottom 12 yards; what was the cost of carting at 15¢ per load (cu. yd.)? *Ans.* $52800.

I

APPENDIX.

THE METRIC SYSTEM

OF WEIGHTS AND MEASURES.

INTRODUCTION.

The old system of weights and measures in our country is irregular, difficult to learn, and inconvenient to apply. The same is true with the old systems of all nations. Originating by chance, rather than by science, they lacked the simplicity of law; and were, therefore, irregular and chaotic.

In 1795, France adopted a system of weights and measures called the Metric System, based upon the decimal method of notation, all the divisions and multiples being by 10. It was regarded as so great an improvement upon the old methods that it has since been introduced into Spain, Belgium, Portugal, Switzerland, Holland, Italy, Germany, Austria, Sweden, Denmark, Greece, Mexico, Brazil, and by most of the South American States, and in the most of these countries its use is compulsory. In 1864, the British Parliament passed an act permitting its use throughout the empire whenever parties should agree to use it.

The introduction of the Metric System into this country had been long recommended by scientific men, and by such statesman as Madison, Jefferson, John Quincy Adams, etc. In 1866, through the influence of Charles Sumner, Congress authorized its use in the United States, and provided for its introduction into the post-offices for the weighing of letters and papers. To facilitate its adoption, a convenient standard of comparison was furnished, by making the new five-cent piece five grams in weight and one fiftieth of a meter, or two centimeters, in diameter. This system will, without doubt, in a few years be in general use in this country.

The *advantages* of the Metric System are numerous and important.

1. It is easily learned; a school-boy can learn it in a single afternoon.

2. It is easily applied, all the operations being the same as in simple numbers.

3. It does away with addition, subtraction, multiplication, division, and reduction of compound numbers.

4. It will facilitate commerce, giving the nations a universal system of weights and measures.

785. The **Metric System** of weights and measures is based upon the decimal system of notation.

786. In this system we first establish the unit of each measure, and then derive the other denominations by taking decimal multiples and divisions of the unit.

787. Names.—We first name the unit of any measure and then derive the other denominations by adding prefixes to the unit name.

788. The *higher denominations* are expressed by prefixing to the name of the unit

Deca,	Hecto,	Kilo,	Myria.
10	100	1000	10,000

The *lower denominations* are expressed by prefixing to the name of unit

Deci,	Centi,	Milli.
$\frac{1}{10}$	$\frac{1}{100}$	$\frac{1}{1000}$

789. Units.—The following are the different units, with their English pronunciation.

Measure.	Unit.	Pronunciation.	Measure.	Unit.	Pronunciation.
LENGTH,	**Meter,**	(meter.)	CAPACITY,	**Liter,**	(leeter.)
SURFACE,	**Are,**	(air.)	WEIGHT,	**Gram,**	(gram.)
VOLUME,	**Stere,**	(stair.)	VALUE,	**Dollar.**	

MEASURE OF LENGTH.

790. The **Meter** is the *unit of length.* It is the ten millionth part of the distance from the equator to the poles, and equals 39.37 inches, or 3.28 feet.

TABLE.

10 millimeters (mm.)	equal	1 centimeter,	cm.
10 centimeters	"	1 decimeter,	dm.
10 decimeters	"	1 meter,	M.
10 meters	"	1 decameter,	DM.
10 decameters	"	1 hectometer,	HM.
10 hectometers	"	1 kilometer,	KM.
10 kilometers	"	1 myriameter,	MM.

NOTES.—1. The *meter* is very nearly 3 feet 3 inches and 3 eighths of an inch in length, which may be easily remembered as the *rule of three threes.*

2. Cloth, etc., are measured by the *meter*; very small distances, by the *millimeter*; great distances, by the *kilometer.*

3. The 5-cent piece of 1866 is very nearly $\frac{1}{50}$ of a *meter* in diameter; hence its diameter is about $\frac{1}{5}$ of a *decimeter*, or 2 *centimeters*. It was ordered to be $\frac{1}{50}$ of a meter in diameter, but owing to the composition of the alloy it was necessary to make its diameter a little greater; 48.6 nickel 5-cent pieces laid side by side measure one meter.

4. A *decimeter* is about 4 inches: a *kilometer*, about 200 rods, or $\frac{5}{8}$ of a mile; a *millimeter*, about $\frac{1}{25}$ of an inch. The *inch* is about $2\frac{1}{2}$ centimeters; the *foot*, 3 decimeters; the *rod*, 5 meters; the *mile*, 1600 meters, or 16 hectometers.

MENTAL EXERCISES.

1. How many centimeters in a meter?
2. How many millimeters in a meter?
3. How many decimeters in a decameter?
4. How many meters in a hectometer?
5. How many meters in a kilometer?

MEASURES OF SURFACE.

791. The **Are** is the *unit of surface* used to measure land. The *are* is a *square decameter.* It equals 119.6 sq. yd., or 0.0247 acre.

TABLE.

10 milliares (ma.)	equal	1	centiare,	ca.
10 centiares	"	1	deciare,	da.
10 deciares	"	1	are,	A.
10 ares	"	1	decare,	DA.
10 decares	"	1	hectare,	HA.
10 hectares	"	1	kilare,	KA.
10 kilares	"	1	myriare,	MA.

NOTES.—1. The *are*, *centiare*, and *hectare*, are the denominations principally used, as these are exact squares. The *centiare* is a square whose side is 1 meter; the *hectare* is a square whose side is 100 meters.

The *are* = 100 square meters. The *centiare* = 1 square meter,
The *hectare* = 10,000 square meters.

2. The *deciare* is not a *square*, it is merely the tenth of an are; the *decare* is not a square, it is merely ten ares.

3. A *hectare* equals nearly $2\frac{1}{2}$ acres; a *centiare* equals nearly $1\frac{1}{5}$ sq. yd. An *acre* is very nearly 40 ares.

MEASURES OF OTHER SURFACES.

792. All surfaces besides land are measured by the *square meter*, *square decimeter*, etc. The measures are shown by the following table:

TABLE.

100 sq. millimeters (mm.^{2})	= 1 sq. centimeter,	cm.^{2}
100 sq. centimeters	= 1 sq. decimeter,	dm.^{2}
100 sq. decimeters	= 1 sq. meter,	M.^{2}

NOTE.—The measures higher than these are not generally used. The usual method of notation is to write *sq.* before the denomination; but I suggest as an abbreviation that we indicate the square by an exponent.

MENTAL EXERCISES.

1. How many centiares in an are?
2. How many ares in a hectare?
3. How many square meters in an are?
4. How many square decimeters in an are?
5. How many ares in 640 square meters?

MEASURES OF VOLUME.

793. The **Stere** is the *unit of volume.* It is a *cubic meter,* and equals 35.3166 cubic ft., or 1.308 cu. yd.

TABLE.

10 millisteres (ms.)	equal	1 centistere,	cs.
10 centisteres	"	1 decistere,	ds.
10 decisteres	"	1 stere,	S.
10 steres	"	1 decastere,	DS.
10 decasteres	"	1 hectostere,	HS.
10 hectosteres	"	1 kilostere,	KS.
10 kilosteres	"	1 myriastere,	MS.

NOTE.—Wood is measured by this measure. The *stere, decistere,* and *decastere* are principally used. 3.6 *steres,* or 36 *decisteres,* very nearly equal the common cord.

MEASURES OF OTHER VOLUMES.

794. Other solid bodies are usually measured by the *cubic meter* and its divisions. The measures are shown by the following table:

TABLE.

1000 cubic millimeters (mm.^{3})	=1 cubic centimeter,	cm.^{3}
1000 cubic centimeters	=1 cubic decimeter,	dm.^{3}
1000 cubic decimeters	=1 cubic meter,	M.^{3}

NOTE.—The higher denominations are not generally used. I indicate the cubic measures with an exponent, instead of writing *cu.* before the denominations.

MENTAL EXERCISES.

1. How many centisteres in a stere?
2. How many decisteres in a decastere?
3. How many decasteres in a kilostere?
4. How many cubic meters in a hectostere?

MEASURES OF CAPACITY.

795. The **Liter** is the *unit of capacity.* It equals a *cubic decimeter;* that is, a cubic vessel whose size is one-tenth of a meter.

796. This measure is used for measuring liquids and dry substances. The *liter* is a cylinder, and holds 2.1135 pints wine measure, or 1.816 pints dry measure.

TABLE.

10 milliliters (ml.)	equal	1 centiliter,	cl.
10 centiliters	"	1 deciliter,	dl.
10 deciliters	"	1 liter,	L.
10 liters	"	1 decaliter,	DL.
10 decaliters	"	1 hectoliter,	HL.
10 hectoliters	"	1 kiloliter,	KL.
10 kiloliters	"	1 myrialiter,	ML.

NOTES.—1. The *liter* is principally used in measuring *liquids*, and the *hectoliter* in measuring grains, etc.

2. The *liter* equals nearly $1\frac{1}{18}$ liquid quarts, or $\frac{9}{10}$ of a dry quart, or nearly $\frac{1}{35}$ of a bushel measure.

3. The *hectoliter* is about $2\frac{5}{6}$ bushels, or $\frac{5}{6}$ of a barrel. 4 *liters* are a little more than a *gallon;* 35 *liters*, very nearly a *bushel*.

MENTAL EXERCISES.

1. How many liters in a hectoliter?
2. How many liters in a kiloliter?
3. How many deciliters in a decaliter?
4. How many liters in a cubic meter? *Ans.* 1000.
5. How many liters in a stere? *Ans.* 1000.

MEASURES OF WEIGHT.

797. The **Gram** is the *unit of weight.* It is the weight of a cubic centimeter of distilled water at the temperature of melting ice. The *gram* equals 15.432 Troy grains.

TABLE.

10 milligrams (mg.)	equal	1 centigram,	cg.
10 centigrams	"	1 decigram,	dg.
10 decigrams	"	1 gram,	G.
10 grams,	"	1 decagram,	DG.
10 decagrams	"	1 hectogram,	HG.
10 hectograms	"	1 kilogram,	KG., or K.
10 kilograms	"	1 myriagram,	MG.

NOTES.—1 The *gram* is used in weighing letters, and mixing and compounding medicines, and in weighing all very light articles. The new 5-cent coin (dated 1866) weighs 5 grams.

2. The *kilogram* is the ordinary unit of weight, and is generally abbreviated into *kilo*. It equals about 2½ pounds avoirdupois. Meat, sugar, etc., are bought and sold by the *kilogram*.

3. In weighing heavy articles, two other weights, the *quintal* (100 kilograms) and the *tonneau* (1000 kilograms), are used. The *tonneau* is between our *short ton* and *long ton*.

4. The *avoirdupois ounce* is about 28 *grams*; the *pound* is a little less than ½ a *kilo*.

MENTAL EXERCISES.

1. How many grams is a kilogram?
2. How many milligrams in a gram?
3. How many decigrams in a kilogram?
4. How many hectograms in a myriagram?

MEASURES OF VALUE.

798. The **Franc** is the French *money unit.* It equals $0.193. The principal gold coin is the 20-*franc* piece; the principal silver coins are the *franc* and the 5-*franc* piece.

TABLE.

10 centimes equal 1 decime.
10 decimes " 1 franc.

NOTE.—It has been suggested that the American dollar and the English pound be so modified that we shall have the following money table:

5 francs = 1 dollar.
5 dollars = 1 pound.

The *franc* equals about 19.3 cents of our present money.

NUMERATION AND NOTATION.

799. In the **Metric System** the decimal point is placed between the unit and its divisions, the whole quantity being regarded as an integer and a decimal. Thus, 3 decagrams, 5 grams, 6 decigrams, 8 centigrams, are written 35.68 grams.

800. The *initials* of the denomination may be placed either before or after the quantity, though they are most frequently placed after it; thus, 27 grams may be written G27, or 27G.

EXERCISES IN NUMERATION.

1. Read 48.05M.

SOLUTION.—This is read 48 and 5 hundredths meters; or it may be read 4 decameters, 8 meters and 5 centimeters.

Read the following:

2. 12.06M.
3. 28.66A.
4. 904.06S.
5. 807.005L.
6. 5062.035G.
7. 20760.508G.

EXERCISES IN NOTATION.

1. Write 6 meters and 5 centimeters.

SOLUTION.—We write the 6 meters with a decimal point to the right, and then, since there are no decimeters, we write a naught in the tenths place, and then write the 5 centimeters in the place of centimeters.

OPERATION.
6.05M

2. Write 17 meters, 4 decimeters, 8 centimeters.
3. Write 7 decameters, 2 decimeters, 5 centimeters.
4. Write 15 ares, 9 deciares, 8 milliares.
5. Write 4 hectares, 8 ares, 5 centiares.
6. Write 12 decasteres, 6 decisteres, 8 centisteres.
7. Write 9 kilosteres, 7 decasteres, 5 centisteres.
8. Write 3 hectoliters, 8 liters, 7 deciliters.
9. Write 16 grams, 4 decigrams, 8 centigrams.
10. Write 8 myriagrams, 7 hectograms, 6 centigrams, and 5 milligrams.

REDUCTION OF THE METRIC SYSTEM TO THE COMMON SYSTEM.

MEASURES OF VALUE.

1. How many dollars in 25 francs? *Ans.* $4.825.
2. How many dollars in 47.50 francs? *Ans.* $9.16¾.
3. How many francs in $15.50? *Ans.* 80.31 fr.
4. How many francs in $37.75? *Ans.* 195.595 fr.

MEASURES OF WEIGHT.

5. How many grains in 12 grams? *Ans.* 185.184 gr.

6. Pounds Troy, in 480.5 grams?
Ans. 1 lb. 3 oz. 8 pwt. $23\frac{19}{250}$ gr.

7. Pounds Av., in 976.25 grams?
Ans. 2 lb. 2 oz. 190.49 gr.

8. Grams in 480 grains? *Ans.* 31.104G.

9. Grams in 12 Troy pounds? *Ans.* 4479.004G.

10. Grams in 12 Av. pounds? *Ans.* 5443.234G.

MEASURES OF LENGTH.

11. How many feet in 24.5 meters? *Ans.* 80.38 ft.

12. Yards in 136.54 meters? *Ans.* 149.3216 yd.

13. Meters in 120 yards? *Ans.* 109.73M.

14. Meters in 2 mi. 120 rd.? *Ans.* 3822.199M.

15. Miles in 4000 meters? *Ans.* 2 mi. 155 rd. 5 ft. 10 in

16. Meters in 3 mi. 272 rd.? *Ans.* 6195.9867M.

MEASURES OF SURFACE.

17. How many ares in 360 sq. yd.? *Ans.* 3.01A.

18. Sq. yd. in 142.5 ares? *Ans.* 17043 sq. yd.

19. Acres in 505.6 ares? *Ans.* 12 A. 78.1312 P.

20. How many ares in 30 acres? *Ans.* 1214.574A.

21. How many ares in 5 A. 104 P.? *Ans.* 228.744A.

MEASURES OF VOLUME.

22. How many cu. ft. in 46 steres? *Ans.* 1624.5636 cu. ft.

23. Cu. ft. in 214.78 steres? *Ans.* 7585.2993 cu. ft.

24. How many steres in 128 cu. ft.? *Ans.* 3.624S.

25. Steres in 16 cu. yd. 8 cu. ft.? *Ans.* 12.458S.

MEASURES OF CAPACITY.

26. How many gallons in 36.08 liters? *Ans.* 9 gal. 2 qt.

27. Gallons in 45.05 liters? *Ans.* 11 gal. 3 qt. 1 pt.

28. How many liters in 24 gallons? *Ans.* 90.844L.

29. How many liters in 36 gal. 2 qt.? *Ans.* 138.16L.

30. How many liters in 6 bu. 2 pk.? *Ans.* 229.07L.

31. Bushels in 65.25 liters? *Ans* 1 bu. 3 pk. 3 qt.

MISCELLANEOUS PROBLEMS.

1. If a letter weighs 2.5 grams, how many such letters will it take to weigh a kilogram? *Ans.* 400.

2. A lady bought 11.5 meters of silk for a dress, at the rate of $4.75 a meter; what did it cost her? *Ans.* $54.625.

3. My butcher's bill one month was 87.5 kilograms of beef, at 18¾ cents a kilo; what was the bill? *Ans.* $16.40⅝.

4. How much must I pay for 56.25 liters of coal oil, at the rate of 18¾ cents a liter? *Ans.* $10.546 +.

5. A kilogram weighs 2.2046 lb.; what is the weight of 56½ tonneaux? *Ans.* 124559.9 lb.

6. A bought 2500 ares of land, at $4.50 an are, and sold it for $525 a hectare; what was the gain? *Ans.* $1875.

7. If 15 steres of wood cost $22.50, what must I pay for 24.5 steres at the same rate? *Ans.* $36.75.

8. If a kilogram of sugar is worth 21¾ cents, how many kilos can I buy for $100? *Ans.* 459.77 +.

9. The height of a pole is 68.325M; how long would it take a worm to climb to its top, at the rate of 15 meters a day? *Ans.* 4.555 days.

10. A kilometer is about ⅝ of a mile; how many kilometers from Lancaster to Philadelphia, 70 miles? *Ans.* 112.

11. How much must I pay for 23¾ meters of silk, at 8 francs 25 centimes a meter? *Ans.* 195.94— fr.

12. What cost 3 kilares, 7 hectares, 6 deciares of land, at $275.25 a hectare? *Ans.* $10185.90.

13. It is about 100 miles from Philadelphia to New York; how many kilometers is it? *Ans.* 160.

14. How much will it cost to excavate 12¾ cubic meters of earth, at $37.25 a cubic meter? *Ans.* $476.80.

15. What is the width of the Atlantic in kilometers, the width being about 3000 miles? *Ans.* 4800 kilometers.

16. What must I pay for 25 steres, 2 decisteres, and 5 centisteres of wood, at the rate of $2.65 a stere? *Ans.* $66.91¼.

17. How long will it take a man to walk from Philadelphia to New York, at 8 kilometers an hour? *Ans.* 20 hours.

18. Two vessels are 432 kilometers apart, and sail toward each other, each at the rate of 18 kilometers an hour; in how many hours will they be together? *Ans.* 12 hours.

19. A block 3.5 meters long, .75 meters wide, and .8 meters thick, cost $12; what would a cubic meter of marble cost, at the same rate? *Ans.* $5.71+.

20. A man bought 7000 grams of jewels, at 40 francs a gram, and sold them at $15 a pennyweight; how much was gained or lost? *Ans.* $13475.

PROBLEMS ON IMPORTS.

1. An importer bought 428.5 meters of silk in France, at 18 francs a meter, sent it to the United States, paying 25 cents a meter shipping and duty, and sold it for $5.25 a meter; what was his gain? *Ans.* $653.89.

2. A man bought a valuable gem in France which weighed 325.75 grams, @ 10.25 francs; the duty on it was $6.25; how must he sell it a gram to clear $150? *Ans.* $2.46.

3. I bought 125.75 liters of wine in France, at 45.25 francs a liter, paid $1.25 a liter duty and freight, and sold it at $12.50 a liter; how much did I gain? *Ans.* $316.48.

4. An importer bought 625.5 liters of French brandy, at 7.55 francs a liter, paid 15 cents a liter duty and freight, and sold it in New York at $1.65 a liter; how much did he gain? *Ans.* $26.80.

5. A man bought 200 meters of cloth in France, at 16.25 francs a meter; he paid 12½ cents a yard duty and freight, and sold it in Boston at $4.62½ a yard; what was the gain? *Ans.* $357.

6. An importer bought 480 grams of jewels, at 12.25 francs a gram, paid $5.25 an ounce shipment and duty, and sold them in Philadelphia at $102.75 an ounce; what was the gain? *Ans.* $369.78.

7. A wine merchant bought 180 liters of brandy in Havre, at 32½ decimes a liter; he paid 2½ decimes a liter shipment, and $2.25 a gallon duty, and sold it in New York at $6.75 a gallon; what was his gain? *Ans.* $92.40.

INSURANCE TABLE FOR COMPUTING SHORT RATES.

Insurances for periods less than one year will be at the following rates. Risks upon Grain, Pork, Wool, and other produce, are sometimes wanted for very short terms.

ANNUAL PREMIUM.	Cts. 30	Cts. 35	Cts. 40	Cts. 45	Cts. 50	Cts. 55	Cts. 60	Cts. 65	Cts. 70	Cts. 75	Cts. 80	Cts. 85	Cts. 90	Cts. 1.00	Cts. 1.10	Cts. 1.25	Cts. 1.50	Cts. 1.75	Cts. 2.00	Cts. 2.25	Cts. 2.50	Cts. 2.75	Cts. 3.00	Cts. 3.25	Cts. 3.50	Cts. 3.75	Cts. 4.00	Cts. 5.00
2 Days or less,	1	1	1	1	2	2	2	2	2	3	3	3	3	3	3	4	5	6	7	8	8	9	10	11	12	12	14	17
5 Days or less,	2	2	3	3	3	4	4	4	5	5	5	6	6	7	7	9	10	12	14	15	17	18	20	22	24	25	27	34
10 Days or less,	3	3	4	4	5	6	6	7	7	8	8	9	9	10	11	12	15	18	20	22	25	27	30	33	35	38	40	50
15 Days or less,	4	4	5	6	6	8	8	8	10	10	10	12	12	14	14	17	20	25	27	30	33	36	40	44	47	50	53	67
20 Days or less,	5	5	6	7	8	9	9	10	11	11	12	13	14	15	17	20	25	30	33	40	42	45	50	55	58	63	66	80
1 Month or less,	6	7	8	9	10	11	12	13	14	15	16	17	18	20	22	25	30	35	40	45	50	55	60	65	70	75	80	1.00
45 Days or less,	8	9	11	12	13	14	16	18	18	19	21	22	24	25	27	31	38	43	50	56	63	68	75	82	88	93	1.00	1.25
2 Months or less,	9	10	12	14	15	16	18	20	21	22	24	25	27	30	33	38	45	52	60	67	75	82	90	98	1.05	1.13	1.20	1.50
75 Days or less,	11	12	14	16	18	20	22	24	25	26	28	30	32	35	38	44	50	60	70	78	88	96	1.05	1.13	1.23	1.31	1.40	1.75
3 Months or less,	12	14	16	18	20	22	24	26	28	30	32	34	36	40	44	50	60	70	80	90	1.00	1.10	1.20	1.30	1.40	1.50	1.60	2.00
4 Months or less,	15	17	20	22	25	27	30	32	35	37	39	42	45	50	55	62	75	88	1.00	1.12	1.25	1.37	1 50	1.62	1.75	1.87	2.00	2.50
5 Months or less,	18	21	24	27	30	33	36	39	42	45	48	51	54	60	66	75	90	1.05	1.20	1.35	1.50	1.65	1.80	1.95	2.10	2.25	2.40	3.00
6 Months or less,	21	24	28	32	35	38	42	46	49	52	55	59	63	70	77	88	1.05	1.22	1.40	1.57	1.75	1.92	2.10	2.27	2.45	2.62	2.80	3.50
7 Months or less,	22	26	30	34	37	41	45	49	52	56	60	63	67	75	82	93	1.12	1.30	1.50	1.68	1.87	2.06	2.25	2.43	2.62	2.81	3.00	3.75
8 Months or less,	24	28	32	36	40	44	48	52	56	60	64	68	72	80	88	1.00	1.20	1.40	1.60	1.80	2.00	2.20	2 40	2.60	2.80	3.00	3.20	4.00
9 Months or less,	26	31	35	40	43	46	51	55	59	64	68	72	76	85	93	1.06	1.27	1.48	1.70	1.91	2.12	2.24	2.55	2.76	2.97	3.17	3.4	4.25
10 Months or less,	27	32	37	42	45	49	54	58	63	67	72	76	81	90	1.00	1.12	1.35	1.57	1 80	2.02	2.25	2.47	2.70	2.9	3.15	3.37	3.60	4.50
11 Months or less,	29	33	39	43	47	52	57	61	66	71	76	80	85	95	1.04	1.18	1 42	1.66	1.90	2.13	2.37	2.61	2.85	3.08	3.32	3.56	3.90	4 75

NOTE.—The upper row of figures shows the rates for the year, from 30 cents up to 500 cents, or 5 per cent., and the rows of figures below them show the price under each, from 2 days to 11 months. Terms of any intermediate number of days or months are taken at the next higher rate; thus, a 40 day policy would command a 45 day rate as above.

TABLE.

Annual Premium Rates for an Insurance of $1000.

LIFE POLICIES. Payable at death only.					ENDOWMENT POLICIES. P'able as indicated, or at death, if prior.			
Age.	Annual Payment.			Single Payment.	Age.	In 10 years.	In 15 years.	In 20 years.
	For Life.	20 years.	10 years.					
16	15.60	23.00	37.34	246.52	16	104.60	65.05	45.90
17	16.00	23.40	38.14	250.78	17	104.65	65.15	45.95
18	16.50	23.80	38.92	255.20	18	104.70	65.20	46.05
19	16.90	24.25	39.73	259.76	19	104.75	65.25	46.15
20	17.30	24.70	40.53	264.50	20	104.85	65.35	46.20
21	17.80	25.20	41.34	269.39	21	104.90	65.40	46.30
22	18.30	25.70	42.17	274.45	22	104.95	65.50	46.40
23	18.70	26.20	43.03	279.68	23	105.05	65.60	46.50
24	19.30	26.75	43.89	285.08	24	105.10	65.70	46.60
25	19.80	27.30	44.78	290.66	25	105.20	65.80	46.75
26	20.30	27.90	45.68	296.43	26	105.30	65.90	46.85
27	20.90	28.50	46.62	302.39	27	105.35	66.00	47.00
28	21.50	29.15	47.57	308.55	28	105.45	66.10	47.15
29	22.10	29.80	48.59	314.91	29	105.55	66.20	47.30
30	22.70	30.45	49.07	321.48	30	105.65	66.35	47.45
31	23.40	31.10	50.44	328.25	31	105.80	66.50	47.60
32	24.10	31.85	51.49	335.25	32	105.90	66.65	47.80
33	24.80	32.60	52.36	342.48	33	106.05	66.80	48.00
34	25.60	33.40	53.56	349.93	34	106.15	66.95	48.25
35	26.50	34.25	54.82	357.63	35	106.30	67.15	48.50
36	27.40	35.10	55.95	365.58	36	106.45	67.35	48.80
37	28.30	36.00	57.26	373.79	37	106.60	67.60	49.10
38	29.30	36.95	59.18	382.27	38	106.80	67.85	49.45
39	30.40	37.95	60.50	391.03	39	107.00	68.15	49.85
40	31.50	39.00	61.68	400.09	40	107.20	68.45	50.25
41	32.60	40.10	63.66	409.46	41	107.45	68.85	50.75
42	33.90	41.25	64.96	419.14	42	107.80	69.25	51.30
43	35.20	42.50	66.43	429.15	43	108.15	69.75	51.90
44	36.50	43.85	68.11	439.44	44	108.55	70.30	52.60
45	38.00	45.20	69.40	450.00	45	109.00	70.85	53.35
46	39.60	46.65	71.64	460.80	46	109.50	71.50	54.20
47	41.20	48.20	73.42	471.82	47	110.05	72.25	55.10
48	43.10	49.85	75.44	483.02	48	110.65	73.05	56.05
49	45.00	51.55	77.77	494.42	49	111.35	73.90	57.15
50	47.00	53.55	80.43	506.01	50	112.05	74.80	58.35
51	49.20	55.25	82.23	517.76	51	112.85	75.85	..
52	51.50	57.25	84.23	529.68	52	113.70	76.95	..
53	53.90	59.40	86.50	541.75	53	114.65	78.20	...
54	56.50	61.65	89.17	553.95	54	115.70	79.55	...
55	59.40	64.05	92.24	566.28	55	116.80	81.00	..
56	62.40	66.60	95.86	578.72	56	118.05	...	...
57	65.60	69.30	97.86	591.26	57	119.40	...	...
58	69.00	72.20	102.04	603.90	58	120.90	...	...
59	72.70	75.30	105.34	616.62	59	122.50	...	...
60	76.40	78.65	108.43	629.41	60	124.30	...	...

TABLES OF WEIGHTS AND MEASURES.

ENGLISH, OR STERLING MONEY.

4 far.	=	1 d.
12 d.	=	1 s.
20 s.	=	1 £.
21 s.	=	1 G.

APOTHECARIES' WEIGHT.

20 gr.	=	1 ℈.
3 ℈	=	1 ʒ.
8 ʒ	=	1 ℥.
12 ℥	=	1 ℔.

TROY WEIGHT.

24 gr.	=	1 pwt.
20 pwt.	=	1 oz.
12 oz.	=	1 lb.

AVOIRDUPOIS WEIGHT.

16 oz.	=	1 lb.
100 lb.	=	1 cwt.
20 cwt.	=	1 T.

LONG MEASURE.

12 in.	=	1 ft.
3 ft.	=	1 yd.
5½ yd., or 16½ ft.	=	1 rd.
320 rd.	=	1 mi.
3 mi.	=	1 lea.
69.16 mi. (69⅛ nearly)	=	1 deg., or °.

SURVEYORS' LINEAR MEASURE.

7.92 in.	=	1 li.
100 li.	=	1 ch.
80 ch	=	1 mi

MARINERS' MEASURE.

6 ft.	=	1 fathom.
880 fath.	=	1 mi.

SURFACE, OR SQUARE MEASURE.

144 sq. in.	=	1 sq. ft.
9 sq. ft.	=	1 sq. yd.
30¼ sq. yd., or 272¼ sq. ft.	=	1 sq. rd.
160 sq. rd.	=	1 acre.
640 acre	=	1 sq. mi.

SURVEYORS' SQUARE MEASURE.

10,000 sq. li.	=	1 sq. ch.
10 sq. ch.,	=	1 A.
640 A.	=	1 sq. mi.
36 sq. mi. (6 mi. sq.)	=	1 T'ship.

CUBIC, OR SOLID MEASURE.

1728 cu. in.	=	1 cu. ft.
27 cu. ft.	=	1 cu. yd.
16 cu. ft.	=	1 cd. ft.
8 cd. ft., or, 128 cu. ft.	=	1 cd.

LIQUID MEASURE.

4 gi.	=	1 pt.
2 pt.	=	1 qt.
4 qt.	=	1 gal.
31½ gal.	=	1 bar.
63 gal.	=	1 hhd.

DRY MEASURE.

2 pt.	=	1 qt.
8 qt.	=	1 pk.
4 pk.	=	1 bu.

NOTE.—A *perch* of masonry=24¾ cu. ft.; a *cubic yard* of earth is called a *load*. A *board foot* is one foot long, 1 foot wide, and 1 inch thick.

NOTE.—For a BUSHEL, in most of the states, it requires 32 lb. of *oats*, 44 lb. of *timothy seed*, 48 lb. of *barley*, 56 lb. of *rye*, 56 lb. of *Indian corn*, 60 lb. of *wheat*, 60 lb. of *clover seed*, 60 lb. of *potatoes*, 60 lb. of *beans*.

APOTHECARIES' FLUID MEASURE.

60 minims (♏)	=1 fluidrachm,	fʒ.
8 fluidrachms	=1 fluidounce,	f℥.
16 fluidounces	=1 pint,	O.
8 pints	=1 gallon,	Cong.

NOTE.—A liquid gallon=231 cu. in.; a bushel=2150.42 cu. in.

CIRCULAR MEASURE.

60 seconds (″)	=1 minute, . .	′
60 minutes .	=1 degree, . .	°
30 degrees .	=1 sign, . . .	S.
12 signs, or 360°	=1 circumference,	C.

TIME.

60 sec.	= .	1 min.
60 min.	= .	1 h.
24 h. .	= .	1 da.
365 da. .	= .	1 yr.
100 yr. .	= .	1 cen.

ALSO,

7 da. .	= .	1 wk.
4 wk. .	= .	1 mo.
12 mo. .	= .	1 yr.

TABLE TO FIND THE TIME IN DAYS.

FROM ANY DAY OF	TO THE SAME DAY OF											
	Jan	Feb	Mar	Apr	May	J'ne	J'ly	Aug	Sep	Oct	Nov	Dec
January	365	31	59	90	120	151	181	212	243	273	304	334
February	334	365	28	59	89	120	150	181	212	242	273	303
March	306	337	365	31	61	92	122	153	184	214	245	275
April,	275	306	334	365	30	61	91	122	153	183	214	244
May	245	276	304	335	365	31	61	92	123	153	184	214
June	214	245	273	304	334	365	30	61	92	122	153	183
July	184	215	243	274	304	335	365	31	62	92	123	153
August	153	184	212	243	273	304	334	365	31	61	92	122
September	122	153	181	212	242	273	303	334	365	30	61	91
October	92	123	151	182	212	243	273	304	335	365	31	61
November	61	92	120	151	181	212	242	273	304	334	365	30
December	31	62	90	121	151	182	212	243	274	304	335	365

MISCELLANEOUS TABLES.

12 units . . .	= 1 dozen.
12 dozen . . .	= 1 gross.
12 gross . . .	= 1 great gross.
24 sheets	= 1 quire.
20 quires	= 1 ream.
480 sheets	= 1 ream.

MISCELLANEOUS WEIGHTS.

The following denominations are frequently used:

25 lb.	of powder	make	1 barrel.
56 "	butter	"	1 firkin.
84 "	"	"	1 tub.
100 "	grain or flour	"	1 cental.
100 "	dry fish	"	1 quintal.
100 "	nails	"	1 keg.
100 lb.	of raisins	make	1 cask.
196 "	flour	"	1 barrel.
200 "	pork, beef or fish		1 barrel.
240 "	lime		1 cask.
280 "	salt at N.Y.S. works		1 barrel.
600 "	rice		1 barrel.

Spotted something that isn't quite right?

Let us know and we'll fix it for future reprints. And as a thank you for your diligence, we'll send you a free Sothis book of your choice.

publisher@sothis.press

This page intentionally left blank

Made in United States
Orlando, FL
04 November 2026

90201499R00252